Global Issues in Water Policy

Volume 13

Global Issues in Water Policy is now indexed in SCOPUS.

Policy work in the water sector has grown tremendously over the past two decades, following the Rio Declaration of 1992. The existing volume of water-related literature is becoming dominant in professional outlets, including books and journals. Because the field of water resources is interdisciplinary in nature, covering physical, economic, institutional, legal, environmental, social and political aspects, this diversification leads in many cases to partial treatment of the water issues, or incomplete analysis of the various issues at stake. Therefore, treating a whole host of a country's water resources issues in one set of pages will be a significant contribution to scholars, students, and other interested public. This book series is expected to address both the current practice of fragmented treatment of water policy analyses, and the need to have water policy being communicated to all interested parties in an integrated manner but in a non-technical language. The purpose of this book series is to make existing knowledge and experience in water policy accessible to a wider audience that has a strong stake and interest in water resources.The series will consist of books that address issues in water policy in specific countries, covering both the generic and specific issues within a common and pre-designed framework.

Upali A. Amarasinghe · Amarnath Giriraj ·
Hasula Wickremasinghe
Editors

Water Governance for Climate Change Adaptation in Sri Lanka

Status, Challenges, and Policy Options

Springer

Editors
Upali A. Amarasinghe
International Water Management Institute (IWMI)
Battaramulla, Sri Lanka

Amarnath Giriraj
International Water Management Institute (IWMI)
Battaramulla, Sri Lanka

Hasula Wickremasinghe
Climate Change Secretariat, Ministry of Environment
Battaramulla, Sri Lanka

ISSN 2211-0631 ISSN 2211-0658 (electronic)
Global Issues in Water Policy
ISBN 978-3-032-04949-0 ISBN 978-3-032-04950-6 (eBook)
https://doi.org/10.1007/978-3-032-04950-6

This Springer imprint is published by the registered company Springer Nature Switzerland AG
The registered company address is: Gewerbestrasse 11, 6330 Cham, Switzerland

Foreword by Mark Smith

Sri Lanka's ancient wisdom on water governance, embodied in King Parakramabahu I's edict to make every drop of water useful, resonates strongly today. The island's legacy of managing water—from tank cascades to major irrigation systems—offers vital guidance as it faces increasing climate and water-related risks.

Increasing competition for water across agriculture, households, industry, and ecosystems is now being compounded by more frequent floods, prolonged droughts, and erratic rainfall patterns. From 2010 to 2024, Sri Lanka experienced 42 major climate-related disasters, impacting nearly 9 million people and resulting in economic losses exceeding USD 2.5 billion. In Sri Lanka, the impacts of climate change are felt most sharply through water—making bold, transformative adaptation an urgent priority.

Encouragingly, Sri Lanka has mobilized over USD 500 million in climate adaptation investments with support from government, development partners, and civil society. Yet infrastructure alone cannot deliver resilience. The success of these efforts depends on effective water governance—encompassing strong institutions, inclusive policies, legal frameworks, coordination across sectors and scales, and the empowerment of communities to participate meaningfully in decision-making and implementation.

This volume highlights five key governance priorities:

- Strengthening institutional coordination to align actions and avoid overlaps;
- Reforming policies and laws to enable integrated water resource management;
- Harnessing digital innovation to enable water-smart and climate-smart services for adaptive planning;
- Promoting equitable water access for vulnerable groups; and
- Empowering local communities, especially women and youth, in decision-making.

IWMI is proud to support Sri Lanka's journey, contributing digital platforms like the GeoGoviya, Climate Smart Governance Dashboard and the AWARE platform, and partnering on localized advisory services that reach farmers through mobile alerts. These innovations demonstrate how technology and inclusive governance can work together to enhance resilience at scale.

Sri Lanka's approach—rooted in history, responsive to climate realities, and forward-looking—offers valuable lessons for other climate-vulnerable nations. This book serves as both a reflection of national experience and a contribution to global efforts aligned with the Sustainable Development Goals and the Paris Agreement.

We are committed at IWMI to supporting water-secure, climate-resilient futures. This volume provides an actionable path forward, showing how good governance can turn water challenges into opportunities for sustainable development.

Mark Smith
Director General
International Water Management
Institute (IWMI)
Battaramulla, Sri Lanka

Foreword by Leel Randeni

Sri Lanka is an island nation situated in the Indian Ocean. A majority of its 103 rivers originate in the Central Highlands and flow radially outwards into the ocean. The country has a long history of effective water management, exemplified by traditional systems such as ancient tank cascade systems. However, Sri Lanka's future water security will depend both on the availability of natural resources and the effective governance of these resources amidst climate change challenges.

As this timely book underscores, water governance in Sri Lanka stands at a critical juncture. It is increasingly shaped by intensifying climate variability, rapid demographic and land-use changes, and institutional arrangements. The core message is clear: effective adaptation to the adverse effects of climate change in the water sector is not merely a technical challenge; it is also one of institutional and governance reform. Without improved coordination across all relevant stakeholders, clearer land and water rights, more inclusive planning processes, and stronger local institutions, even well-funded adaptation initiatives may fall short of intended outcomes.

This book, *Water Governance for Climate Change Adaptation in Sri Lanka: Status, Challenges, and Policy Options*, presents a framework to reconceptualize water governance as a dynamic, context-specific process—one that must consider power dynamics, knowledge systems, and institutional legacies. It comes at a crucial juncture in policy. As Sri Lanka revises its National Adaptation Plan, updates its Nationally Determined Contributions (NDCs), and introduces Provincial Adaptation Plans encouraging sub-national entities to develop localized resilience strategies, the insights offered here provide a roadmap for evidence-based, locally tailored decision-making. The emphasis on polycentric and participatory governance approaches is particularly relevant for ensuring that adaptation planning reflects both the diversity of water systems and the communities that rely on them.

Moreover, this book makes a valuable contribution to bridging the gap between research and policy. It offers practical tools, institutional insights, and a progressive agenda that will benefit decision-makers, scholars, and development practitioners working to build a climate-resilient Sri Lanka.

We are grateful to the International Water Management Institute (IWMI) for leading the development of these comprehensive chapters, which lay a strong foundation for the future of water governance in Sri Lanka.

Mr. Leel Randeni
Director, Climate Change Secretariat
Ministry of Environment
Battaramulla, Sri Lanka

Acknowledgements

We gratefully acknowledge the financial support provided by the Climate Resilience (CLIMBER) initiative and the Climate Action Program of the CGIAR, and the Ministry of Agriculture, Forestry, and Fisheries (MAFF) of Japan, which enabled the research underpinning this book. We extend our sincere thanks to the Editorial Board Members of *Global Issues in Water Policy*—Prof. Ariel Dinar, Editor-in-Chief, and Drs. José Albiac, Guillermo Donoso, and Stefano Farolfi—for their critical review of the book proposal.

We are especially grateful to Dr. Rathinasamy Maria Saleth for his encouragement and insightful suggestions, which contributed significantly to our involvement in this book series. This volume, a timely contribution to water policy discourse in Sri Lanka, is the result of collaborative efforts by numerous researchers from various Sri Lankan institutions, to whom we express our deep appreciation.

The book was further enriched by the editors' extensive discussions with national and provincial water professionals, as well as other experts and farmers, during the CLIMBER project research in Sri Lanka. We sincerely acknowledge their valuable contributions.

Finally, we would like to thank the International Water Management Institute (IWMI) and the Climate Change Secretariat of the Ministry of Environment for their support in bringing this book to fruition.

Battaramulla, Sri Lanka

Upali A. Amarasinghe
Amarnath Giriraj
Hasula Wickremasinghe

Contents

Contributors

Mohamed Aheeyar International Water Management Institute, Colombo, Sri Lanka

Niranga Alahacoon International Water Management Institute, Colombo, Sri Lanka

Upali A. Amarasinghe International Water Management Institute, Colombo, Sri Lanka;
Department of Zoology and Environmental Management, University of Kelaniya, Kelaniya, Sri Lanka

Saman B. Athauda Department of Animal Science, Faculty of Agriculture, University of Peradeniya, Peradeniya, Sri Lanka

Emmanuel Attoh International Water Management Institute, Colombo, Sri Lanka

S. S. K. Chandrasekara Department of Agricultural Engineering, Faculty of Agriculture, University of Peradeniya, Peradeniya, Sri Lanka

Chandra Embuldeniya Founder and Former Vice Chancellor, Uva Wellassa University of Sri Lanka, Badulla, Sri Lanka;
Past Chairman, Technology Development and Innovations Arm, National Science Foundation, Colombo, Sri Lanka

Surajit Ghosh International Water Management Institute (IWMI), Colombo, Sri Lanka

Amarnath Giriraj International Water Management Institute, Colombo, Sri Lanka

S. W. Goonathilake Department of Civil Engineering and Construction, Skills College of Technology, Nugegoda, Sri Lanka

Upali Imbulana Water Resources Engineer, Colombo, Sri Lanka

Shiromani Jayawardane Climate Change Secretariat, Ministry of Environment, Battaramulla, Sri Lanka

Thilina Madushanka International Water Management Institute, Colombo, Sri Lanka

Herath Manthrithilake International Water Management Institute, Colombo, Sri Lanka

Thilani P. Munaweerage Griffith University, Gold Coast, Australia

Lal Mutuwatte International Water Management Institute, Colombo, Sri Lanka

Erandathie Pathiraja Institute of Policy Studies, Colombo, Sri Lanka

Madhusha Perera International Water Management Institute, Colombo, Sri Lanka

Muditha Prassanajith Perera Department of Geography, University of Peradeniya, Peradeniya, Sri Lanka

P. H. Sarath Gamini Independent Consultant and Former Project Director, National Water Supply & Drainage Board, Kandy, Sri Lanka

Kaushika Seelanatha International Water Management Institute, Colombo, Sri Lanka

Piriyanka Sivananthan International Water Management Institute, Colombo, Sri Lanka

Manoj Thibbotuwawa Institute of Policy Studies, Colombo, Sri Lanka

Sachini Ukwattage International Water Management Institute, Colombo, Sri Lanka

Hasula Wickremasinghe Climate Change Secretariat, Ministry of Environment, Battaramulla, Sri Lanka

Abbreviations

AEZ	Agroecological zones
CBF	Culture-Based Fisheries
CC	Climate Change
CCD	Coast Conservation Department
CEA	Central Environmental Authority
CGIAR	Consultative Group of International Agricultural Research
CHIMP	Coupled Model Intercomparison Project
CMC	Cascade Management Committee
CRIWMP	Climate Resilient Irrigated Water Management Project
CSA	Community Supported Agriculture
CSA	Climate Smart Agriculture
CSG	Climate Smart Governance
CS-GGS	Climate Smart Green Growth Strategy
CSIAP	Climate-Smart Irrigated Agriculture Project
DAD	Department of Agrarian Development
DCS	Department of Census and Statistics
DMC	Disaster Management Center
DS	Divisional Secretariat
EFFRF	Effective rainfall
ENSO	El Niño–Southern Oscillation
Eta	Actual evapotranspiration
FAO	Food and Agriculture Organization
FIM	First Inter monsoon
FO	Farmer Organizations
GCF	Global Climate Fund
GDP	Gross Domestic Product
GN	Grama Niladari
GoSL	Government of Sri Lanka
GSMB	Geological Survey and Mines Bureau
HCR	Headcount ratio
ID	Irrigation Department

IMD	Irrigation Management Division
INMAS	Integrated Management of Agricultural Settlements
IRCWU	Irrigation consumptive water use
IWMI	International Water Management Institute
JVP	Janatha Vimukthi Peramuna (Peoples Liberation Front)
LCGD	Land Commissioner General's Department
LTTE	Liberation Tigers of Tamil Eelam
MASL	Mahaweli Authority, Sri Lanka
MC	Municipal Councils
NAP	National Adaptation Plan
NAP	National Agriculture Policy
NCCP	National Climate Change Policy
NDC	Nationally Determined Contributions
NEAP	National Environment Action Plan
NEM	Northeast monsoon
NEP	National Environmental Policy
NWSDB	National Water Supply and Drainage Board
OPHI	Oxford Poverty and Human Development Initiative
PC	Provincial Council
PS	Pradeshiya Saba
RFCWU	Rainfall consumptive water use
SIM	Second inter-monsoon
SSP	Shared Socioeconomic Pathways
SWM	Southwest monsoon
TCWU	Total consumptive water use
UNICEF	United Nations Children's Fund
UNOCHA	United Nations Office for the Coordination of Humanitarian Affairs
WASSIP	Water Supply and Sanitation Improvement Project
WFP	World Food Programme
WRB	Water Resources Board

List of Tables

List of Figures

Chapter 1
Water Governance for Climate Change Adaptation in Sri Lanka: Overview and Synthesis

Upali A. Amarasinghe, Amarnath Giriraj, and Hasula Wickremasinghe

Abstract This chapter provides a concise overview of water-related climate risks and governance in Sri Lanka, along with a synthesis of the governance challenges associated with managing water resources and adapting to the increasing impacts of climate change across various scales and sectors. The discussion is guided by an analytical framework that evaluates the effects of transformative adaptation interventions designed to address climate change and related disasters. These interventions focus on strengthening water, food, livelihood, energy, and environmental security. The framework comprises six exogenous governance components: (a) water-related institutions, (b) agricultural and the environmental institutions, (c) legal and policy framework, (d) market mechanisms, (e) structural systems, and (f) information, extension, and local support services. The review highlights the urgent need to develop transformative adaptation strategies and to foster institutional collaboration across sectors and governance levels to achieve systemic and sustained resilience outcomes.

Keywords Climate change · Climate risks · Water governance · Impact transmission · Transformative adaptation · Policies · Institutions

---

U. A. Amarasinghe (✉) · A. Giriraj
International Water Management Institute, Colombo, Sri Lanka
e-mail: u.amarasinghe@cgiar.org

A. Giriraj
e-mail: a.giriraj@cgiar.org

H. Wickremasinghe
Climate Change Secretariat, Ministry of Environment, Battaramulla, Sri Lanka
e-mail: hasulaw@yahoo.com

U. A. Amarasinghe et al. (eds.), *Water Governance for Climate Change Adaptation in Sri Lanka*, Global Issues in Water Policy 13,
https://doi.org/10.1007/978-3-032-04950-6_1

1.1 Introduction

> Let not even a drop of water flow into the ocean without being made useful for the benefit of all Earth.
>
> —King Parakramabahu I ruled Sri Lanka from 1153 to 1186.

King Parakramabahu's famous dictum highlights how decisively Sri Lanka's ancient monarchs shaped the island's water governance. Its relevance is even greater today, as growing competition among domestic, agricultural, industrial, and environmental demands strains river-basin supplies already stressed by increasingly erratic weather. Worldwide, climate change is altering rainfall patterns, intensifying extreme events, and raising temperatures—undermining water and food systems and damaging ecosystems (Browder et al., 2021; Walsh & Hallegatte, 2019). Sri Lanka, a tropical island influenced by both the northeast and southwest monsoons, is no exception. Although floods, droughts, and landslides have long been familiar hazards, their frequency and severity have risen markedly in recent decades.

Between 1995 and 2010, Sri Lanka experienced 14 major natural disasters, which affected approximately 3.58 million people and caused an estimated USD 48 million in damages. However, from 2010 to mid-2024, the country faced 42 major disasters, affecting 8.77 million people and resulting in over USD 2.57 billion in losses (EM-DAT, 2024). One notable example of multiple hazards occurred between 2016 and 2017, when a tropical storm triggered severe floods and landslides, disrupting 22 of the country's 25 districts, including Colombo (DMC, 2016; UNOCHA 2016). This was followed by a prolonged drought that began in August 2016, which was exacerbated by weak monsoon rains. The drought led to widespread water shortages, severely affecting agriculture, coastal areas, public health, and ecosystems. By mid-2017, over 848,000 people were impacted, and paddy production had declined by 53% (WFP, 2018). At the same time, heavy rains in the southern districts caused further devastation, affecting 415,000 people, claiming over 200 lives, and inflicting extensive economic damage (DMC, 2017).

Although databases like EM-DAT capture major disasters, many smaller, localized climate-related events, particularly in rural areas, go undocumented. As such events become more frequent, the cost of post-disaster compensation is expected to rise unless Sri Lanka adopts more proactive adaptation and mitigation strategies. The country's continued reliance on reactive responses is increasingly inadequate in the face of escalating climate risks. Notably, the impacts on water-scarce river basins in the northern and eastern regions, which are vital to national food production, highlight the urgent need to accelerate climate adaptation efforts.

In response to these growing challenges, the Government of Sri Lanka (GoSL) has launched large-scale, donor-funded climate adaptation projects totaling over USD 380 million since 2018 (e.g., GCF 2016; World Bank, 2014, 2019). In parallel, central and provincial governments continue to allocate substantial resources toward addressing water and climate challenges. The private sector and non-governmental organizations also contribute to these efforts (https://www.sdb.lk/en/2024/sdb-bank-and-usaid-in-a-landmark-partnership-to-drive-climate-adaptation-in-sri-lanka), although many of their initiatives remain underreported.

Governance plays a crucial role in addressing Sri Lanka's water-related challenges, particularly in the face of increasing climate variability and the need for large-scale adaptation efforts. While some of these initiatives address certain governance aspects, their long-term success and systemic benefits for vulnerable communities hinge on stronger institutional coordination and improvements in legal and regulatory frameworks (Saleth et al., 2022). Given the scale of climate-related investments, it is important to examine water governance across sectors and levels. This need is reflected in national policy documents such as the National Adaptation Plan (NAP) 2016–2025 (GoSL, 2016), Sri Lanka's Nationally Determined Contributions (GoSL, 2021), the National Environmental Policy (GoSL, 2022), National Climate Change Policy (2023), Provincial Adaptation Plans (PAPs) for 2025–2034, the Climate Smart Green Growth Strategy (CS-GGS), and the upcoming NDC 3.0 (2025–2034).

This chapter outlines the key cross-sectoral and multi-scale aspects of governance that are essential for planning effective adaptation strategies. It highlights the importance of aligning governance systems to ensure that interventions yield positive outcomes, particularly in the areas of water, food, and livelihood security. It also connects insights from other chapters in this book, directing readers to explore governance across related sectors in greater detail.

Governance challenges in managing water-related climate risks are complex and multifaceted. Physical interventions, such as enhanced storage capacity or modern irrigation technology, often receive attention, but governance structures (institutions, policies, laws, and regulations) are frequently overlooked. Without robust governance, these technical solutions offer only limited incremental benefits (Saleth et al., 2022). Effective governance requires improved coordination across sectors and scales, enabling transformative interventions that address root causes and foster resilience.

However, transformative change must be supported by stronger institutional coordination while respecting the autonomy of local governance. An example is Sri Lanka's tank cascade system, where fully functioning management committees facilitate cross-sector coordination and support interventions such as crop diversification and improved irrigation (Amarasinghe et al., 2024). Similarly, the "Mahaweli Water Management Panel," chaired by the Director General of the Mahaweli Authority of Sri Lanka and attended by all relevant line departments and District Secretaries, is close to a polycentric governance model that demonstrates successful collaboration across scales and institutions.

Recent climate adaptation projects funded by the World Bank and the Green Climate Fund (GCF) incorporate elements of transformative adaptation, which aim to bring about systemic change through innovative technologies, policies, and institutional reforms (GCF, 2016; World Bank, 2014, 2019). Unlike incremental approaches that merely boost coping capacity, transformative adaptation, combined with robust governance strategies, seeks long-term solutions by altering the structure and function of systems (Fedele et al., 2019). These transformative adaptations addressing disparities across sectors and regions are essential for enhancing systemic resilience.

This chapter summarizes the 15 chapters in this book that address governance challenges in climate adaptation through the lens of water governance in Sri Lanka. The next section of this introductory chapter provides an overview of water-related climate risks, and the third section synthesizes governance challenges highlighted in the following chapters.

1.2 Water-Related Climate Risks

El Niño and La Niña affect seasonal rainfall patterns and agricultural production in Sri Lanka (Fig. 1.1). El Niño and La Niña conditions are the primary factors driving extreme variations in rainfall patterns in the tropics. El Niño–Southern Oscillation (ENSO) is a shift in patterns of oceanic warming and atmospheric circulation in the Pacific Ocean. La Niña is the cold analog of El Niño. El Niño and La Niña conditions appear when the sea surface temperature in the Eastern Pacific Ocean (5°N–5°S, 120–180°W) exceeds or falls below 0.4 °C of the normal sea surface temperature for three consecutive months (Trenberth, 1997).

Figure 1.1 illustrates the annual growth of agricultural GDP and the variation in rainfall during the Maha and Yala seasons. The text within the figure displays the years of floods and droughts associated with El Niño and La Niña events.

Between 1991 and 2022, there were four strong to very strong El Niño conditions, three moderate conditions, and four weak conditions. In the same period, there were four strong, five moderate, and six weak La Niña events (http://ggweather.com/enso/oni.htm). The 2015–2016 El Niño was one of the strongest to occur since the 1997–1998 event. According to the EM-DAT database, 28 of the 32 flood events and seven of the eight drought events that occurred between 1990 and 2022 occurred during El Niño and La Niña events.

Catastrophic agricultural losses (more than one standard deviation above the mean) caused by water-related risks—particularly floods and droughts—can occur approximately once every ten years. Severe agricultural losses (0.33 to 1 standard deviation above the mean) can happen every three to four years, primarily due to El Niño and La Niña conditions (Fig. 1.1). Table 1.1 summarizes these rainfall variations by climatic zone, illustrating, for example, about a 15% increase in early-season rainfall in dry zones during El Niño years. During El Niño years, rainfall is significantly above normal in the first half of the Maha season (October to December) and markedly lower than normal in the second half. In La Niña years, rainfall decreases in the first half of the Maha season and increases in the second half (July to September) of the Yala season.

Climate change is expected to increase both rainfall variability and temperature in Sri Lanka. Many of the country's 103 river basins flowing to the sea from the dry zone already experience little or no flow during the dry season. While this is a natural

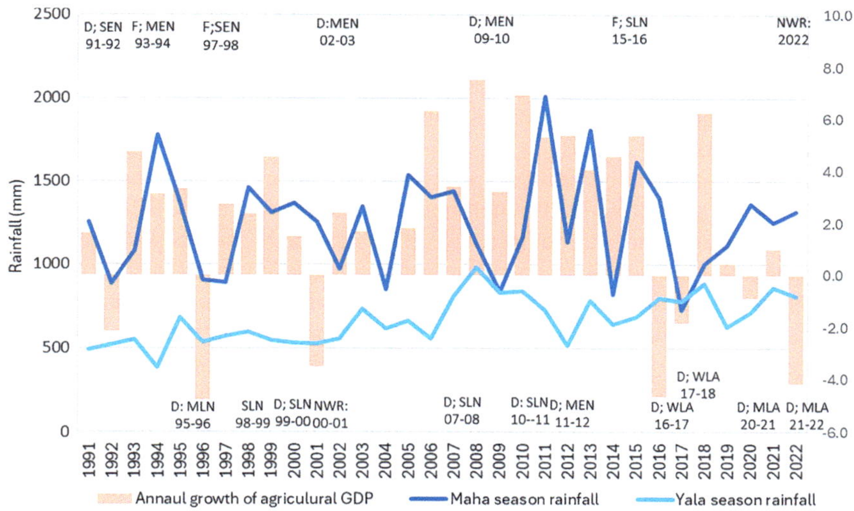

Fig. 1.1 Annual growth of agricultural GDP (bar graph) and variation of seasonal rainfall (line graphs). *Source* World Bank (2024), Department of Meteorology, Sri Lanka, and http://ggweather.com/enso/oni.htm. *Notes* F—major floods; D—major droughts; SEN, MEN—strong and moderate El Niño events; SLN, MLN, and WLN—strong, moderate, and weak La Niña events; NWR—non-water related (internal conflicts or economic conditions)

Table 1.1 Average rainfall in the wet and dry zones in El Niño and La Niña years between 1991 and 2022

Climatic zone	Seasons	Months	Average 1980–2015	El Niño years	% Change	La Niña years	% change
Dry zone	Maha	ONDJFM	1036	1046	0.9	1096	5.8
		OND	743	854	14.9	694	−6.6
		JFM	293	192	−34.7	402	37.2
	Yala	AMJJS	404	401	−0.9	384	−5.0
		AMJ	236	249	5.2	209	−11.7
		JAS	168	152	−9.5	175	4.4
Wet zone	Maha	ONDJFM	1306	1350	3.4	1343	2.8
		OND	920	1046	13.7	841	−8.5
		JFM	386	304	−21.3	501	29.8
	Yala	AMJJS	1473	1419	−3.6	1559	5.9
		AMJ	838	798	−4.8	889	6.1
		JAS	635	622	−2.1	670	5.5

Source Authors' estimation based on Meteorological Department data
Notes JFM: January, February, and March; AMJ: April, May, and June; JAS: July, August, and September; OND: October, November, and December

condition for some basins, it is increasingly the result of a mismatch between water demand and availability. Growing urban and industrial demand will likely intensify water stress, particularly in closed basins that are unable to meet the additional water commitments required for domestic, industrial, agricultural, and environmental use throughout the year.

With only limited opportunities to develop new large surface-water storage, climate-induced rainfall variability will heighten agricultural risks. Effective water storage and management—through surface reservoirs, tanks, groundwater recharge, or soil moisture conservation—will become increasingly vital as the climate continues to change. These are critical for agricultural production in the dry zone, which is especially vulnerable to droughts, and are crucial for national food security.

Droughts in the dry zone are most severe when rainfall during the Maha season is substantially below normal. Such events have occurred under both El Niño (1987, 1988) and La Niña (2001, 2012) conditions. During El Niño years, excess rainfall in the early Maha season often exceeds the dry zone's storage capacity, while the latter half of the season and the following Yala season bring reduced rainfall, leading to drought. Under La Niña, the dry zone receives lower rainfall from the start of the Maha season through the Yala season, similarly increasing drought risk. In both cases, improved soil and water conservation practices can help reduce these risks.

Conversely, floods can also impact dry zone agriculture when rainfall during the Maha season is abnormally high. This usually occurs from October to December under El Niño conditions and from January to March under La Niña conditions. Effective agricultural planning and the use of improved seed varieties play a crucial role. Together with climate-resilient infrastructure and practices, these strategies can enhance the resilience of dry zone agriculture to both droughts and floods in the context of increasing climate variability. Water governance is an essential intervention in this process.

1.3 Governance Challenges in Managing Climate Risks

1.3.1 Analytical Framework for Assessing Governance Challenges

The governance challenges associated with adapting to climate change impacts are multidimensional. In the realm of irrigation water management, infrastructural interventions—such as increasing water storage and deploying advanced irrigation techniques, including drip, sprinkler, and underground pipe systems—often receive substantial attention. However, other governance components, including water-related institutions, policies, rules, and regulations, as well as their interactions with governance issues in other water use sectors, are frequently overlooked in risk mitigation strategies.

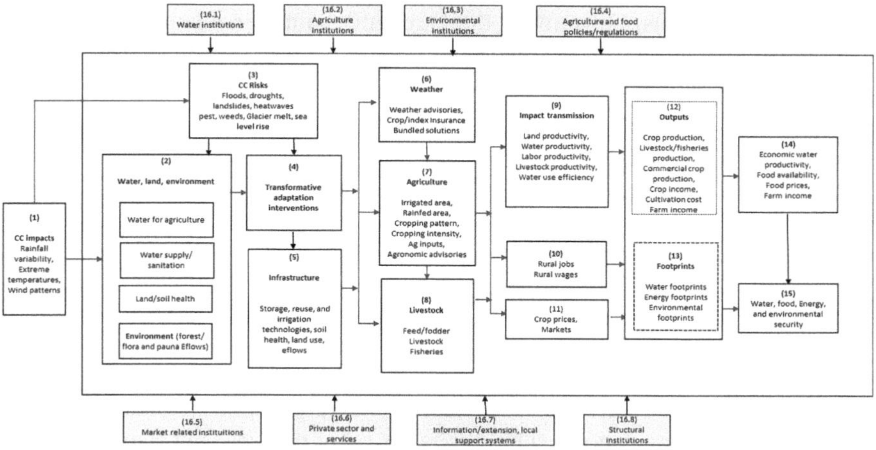

Fig. 1.2 Climate change, transformative adaptation interventions, and impact transmission factors to enhance water, food, livelihood, and environmental security. *Source* Authors, based on the climate change adaptation analytical framework of Saleth et al. (2022)

Neglecting these governance elements often limits the effectiveness of physical interventions, resulting in only incremental benefits (Saleth et al., 2022). In contrast, improved coordination of governance across multiple scales and sectors enables the development of effective, transformative adaptation strategies. These strategies generate systemic impacts that not only mitigate climate change risks but also enhance overall resilience.

Addressing multi-scale and cross-sectoral governance challenges is essential for the planning and implementation of transformative adaptation interventions. Such interventions support sustainable outcomes from the sequence of actions or events that follow, contributing to greater water, food, livelihood, energy, and environmental security, ultimately strengthening climate resilience (Fig. 1.2).

In Fig. 1.2, increased rainfall variability, higher temperatures, and altered wind patterns are examples of climate change impacts (Box 1). These changes have direct impacts on water, land, and ecosystems, as well as indirect effects through climate-related risks such as floods, droughts, and landslides. Transformative adaptation interventions address these effects across multiple domains, including climate, water, agriculture, livestock, and the environment.

Impact transmission factors, such as productivity, rural employment, and market dynamics, influence agricultural output, crop production, and farm income, while also reducing water footprints, environmental degradation, and ecological footprints. These improvements lead to higher economic water productivity, increased food availability, higher farm incomes, and lower food prices, which collectively contribute to water, food, energy, and environmental security.

Throughout the impact pathway, the governance components—which are often overlooked—are essential for achieving sustainable improvements in productivity, outputs, and footprints. Collectively, they strengthen climate resilience by enhancing water, food, energy, and environmental security.

The governance components under consideration include:

- **Water-related institutions**—such as irrigation departments, the Mahaweli Authority, water user associations, and farm organizations; as well as water infrastructure, including storage systems and the water supply and distribution network.
- **Agriculture- and environment-related institutions**—such as the Ministry of Agriculture, the Departments of Agriculture, Forests, and Wildlife, agricultural credit and investment institutions, agricultural input suppliers, science and technology research and extension departments, and climate information and decision support systems.
- **Legal and policy-related institutions**—encompassing price regulation policies, water laws and regulations, subsidy and tax relief policies, and social safety net policies.
- **Market-related institutions**—including agricultural value chains, production and marketing corporations, market price monitoring, and crop and employment insurance schemes.
- **Private sector institutions**—including international donors, investors, and corporate sector agencies.
- **Information, extension, and local support systems**—including decision support systems, extension services, and rural service providers.
- **Structural institutions**—including land tenure systems and customary institutions.

We utilize the above framework to synthesize the governance challenges identified throughout the rest of the chapters in this book, which is structured into four sections comprising 15 chapters that examine governance mechanisms across various sectors and scales.

Section 1.1 has four chapters. Chapter 2 examines spatial variations in renewable water supply, usage, and depletion (consumptive water use) across various sectors in Sri Lanka, including agriculture, domestic, industrial, and environmental sectors. It also discusses potential water scarcity, both with and without environmental flow allocations, as well as the challenges of meeting future water needs in the face of climate change. Chapter 3 provides an overview of Sri Lanka's physical setting, as well as the evolution of its political, socioeconomic, and institutional contexts. Chapter 4 synthesizes climate change projections for Sri Lanka, highlighting their implications for the water sector and climate adaptation. Chapter 5 explores climate variability-induced disasters, along with the related vulnerabilities and risks.

Section 1.2 comprises two chapters that focus on the National Adaptation Plan (NAP), current governance, and the challenges of NAP implementation. Chapter 6 outlines the current state of governance and the challenges faced in water resources management. Chapter 7 summarizes the NAP and Nationally Determined Contributions (NDCs), with a focus on the water-related difficulties associated with their implementation.

Section 1.3 of the book comprises four chapters that address water governance challenges and climate change adaptation across various water-use sectors, including irrigation, fisheries, livestock, domestic, industrial, and power sectors. Chapter 8 identifies governance challenges in the major and medium canal irrigation sector, as well as groundwater use. (Major and medium irrigation systems are those with culturable command areas of more than 80 hectares, where reservoirs and major diversions serve as the primary sources of water. Groundwater use, especially in and around these irrigation areas, is primarily recharged by canal irrigation and rainfall.) Chapter 9 assesses governance challenges in the minor irrigation sector, including small tanks and tank cascades, which are widely distributed, particularly in the dry and intermediate zones. These systems play a substantial role in mitigating the impacts of floods and droughts. Chapter 10 examines water governance issues in the fisheries and livestock sectors, which are increasingly recognized as helpful in building resilience in rural populations. Chapter 11 discusses governance challenges in non-agricultural sectors, including domestic, industrial, and power sectors.

Section 1.4 of the book comprises five chapters that address the policy, institutional, and technical dimensions of climate change adaptation and water governance challenges. Chapter 12 discusses the challenges within environmental and agricultural policy and action plans. Chapter 13 focuses on land use policy, specifically land and water conservation strategies aimed at mitigating erosion through agroforestry and other agricultural practices. Chapter 14 explores traditional water management practices and the governance challenges they present. These practices are gaining renewed attention for their potential to contribute to the sustainable management of water resources. Chapter 15 identifies the climate information and service needs essential for effective climate change adaptation. Many large adaptation projects prioritize enhancing climate information services, as they are crucial for optimizing reservoir and tank operations and improving on-farm water management. Chapter 16 concludes the book by outlining future research needs related to water governance and climate change. The following sections synthesize the governance challenges in managing climate risks identified throughout the book.

1.3.2 Climate Change Impacts and Disaster Risks

Climate change impacts (see Box 1; Fig. 1.2) directly affect water, food, and environmental systems, while related disasters (Box 3) indirectly affect these systems.

Climate variability is expected to increase with climate change. Projections show an increasing frequency of drier “dry” periods and wetter “wet” periods (Chap. 3). Drier dry periods will exacerbate droughts, and wetter wet periods will increase the frequency of floods.

Due to excessive water depletion relative to available resources, many river basins already experience significant seasonal water scarcity, especially during the dry season (Chap. 2). These basins include those draining to the sea from the eastern, northern, and northwestern parts of the country. Additionally, major basins such as Yan Oya, Kala Oya, and Malwathu Oya, as well as several smaller basins, experience year-round severe water stress. If environmental flow allocations are also considered, these basins experience very severe water stress. Without augmenting the water supply from outside, climate change impacts can only exacerbate water stress in these basins.

Increased industrial activities are already creating water quality issues in major river basins (Chap. 11). Increasing water scarcity and quality issues threaten water, food, livelihood, and environmental security in some river basins. Since these basins are primarily located in the dry zone, climate change is expected to exacerbate the impacts of water scarcity further.

Projected temperature increases and changes in wind patterns may increase evaporation and transpiration rates (Chap. 3), which would negatively affect surface storage, reduce surface runoff, and impair groundwater recharge. These changes in runoff and storage may increase conflicts between agricultural and other sectors (domestic, industrial, and power) over reservoir releases. Increased rainfall variability and temperature fluctuations will also impact crop yields.

Therefore, all governance components, particularly those focused on water, should play a crucial role in mitigating the vulnerabilities associated with water risks (scarcity and quality issues) and agricultural risks, which are likely to be exacerbated by climate change (Chaps. 2 and 6). Properly designed adaptation measures should immediately target the vulnerable hotspots identified in Chap. 3, especially those affecting women, children, youth, and older people. These interventions can alleviate factors contributing to multidimensional poverty (as highlighted in Chap. 3), especially the most significant ones, such as lack of access to gas cooking fuel and clean drinking water. These deficiencies primarily affect women and children, who spend a considerable amount of time collecting firewood and fetching water.

1.3.3 Water Infrastructure: Storage and Water Conveyance Facilities

The major and minor irrigation systems contribute significantly to food production, particularly of staple foods such as paddy. Therefore, water storage and conveyance facilities play a crucial role in mitigating climate change risks and water scarcity.

Small tanks, tank cascades, and major and medium reservoir storages can protect against floods in the wet season and meet water needs during dry periods and droughts. These storages benefit multiple sectors, supporting crops (Chaps. 2, 6 and 8), livestock and inland fisheries (Chap. 11), and domestic and industrial water supply (Chap. 12) during times of water scarcity (Chaps. 6, 8, 9, 11 and 12).

Major and medium reservoir storages in the dry zone have immeasurable value during extended droughts, as these storages recharge groundwater resources that support drinking water supplies from wells, as well as agricultural and industrial water needs through increased groundwater use (Chap. 8). The NAP and NDCs, as well as climate, agricultural, and environmental policies, have underscored the importance of small tanks, tank cascades, reservoir storage, and groundwater in managing future disaster risks associated with climate change (Chaps. 6 and 10).

Lakes and wetlands support invaluable ecosystem services. Many of the national parks that span the country are key attractions for both domestic and international tourists, a fast-growing sector of the economy.

Soil moisture storage is an essential resource for agriculture. Increased soil moisture storage would enhance rainfall-use efficiency in all crop areas and reduce irrigation requirements in irrigated areas (Chap. 2). This is particularly important for most highland and plantation crops, as rainfall is the primary source for meeting crop water requirements. These crops play a crucial role now and in the future in enhancing the food, nutritional, and livelihood security of rural, agriculture-dependent populations.

Moreover, high-value crops—such as plantation crops, spices, and vegetables—are already making substantial contributions to agricultural GDP and exports. Their role is expected to expand in the future, further enhancing economic water productivity, rural incomes, livelihoods, and nutritional security.

The dilapidated state of conveyance infrastructure, including main canals, distributaries, and field channels, poses a significant constraint to the introduction of innovative water management techniques. These include canal automation and bulk water supply, which require a reliable water supply delivered according to agreed-upon schedules (Chap. 8).

Additionally, the lack of accurate flow measurement devices in canals and the absence of effective climate information systems to assess supplementary irrigation supply are significant constraints for operations and management fee collection, as farmers lack confidence in the water supply and the duty (water allowance) estimated for their farms.

Groundwater is a resource that can mitigate the impacts of drought in and around canal command areas. There is substantial groundwater recharge from reservoirs, tanks, and canal flows. Within command areas, groundwater can serve as a supplementary supply to meet water deficits in critical crop water requirements during normal rainfall years. It can also be a primary irrigation source in drought years.

Groundwater use should be encouraged, but proper policies must be introduced to prevent overexploitation of the resource and to safeguard water levels in nearby wells that supply drinking water (Chaps. 2 and 8). Increased groundwater usage would contribute to equitable water use in canal irrigation systems.

1.3.4 Water Institutions and Coordination with Other Institutions

There is a severe shortage of technically skilled personnel to manage canal water systems. The existing technical staff have many non-technical responsibilities and find little time to examine technical failures or implement improvements required to operate systems at optimal levels (Chaps. 8 and 9).

More than 40 institutions manage water across different sectors and scales, with many overlapping mandates. With no coherent water policy in place, the water sector faces numerous conflicts and accountability issues. With increasing climate variability and scarcity, a coherent water policy and a policy action plan outlining the responsibilities and accountabilities of various institutions are top priorities (Chaps. 6, 7, 8 and 9).

Improving rainfall and irrigation water-use efficiency can mitigate the scarcity risks that many river basins currently face and that are expected to intensify with future climate change. Moreover, drought-tolerant, short-duration, high-yielding crop varieties can be used to reduce consumptive water use and enhance irrigation water-use efficiency (Chaps. 8 and 9). Although these innovations have water-centric benefits, they require substantial support and coordination not only from the agricultural research and development institutions that developed these varieties, but also from agricultural extension departments that provide training and improved agronomic practices, from the Meteorological Department, which offers reliable local-level weather forecasts, and from the private sector to increase seed production and distribution to suit different agro-ecological regions of the country.

The tank cascade system is a robust first line of defense against floods and provides relief during droughts. However, the interconnectivity of tanks in most tank cascade systems is currently dysfunctional; these tanks often operate without a proper understanding of the benefits of return flows from one tank to another (Chap. 9). Moreover, the farmer organization is the dominant decision-maker within a tank command, the fisheries organization coordinates fishing activities in the tank, and the women's organization primarily manages cultivation in the homestead, which benefits from groundwater recharged by the tank's water resources (Chap. 10). The water governance challenge is to develop an institutional mechanism that ensures an equitable water supply and shared benefits for all groups, including fisheries and women, both within a single tank and between tanks.

1.3.5 Policies, Rules, and Regulations

Increasing water productivity is a potential response in areas where water scarcity is a major issue. While physical water productivity—measured as "more crops per drop"—enhances production, economic water productivity—measured as "more value per drop"—enhances crop income (Chaps. 2 and 8). There is also potential to reduce water depletion. However, past policies and regulations constrain crop diversification in major irrigation systems. These policies, laws, and regulations must be addressed to support crop diversification in canal commands.

Fragmented farmlands, many of which are smaller than one acre, constrain the introduction of efficient water management techniques in Sri Lanka's agriculture sector (Chap. 13). Laser land leveling, underground pipe irrigation, and agricultural mechanization have significant potential for enhancing water-use efficiency and improving water productivity. Yet, operationalizing these technologies requires larger farm parcels, which presents a considerable governance challenge due to increasing fragmentation resulting from population growth.

Unclear land ownership and overlapping formal and traditional tenure systems hinder farmers from making long-term investments to improve irrigation and land-use practices. These factors are barriers to enhancing water-use efficiency, improving soil health, and preventing land degradation. Land ownership also grants individuals the ability to engage in practices that have long-term effects on water quality, soil health, and downstream ecosystems. Governance reforms should prioritize formalizing land titles and integrating traditional landholding systems into the official legal framework, along with land-use regulations, to prevent unsustainable practices that threaten land, water, and ecosystem services.

1.3.6 Market-Related Institutions

Underdeveloped value chains and inadequate corporate sector participation are major challenges to productive agriculture in rural areas. Access to seeds, agronomic inputs, and markets is a significant constraint that farmers currently face. Policy uncertainty and political interference also hinder private sector investments in public–private partnerships, particularly in the agriculture sector and more specifically in adaptation interventions.

1.3.7 Information Systems, Extension Services, and Local Support Systems

A lack of proper information systems is a major challenge in mitigating disaster risks. It is imperative that decision support systems, such as the AWARE platform (Chaps. 4 and 15), be available to anticipate and plan for disasters before they occur. These early warning systems should be available at all scales and across all sectors, and they require institutional collaboration to achieve optimal results.

A lack of advanced decision support systems, such as the Climate Smart Governance (CSG) Dashboard (Chap. 4), is a constraint to prioritizing and monitoring adaptation investments for vulnerable communities. These systems should help identify hazards that affect vulnerable locations, document the types of past adaptation investments made, and monitor their impacts. They help reduce duplication of adaptation efforts and streamline investments to address multiple hazards that affect many vulnerable locations across the country.

Local communities require implementation and maintenance support to operationalize and benefit from advanced, high-tech climate-smart adaptation interventions. Many of these communities are in rural areas with limited access to support systems available in urban centers. Properly trained and equipped rural service providers could accelerate the upscaling of high-tech adaptation technologies.

1.3.8 Structural Institutions

Unclear land ownership and overlapping formal and traditional tenure systems hinder farmers from making long-term investments to improve irrigation and land-use practices. These factors serve as barriers to enhancing water-use efficiency, improving soil health, and mitigating land degradation (Chap. 13). The challenge is to develop mechanisms to consolidate small farms into larger units to achieve hydro-economic benefits of scale.

Traditional water management practices and the maintenance of irrigation tanks are integral to tank cascade systems. However, given the individualistic operation of these tanks at present, the governance challenge is to organize communities to manage tank cascades collectively and to maintain traditional systems such as Kuluwewa and Olagam Wewa (different forms of tanks), as well as traditional land management systems like the Bethma system and the Puranawela–Akkarawela systems (Chap. 14).

Kuluwewa is a small tank situated in the immediate catchment of medium- to large-scale tanks, where it traps sediments that reach it through ephemeral streams. *Olagam wewa* is a tank built in forest areas to cultivate crops only during the Maha season, as the village tanks cannot meet the food deficit caused by the expanding population. These tanks are left for wildlife to use during the Yala (dry) season.

The Bethma system, used in periods of scarce water availability, shares water equitably among all farmers by cultivating only a portion of the command area—either in a selected canal, near a tank, or on the most fertile land within the command area.

1.4 Conclusions

Sri Lanka faces numerous water governance challenges, including increasing weather variability associated with climate change. The country must prioritize investments to address these governance challenges, as they are critical to ensuring water, food, and environmental security. This prioritization is crucial as the country has been recovering from severe economic challenges since 2022. Some solutions proposed in different chapters (summarized in Chap. 16) to address these governance challenges will require investments. In contrast, others require minimal financial investment and instead advocate for changes in policies, laws, and regulations, as well as enhanced coordination among existing institutions. These measures alone can help change attitudes, enhance corporate sector participation, and increase private investment in improving water-use efficiency and water productivity.

Future climate adaptation measures supported by government and donor investments should consider transformative adaptation interventions that shift the focus from achieving short-term incremental benefits to attaining long-term sustainable outcomes. These transformative measures, backed by improved coordination among institutions and complemented by coherent policies and laws, will lead to systemic impacts and help ensure water, food, livelihood, and environmental security, thereby enhancing resilience to climate change.

References

Amarasinghe, U. A., Ekanayake, C., Amarnath, G., & Alahacoon, N. (2024). *Improving water governance in climate adaptation for Sri Lanka's tank cascade systems.* Policy Brief. International Water Management Institute, Colombo, Sri Lanka.

Browder, G., Nunez Sanchez, A., Jongman, B., Engle, N., van Beek, E., Castera Errea, M., & Hodgson, S. (2021). *An EPIC response: Innovative governance for flood and drought risk management.* © World Bank. http://hdl.handle.net/10986/35754

DMC. (2016). *Situation Report—Floods (Special Issue).* Retrieved June 2017, from http://www.dmc.gov.lk/attachments/Situation_19May_2016-1.pdf

DMC. (2017). *Situation Report—Drought.* Retrieved June 2017, from http://www.dmc.gov.lk/index_english.htm

GCF (Green Climate Fund). (2016). *Strengthening the resilience of smallholder farmers in the dry zone to climate variability and extreme events through an integrated approach to water management.* https://www.greenclimate.fund/sites/default/files/document/funding-proposal-fp016-undp-sri-lanka.pdf

GoSL. (2016). *National adaptation plan for climate change impacts in Sri Lanka 2016–2025.* Climate Change Secretariat (CCS), Ministry of Environment, Government of Sri Lanka, May 2016. ISBN: 978-955-0033-98-0. http://climatechange.lk/NAP/NAP%20For%20Sri%20Lanka_2016-2025.pdf

GoSL. (2021). *Sri Lanka's updated NDCs.* Ministry of Environment, Government of Sri Lanka, July 2021. https://www.climatechange.lk/CCS2021/UpdatedNDCsSriLanka2021.pdf

GoSL. (2022). *National environmental policy.* Ministry of Environment, GoSL, June 2022. http://env.gov.lk/web/images/pdf/policies/National_Environment_Policy_-_English.pdf

Fedele, G., Donatti, C. I., Harvey, C. A., Hannah, L., & Hole, D. G. (2019). Transformative adaptation to climate change for sustainable social-ecological systems. *Environmental Science & Policy, 101*, 116–125.

Saleth, R. M., Amarasinghe, U., Amarnath, G., Ait El Mekki, A., & Brouziyne, Y. (2022). *Climate change, transformative adaptation options, multiscale polycentric governance, and social welfare: Towards an empirical evaluation in the context of Oum Er Rbia Basin, Morocco.*

Trenberth, K. E. (1997). The definition of El Niño. *Bulletin of the American Meteorological Society, 78*(12), 2771–2777.

UNOCHA (United Nations Office for the Coordination of Humanitarian Affairs). (2016). *Sri Lanka: Floods and Landslides Situation Report No. 1 (as of 22 May 2016).* Retrieved June 2017, from http://reliefweb.int/report/sri-lanka/sri-lanka-floods-and-landslides-situation-report-no-1-22-may-2016

Walsh, B., & Hallegatte, S. (2019). *Socioeconomic resilience in Sri Lanka: Natural disaster, poverty, and wellbeing impact assessment.* Policy Research Working Paper No. 9015. © World Bank. http://hdl.handle.net/10986/32423

WFP (World Food Programme). (2018). *Emergency assistance to the most vulnerable drought-affected households.* Operations Document. Retrieved February 2018, from http://www1.wfp.org/operations/201072-emergency-assistance-most-vulnerable-drought-affected-households
World Bank. (2014). *Sri Lanka—Improving Climate Resilience Project.* World Bank Group, Washington, D.C. http://documents.worldbank.org/curated/en/256411468334291922/Sri-Lanka-Improving-Climate-Resilience-Project
World Bank. (2019). *Climate Smart Irrigated Agriculture Project (P163742).* https://documents1.worldbank.org/curated/en/099061223005066914/pdf/P16374209fc67b06e09a9300f711645bd07.pdf
World Bank. (2024). *World Development Indicators.* Retrieved 2024, from http://databank.worldbank.org/data/home.aspx

Dr. Upali A. Amarasinghe Emeritus Scientist at the International Water Management Institute, has over 33 years of experience as a researcher and project leader in water sources management, as well as ten years of teaching experience as a university lecturer. He holds a Ph.D. degree in Statistics from the University of Illinois, Urbana-Champaign, USA. His research focused on water management issues at different spatial scales; which included estimating water productivity and footprints of crops to river basins; food and water supply and demand and projections of river basins to countries; benchmarking of performance of canal irrigation systems; poverty and vulnerability assessment; poverty mapping using small area estimation; water security assessment of industries; impact assessments of interventions in water resources management. He has extensive experience managing research projects in India and Sri Lanka, as well as short-term work experiences in other countries, including Pakistan, China, Malaysia, Thailand, and Vietnam. Dr. Amarasinghe has published over 130 research articles and edited four books.

Dr. Amarnath Giriraj has a Ph.D. in Remote Sensing. He is the Research Group Leader for Water Risks and Disaster at IWMI, Sri Lanka, a Steering Committee member of Sentinel Asia by JAXA, and Chair of the Working Group on Water-related Disasters at the Asia Pacific Space Agency Forum. Dr. Giriraj's primary expertise includes disaster risk assessment, hydrological process research, agricultural disaster risk management, climate services, and food security, with a focus on developing solutions for pressing climate risk challenges. Many of his research and development projects have demonstrated the impact of climate risks, utilizing advanced science and technology in various hydro-climatic regions worldwide. He is keen on translating science-based knowledge into effective policies and strategies contributing to environmental, economic, and societal sustainability. He established the Water Risks and Disasters mission at IWMI and developed regional and national water-risk data systems and reports to inform policy and decision-makers on key disaster, agricultural, and water challenges and solutions.

Ms. Hasula Wickremasinghe works at the Ministry of Environment. She has worked at the Biodiversity Secretariat for ten years and coordinated the preparation of the National Red List (fauna groups) in 2012. She has worked on the Mitigation and Adaptation sectors at the Climate Change Secretariat for the past seven years. She assisted in submitting the Updated Nationally Determined Contributions (NDCs) in 2021 and is currently involved in developing the NDC Implementation Plan. She was also instrumental in designing a Climate Smart City for Kurunegala and assisted in implementing the National Adaptation Plan Readiness Support Project. Hasula has a degree in Biosciences from the Open University of Sri Lanka, an MSc in Environment Sciences from the University of Colombo, and a Postgraduate Diploma in Archaeology from the University of Kelaniya.

Part I
Water Sector in Sri Lanka: Physical, Political, and Socioeconomic Setting

Chapter 2
Temporal and Spatial Patterns of Water Supply and Consumption in Sri Lanka

Upali A. Amarasinghe, Madhusha Perera, Niranga Alahacoon, and Thilina Madushanka

Abstract This chapter examines spatial and temporal variations in water supply and consumption across 103 river basins in Sri Lanka. Water accounting of the surface runoff, a crucial initial step in evaluating current water resource use, shows that several basins in the dry zone are already experiencing seasonal or year-round water stress. In these areas, 12 total process water depletion exceeds 60% of the utilizable water resources. Process water depletion encompasses the evaporation and transpiration (or consumptive water use) from the irrigation, domestic, and industrial sectors, as well as non-process evaporation from water bodies. Many river basins in the northern regions face severe seasonal water stress during the dry season. This stress intensifies when environmental flows are allocated from the utilizable water resources. The increasing water shortages underscore the necessity of providing accurate weather predictions, enhancing rainfall and irrigation water use efficiency, improving storage management, promoting crop diversification, and other measures to address the challenges posed by population growth, economic development, environmental flow allocations, and climate change.

Keywords Sri Lanka · River basins · Water supply · Consumptive water use · Spatial variation · Water stress

---

U. A. Amarasinghe (✉) · M. Perera · N. Alahacoon · T. Madushanka
International Water Management Institute, Colombo, Sri Lanka
e-mail: u.amarasinghe@cgiar.org

M. Perera
e-mail: m.perera@cgiar.org

N. Alahacoon
e-mail: n.alahacoon@cgiar.org

T. Madushanka
e-mail: t.madushanka@cgiar.org

U. A. Amarasinghe et al. (eds.), *Water Governance for Climate Change Adaptation in Sri Lanka*, Global Issues in Water Policy 13,
https://doi.org/10.1007/978-3-032-04950-6_2

2.1 Introduction

Sri Lanka encompasses 103 river basins that span a total land area of 65,610 km^2 and generate an annual surface runoff of approximately 52 billion cubic meters (Bm^3). In 2020, this equated to about 2,438 cubic meters of surface runoff per person. With an increasing population, per-person runoff will decrease to approximately 2,230 m^3 per person by 2050. Despite this reduction, the country's per capita surface runoff will remain above the widely accepted national water scarcity threshold of 1,700 m^3 per person (Falkenmark & Lundqvist, 1998).

The monsoonal rainfall patterns have a significant influence on the spatial and seasonal variations in water supply across Sri Lanka. The wet zone districts in the central, western, and southwestern regions, comprising only 25% of the country's land area, contribute 51% of the annual surface runoff, with over 80% occurring between April and September. The 12 major perennial rivers, which cover 46% of the land area, generate as much as 72% of the total surface runoff. Consequently, large portions of the dry and intermediate zones, which include many non-perennial river basins, experience severe water stress (Amarasinghe et al., 1999).

In addition to limited water availability, water use patterns contribute to seasonal and regional water stress (Amarasinghe et al., 2010). In 2000, over 90% of water withdrawals were for the agricultural sector, primarily in the dry zone. A significant portion of these withdrawals meets the water requirements of paddy cultivation. The dry zone, which accounts for most of the irrigated paddy area, made up 90% of the total withdrawals, despite having only 44% of the total population. This limited water availability and high irrigation demand make the dry zone particularly vulnerable to seasonal and annual water stress.

The growing population and increasing food demand necessitate higher agricultural water use across various regions. The expanding domestic and industrial sectors also elevate the demand for water resources. These two sectors now account for approximately 90% of the gross domestic product, compared to a significantly smaller share in the early 1990s. The rising water demand from non-agricultural sectors further puts pressure on the available water resources. If environmental water needs, such as minimum river flows, are also considered, many river basins may face severe water stress.

This chapter examines process consumptive water use (PCWU), also known as evapotranspiration, across different sectors and the supply and demand gaps that contribute to regional water stress. The analysis also assesses water stress in river basins and explores the challenges of managing these pressures amid increasing weather variability due to climate change, as well as demographic and lifestyle changes.

Following this introduction, the next section outlines the methodology and data used to estimate the CWU and water stress of river basins. Section three presents the temporal and spatial variations in water supply across river basins, including rainfall and surface runoff. Section four discusses the CWU of different sectors and the variations across river basins. Section five explores the water stress of river basins with and without environmental flows. The chapter concludes by discussing the challenges of climate adaptation in relation to water demand and supply, as well as the existing gaps in river basins.

2.2 Methodology and Data

2.2.1 Agriculture CWU Estimation

In agriculture, the total consumptive water use (TCWU) represents evapotranspiration from the cropped area during production, which is the water depleted from the system and unavailable for reuse during the production process. The TCWU comprises two components: evapotranspiration from the effective rainfall and irrigation. We refer to these two components as rainfall CWU (RFCWU) and irrigation CWU (IRCWU), respectively. In the agricultural sector, TCWU refers to the total crop water requirement of crops, which is the sum of the actual evapotranspiration (ETa) of four growth stages: initial, development, middle, and late.

For crop j

$$\mathrm{TCWU}_j = \sum_{k=1}^{4} \mathrm{ETa}_k = A_j \sum_{k=1}^{4} C_{jk} \times \sum_{\mathrm{months} \in l} \mathrm{ETP}_{jkl} \times n_{jkl} \tag{2.1}$$

where A_j is the cropped area, C_{jk} is the crop coefficient of the kth growth period, and n_{jkl} and ETP_{jkl} are the number of days and daily potential evapotranspiration (mm/day) of the lth month in the kth growth period.

For irrigated crops, the IRCWU is the difference between TCWU and RFCWU, which is the effective rainfall at the rootzone. i.e., for crop j,

$$\mathrm{IRCWU}_j = \mathrm{TCWU}_j - \mathrm{RFCWU}_j \tag{2.2}$$

where the RFCWU_j is the effective rainfall EFFRF_j of the crop area

$$\mathrm{RFCWU}_j = \mathrm{EFFRF}_j = A_j \sum_{k=1}^{4} \sum_{\mathrm{months}\ l} \frac{n_{jkl}}{N_{kl}} \mathrm{EFFRF}_{jkl} \tag{2.3}$$

And EFFRF_{jkl} and N_{kl} are the effective rainfall and the number of days of the lth month of the kth growing period. The EFFRF_{jkl} is estimated from the United States Geological Survey method and is given by

$$\mathrm{EFFRF}_{kl} = \begin{cases} \mathrm{RF}_{kl} \times (125 - 0.2\,\mathrm{RF}_{kl})/125 & \text{if } \mathrm{RF}_{kl} \leq 250\,\mathrm{mm} \\ 125 + 0.1 \times \mathrm{RF}_{kl} & \text{if } \mathrm{RF}_{kl} > 250\,\mathrm{mm} \end{cases} \tag{2.4}$$

For rainfed crops, the RFCWU is the minimum of TCWU and the effective rainfall at the root zone, i.e., for crop *j*,

$$\mathrm{TCWU}_j^{RF} = \min\{\mathrm{TCWU}_j, \mathrm{EFFRF}_j\} \tag{2.5}$$

The CWU of a district is the sum of the total CWUs from agriculture, domestic, and industrial sectors.

$$\mathrm{CWU}_j = \sum_{j=1}^{J} \mathrm{TCWU}_j + \mathrm{CWU}_{\mathrm{domestic}} + \mathrm{CWU}_{\mathrm{Industrial}}$$

TCWU of agriculture in a river basin is estimated as the weighted average of the DS divisions area that intersect a river basin, as weights, i.e., the CWU of crop *j* in the river basin is

$$\mathrm{CWU}_j^t = \sum_{l=1}^{L}\left(\sum_{j=1}^{J} \mathrm{TCWU}_{jl}^t\right) \times \frac{A_l}{A} \tag{2.6}$$

where A is the river basin area, and A_l is the area common to the river basin and the lth district.

2.2.2 Domestic and Industrial CWU Estimation

Official records from the Water Supply and Drainage Board (2024) indicate that piped water supply is available in the industrial and domestic sectors. However, only 33.8% of the population has access to a piped water supply in the domestic sector, with district-level access ranging from 0.2 to 77.8%. To estimate total domestic water withdrawals in each district, we assume that individuals without a piped water supply use 50 L per day per person for domestic purposes. It is further assumed that 40 and 25% of the water withdrawn are used to estimate the process CWU in the domestic and industrial sectors, respectively, i.e., CWUDomestic and CWUIndustrial.

2.2.3 River Basin Water Stress

We assess the water stress in river basins by accounting for the depletion of surface runoff resulting from various production or service activities. The water depletion ratio, defined as the ratio of total water depletion to utilizable water resources, is used to assess water stress. Total depletion encompasses both process and non-process depletion. Process depletion refers to the consumptive water use (CWU) for irrigation, industrial, and service sectors, while non-process depletion pertains to evaporation from reservoir surfaces. The utilizable supply is the dependable surface runoff, corresponding to the surface runoff with a 75% exceedance probability. Annual water stress is evaluated under two scenarios: one where the utilizable water supply is the dependable surface runoff, and another where it is the dependable runoff minus the environmental flow requirement of the rivers (Eriyagama et al., 2015). Seasonal environmental flow requirements are not available for this analysis.

2.2.4 Data

The study analyzes time series data from 1999 to 2022 to examine trends and uses average data from 2020, 2021, and 2022 to explore spatial variations.

The source for data on sown and harvested crop areas, as well as crop production, is the Department of Census and Statistics of Sri Lanka (DCS, 2024). The DCS publishes district-level data on cropped area and production for the Maha (October to March) and Yala (April to September) seasons, available at https://www.statistics.gov.lk/Agriculture/StatisticalInformation/new. This study includes data for various crops, such as paddy (major and minor irrigation, and rainfed areas), maize, other cereals (kurakkan, sorghum, and small millet), pulses (green gram, black gram, cowpea, and dhal), oilseeds (groundnut, soybean, mustard, and gingelly), roots and tubers (manioc, sweet potatoes, potatoes, and onions), upcountry vegetables (lady's fingers, brinjal, bitter gourd, snake gourd, cucumber, ash and red pumpkin, ash plantain), low country vegetables (tomatoes, cabbages, carrots, knolkhol, beetroot, radish, beans, leeks, and capsicum), spices (cinnamon, pepper, cardamom, and cloves), fruits (oranges, lime, mangoes, plantains, papaw, and pineapple), as well as ginger, turmeric, sugarcane, tobacco, and cigars.

Monthly rainfall data is sourced from the Department of Meteorology, while evapotranspiration (ETP) estimates at the district or Divisional Secretariat (DS) level are derived from global ETP grids provided by the Climate Research Unit at the University of East Anglia (Harris et al., 2020; data available at https://crudata.uea.ac.uk/cru/data/hrg/).

Crop coefficients and growth periods are referenced from the FAO 56 paper (Allen et al., 1998), along with crop calendars for different crops available through the Department of Agriculture, Sri Lanka (DOA, 2024) at https://doa.gov.lk/naicc-publications-crop-calender/.

Water withdrawal data for the industrial and domestic sectors for 2022–2023 was provided by the Ministry of Water Supply and Sanitation.

2.3 Water Supply Variations

2.3.1 Surface Runoff

Monsoonal rainfall variations (see Chap. 1) result in significant fluctuations in river basin surface runoff. Of the 103 river basins (Fig. 2.1), 21 (including the Jaffna Peninsula) account for 78% of the surface runoff and have drainage areas exceeding 750 km^2 (Annex Table 2.4). These basins, which cover 76% of the country's total land area, are vital for both agricultural production and energy generation. Seven of these basins drain to the sea from the wet zone and contribute substantially to hydroelectric power generation. The other 16 basins, draining from the dry or intermediate zones, support much of the seasonal agricultural production.

In total, 90 river basins drain to the sea from the dry zone. These basins cover 80% of the country's land area but contribute only 52% of the surface runoff. Excluding the Mahaweli River Basin, this group accounts for just 33% of the surface runoff. In 61 of these river basins, over 80% of the runoff is generated during the Maha season (September to February), while many rivers originating in the dry or intermediate zones experience little or no flow during the Yala season (March to August). These basins are expected to experience increased seasonal runoff variability, as projected by climate change models, with drier conditions in the dry zones and wetter conditions in the wet zones (see Chap. 4).

2.3.1.1 Mahaweli River Basin

The Mahaweli River is the largest river basin in Sri Lanka. Originating in the Central Province of the wet zone, it flows for 384 km before draining into the sea in the Trincomalee District of the Eastern Province. Key characteristics of the Mahaweli River Basin include:

- It encompasses 17% of the country's land area, contributes 19% of the total surface runoff, and is home to 17% of its total population.
- The basin spans three climatic zones: the upper, middle, and lower reaches correspond to the wet, intermediate, and dry zones, respectively. The upper reach, covering the Kandy and Nuwara Eliya Districts, occupies 30% of the basin's area; the middle reach, mainly covering the Matale and Badulla Districts, accounts for 20%; and the lower reach, spanning the Polonnaruwa, Trincomalee, and Batticaloa Districts, makes up 50%. These areas receive average annual rainfall of approximately 2,400 mm, 2,025 mm, and 1,550 mm, respectively. Most of the Mahaweli's surface runoff is generated in the upper and middle reaches, which lie in the wet and intermediate zones.

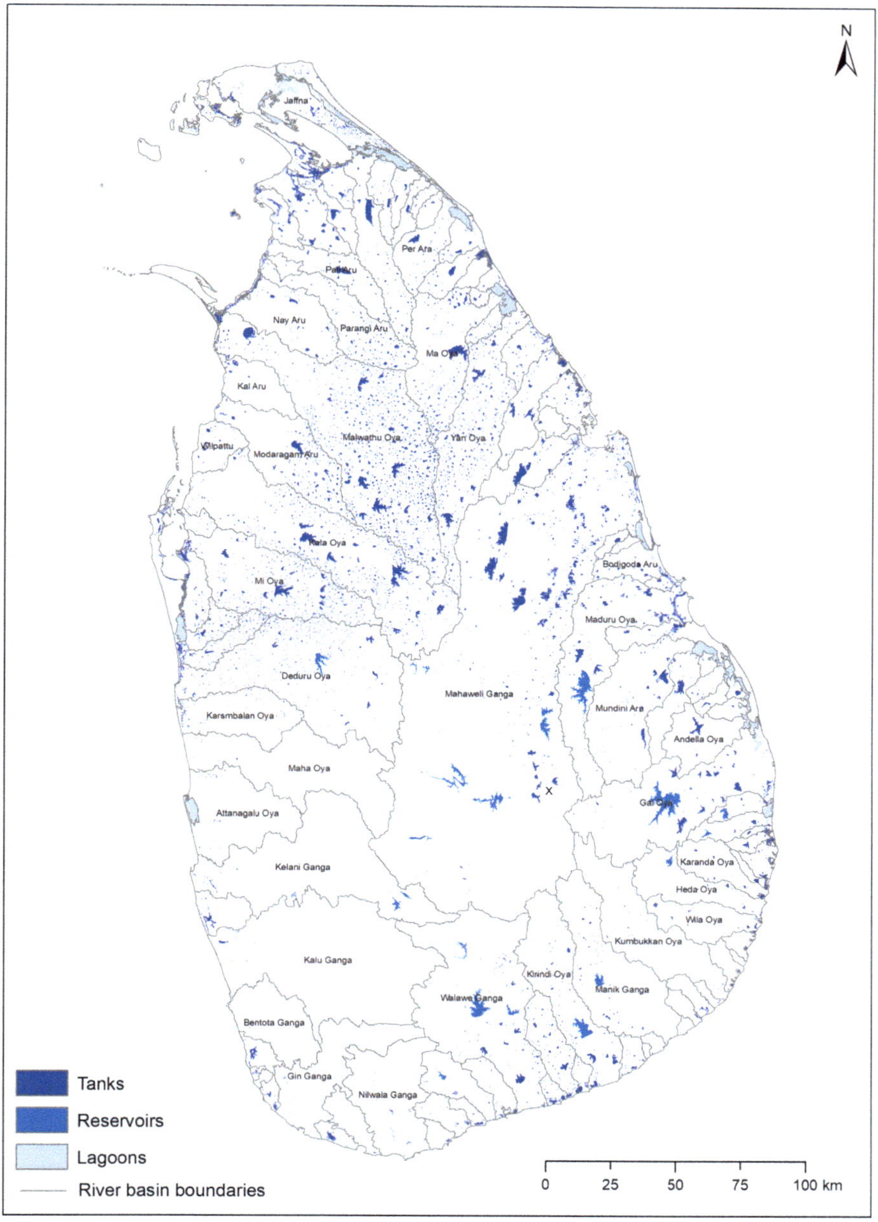

Fig. 2.1 River basins and storage spread

- The basin has the largest storage capacity in the country (Annex Table 2.5), with most of the major storages being multi-purpose reservoirs or tanks. These storages primarily meet the irrigation needs of the Kandy, Matale, Anuradhapura, Polonnaruwa, Batticaloa, and Trincomalee Districts.
- The Mahaweli River also supplies water to other basins through inter-basin transfers, including to the Malwathu Oya, Yan Oya, Maduru Oya, and Pettiyan Oya (Kantale Oya).
- The basin accounts for 19% of the total paddy area in Sri Lanka, which was approximately 1.185 million hectares between 2019 and 2022, of which about 629,000 hectares are under major irrigation, 300,000 hectares under minor irrigation systems, and 9% (approximately 255,000 hectares) are rainfed paddy areas (Annex Table 2.6).
- The major and minor irrigated areas, along with the rainfed areas, contribute significantly to the country's paddy production. Paddy production in this basin exceeds local consumption needs, contributing significantly to national production.
- The basin has 962 GWh of installed hydropower generation capacity and generates about 2,619 GWh per year.

Among other prominent river basins important for agriculture and water management is the Malwathu Oya, which is the second-longest river at 164 km, with the second-largest drainage area. However, its surface runoff accounts for only 2% of the total runoff (about 10% of the runoff from the Mahaweli basin). The Malwathu Oya basin is primarily located in the dry zone, with its source in the North-Central Province, and it drains to the sea in the Northern Province. Key features of the Malwathu Oya Basin include:

- Most of the runoff in the basin is generated during the Maha season, accounting for over 85% of the total runoff. This makes it the 14th largest basin in terms of runoff.
- Storage is critical for maintaining year-round agriculture in the basin, with village tank-based storage systems being a prominent feature. Many small tanks are scattered across the basin's landscape.
- Minor irrigation systems, including tanks and anicuts, support 50% of the basin's paddy area, making it the second-largest paddy-producing area in the country.

Kala Oya, the third-longest river basin at 148 km, has the third-largest drainage area and drains into the sea from the west. Key features of the Kala Oya basin include:

- It ranks as the 16th largest basin in terms of surface runoff, with 82% of the runoff generated by the Southwest Monsoon (SWM) and Northeast Monsoon (NEM) rainfall during the Maha season.
- The basin has the third-largest irrigated paddy area, with 47% of it served by minor irrigation systems.
- The Wilpattu National Park, which covers a large part of the lower reach of the basin, makes it an important eco-tourism destination.

The Deduru Oya and Yan Oya basins are approximately 142 km long, ranking 5th and 10th in terms of drainage area, respectively. These basins also rank 11th and 21st in terms of surface runoff. Large portions of these basins lie in the dry zone, where agriculture is the predominant economic activity.

The Walawe Ganga basin ranks 6th in terms of drainage area and 5th in surface runoff. A significant portion of the headwaters lies in the wet zone, and 57% of the runoff is generated during the Maha season. In addition to agriculture, the Udawalawe National Park is an important eco-tourism attraction. The basin has 126 GWh of installed hydropower generation capacity, generating approximately 388 GWh annually.

Gal Oya has the 8th largest drainage area and the 10th largest surface runoff. The basin is home to the largest reservoir in Sri Lanka, known as "Senanayake Samudraya," with a gross storage capacity of 956 million cubic meters, which provides irrigation water to a significant portion of the Eastern Province. The gross storage capacity of Senanayake Samudraya accounts for 75% of the surface runoff from the Gal Oya River basin, and it generates 20 GWh of hydropower.

The Kalu, Kelani, Bentota, Gin, and Nilwala rivers, along with the Maha Oya and Attanagalu Oya, predominantly lie in the wet zone and drain into the sea from the west and southwest. Kalu Ganga, with the 4th largest drainage area and the 2nd largest runoff, and Kelani Ganga, with the 7th largest drainage area and the 3rd largest surface runoff, together illustrate the concentration of water resources. Altogether, these seven basins contribute 44% of the total surface runoff, while covering only 16% of the country's drainage area. The Kalu and Kelani rivers have 79 and 289 GWh of installed hydropower capacity, generating approximately 338 and 1,595 GWh annually.

2.3.2 Storage

Given the seasonal variability of rainfall, water storage plays a critical role in managing water resources. Water is stored in various forms, including soil moisture, groundwater, wetlands, lakes, man-made reservoirs, tanks, and ponds. The reservoirs supporting major and medium irrigation systems, along with minor tanks supporting smaller irrigation systems, as well as ponds and lakes (Fig. 2.2), especially in the dry and intermediate zones, provide vital resilience against fluctuations in rainfall and runoff. The gross storage capacity of major and medium reservoirs and tanks is approximately 8.459 billion cubic meters (Bm^3), or 16% of the total surface runoff (Table 2.1).

These major and medium reservoirs support irrigation for a 544,000-hectare paddy area, while minor irrigation systems currently serve an additional 386,000 hectares of paddy. Due to variability in seasonal rainfall, water storage is essential for effective water resource management.

Additionally, small tanks, including those in tank cascade systems, provide approximately 0.42 Bm^3 of extra storage. Many of these tanks in the dry zone are

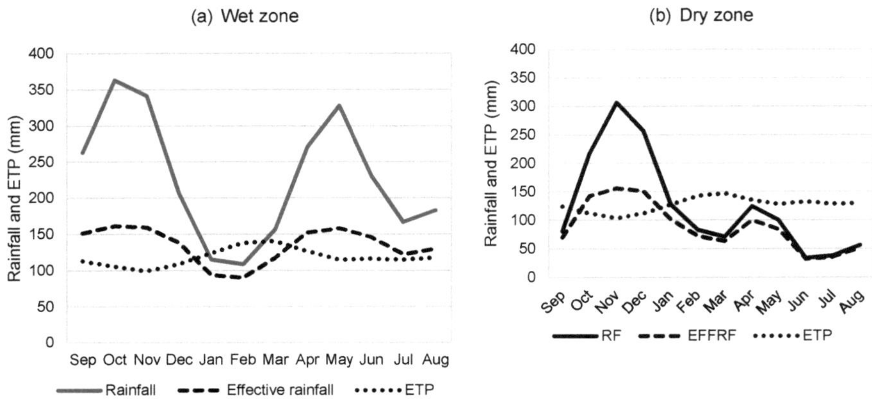

Fig. 2.2 Average monthly rainfall and ETP in the **a** wet zone, **b** dry zone. *Source* Authors' estimation based on rainfall data from the Department of Meteorology and ETP data from the Climate Research Unit, East Anglia

part of tank cascade systems, with an estimated total of up to 16,000 tanks nationwide (Panabokke, 2004). These minor irrigation systems primarily support paddy cultivation during the Maha season, while a more diversified cropping pattern in the Yala season uses any remaining water in the tanks. Currently, tank storage enables supplemental irrigation for approximately 386,000 hectares of paddy fields.

2.3.3 Groundwater

Traditionally, groundwater has been the primary source of safe drinking water for both rural and urban populations. The 2019 Household Income and Expenditure Survey indicates that 42% of the population still relies on groundwater (via dug wells and tube wells) for their drinking water needs (DCS, 2020). The volume of groundwater extracted by individual domestic users without access to piped water is estimated at 400 million cubic meters per year (Herath, 2006). Additionally, nearly 30% of both rural and urban piped water supply systems rely on groundwater resources, which collectively extract over 16 million cubic meters of water annually (Panabokke & Perera, 2005).

Before the 1980s, the use of groundwater for agriculture was mainly restricted to a few areas, such as Jaffna and Kalpitiya, where surface water sources were scarce. However, since the mid-1980s, the extraction of groundwater has increased significantly to mitigate the effects of droughts and improve the livelihoods of people in marginal areas. By 2003, over 50,000 agro-wells had been established in Sri Lanka's dry zone (IWMI, 2003), and this number has continued to rise (Sood et al., 2015). According to Shah et al. (2013), the number of agro-wells in use in Sri Lanka had reached approximately 200,000 to 250,000 by 2013. Currently, groundwater

Table 2.1 Distribution of storage capacity of major/medium reservoirs and tanks (in Mm^3)

District	Managed by the irrigation department	Managed by the Mahaweli development authority	Total
Major/medium reservoirs			
1. Ampara	1,330	167	1,497
2. Anuradhapura	828	104	932
3. Badulla	103	62	165
4. Batticaloa	193	596	789
5. Colombo	2		2
6. Galle	9		9
7. Hambantota	483	17	500
8. Kandy	41	1,534	1,575
9. Kurunegala	196		196
10. Mannar	94		94
11. Matale	34	876	910
12. Matara	4		4
13. Monaragala	91		91
14. Nuwara Eliya		171	171
15. Polonnaruwa	4352		4,352
16. Puttalam	108		108
17. Rathnapura		580	580
18. Trincomalee	249		249
19. Vavuniya	33		33
Minor tanks			422
Total	8,150	4,107	12,257

Source Irrigation Department, Mahaweli Authority (Irrigation Department, 2022, MA, 2022)
Notes The major and medium reservoirs are those with an irrigated command area of more than 400 ha and between 200 and 400 ha, respectively. The minor irrigation systems (including tank cascade systems) are those with a command area of up to 200 ha

is also being used to cultivate areas and increase the incomes of farmers in canal command areas (Amarasinghe et al., 2021).

Numerous studies have demonstrated that excessive groundwater extraction can have a negative impact on overall agricultural production, with significant long-term consequences (IWMI, 2003). Most of Sri Lanka's agricultural wells are in the dry zone, where groundwater recharge and aquifer replenishment are under severe stress (Kikuchi et al., 2003). The development of agro-wells in the country has been haphazard (De Silva et al., 1996), lacking a comprehensive assessment of the hydrogeological properties of aquifers, safe yield, well spacing, and recharge levels.

2.4 Consumptive Water Use

2.4.1 Agricultural Sector CWU

The two components of agricultural consumptive water use (CWU)—irrigation CWU and rainfall CWU—are estimated for paddy and other major crops at the administrative unit and river basin levels from 1999 to 2022. While district-level data on paddy and non-paddy cropped areas and production are available for all years, data on the sown and harvested paddy areas at the Divisional Secretariat (DS) level are only available for select years. Changes in CWU over time are analyzed using district-level data. Spatial variation in paddy crop CWU is examined for DS divisions from 2019–20 to 2021–22. Reliable paddy yield data, based on crop-cut experiments, as well as yield and production data for other crops, are only available at the district level. Area and production data for all crops are available for both the Maha and Yala seasons.

Weather patterns generally show that effective rainfall exceeds crop water requirements during the Maha season in both the wet and dry zones, as well as during the Yala season in the wet zone (Fig. 2.2a, b). For example, between October and February (Maha season), effective rainfall in the wet and dry zones is 640 mm and 620 mm, respectively, while total evapotranspiration potential (ETP) during the same period is 559 mm and 579 mm, respectively. However, between March and August (Yala season), the effective rainfall of 372 mm in the dry zone is less than half the ETP. As a result, irrigation is essential for crop growth during the Yala season in the dry zone. Due to intra-seasonal variability, both major and minor irrigation systems also provide water for paddy crops during the Maha season (Table 2.2).

The total cropped and irrigated areas used to estimate agricultural CWU are approximately 2,412,000 ha and 968,000 ha, respectively (see Chap. 1, Table 1. 3). The irrigated area encompasses both the Maha and Yala season paddy areas, utilizing both major and minor irrigation systems. The other cropped area includes part of the Yala season, featuring maize, vegetables, and annual crops such as ginger and sugar in the dry-zone districts. The total and irrigation CWU for all crops are shown in Table 2.2, while the spatial variation of CWU for paddy and other crops across river basins is presented in Table 2.2.

Seasonal crops account for 6,238 Mm^3, or 34% of the total CWU. Approximately 80% of the CWU for seasonal crops is in the dry zone, with 70% of this occurring during the Maha season.

Annual crops account for approximately 12,000 Mm^3, or 66% of the total CWU. Among these, plantation crops—tea, rubber, and coconut—account for more than 11,000 Mm^3, or 93% of the CWU for annual (permanent) crops. Other annual or permanent crops account for only 795 Mm^3 (Table 2.2). The wet zone accounts for 53% of the CWU for annual or permanent crops.

- The irrigation CWU for all crops totals 1,930 Mm^3, which represents just 10% of the total CWU.

Table 2.2 Total and irrigation crop CWU

Crops	Total CWU (Mm^3)							Irrigation CWU (Mm^3)						
	Wet zone		Dry zone		Sri Lanka			Wet zone		Dry zone		Sri Lanka		
	Maha	Yala	Maha	Yala	Maha	Yala	Total	Maha	Yala	Maha	Yala	Maha	Yala	Total
Seasonal crops														
Paddy	925	249	2927	1404	3852	1653	5505	68	26	1122	849	1190	875	2065
Major irrigation	69	53	1494	988	1563	1041	2605	26	12	726	669	752	681	1433
Minor irrigation	118	73	840	326	958	400	1357	42	14	396	180	438	194	632
Rainfed	737	123	593	90	1330	213	1543	0	0	0	0	0	0	0
Maize	2	2	192	23	194	25	219	0	0	0	1	0	1	1
Other cereals	1	1	8	1	9	2	10	0	0	0	0	0	0	0
Pulses	1	1	26	18	26	19	45	0	0	0	0	0	0	0
Oilseeds	1	1	27	40	28	41	68	0	0	0	0	0	0	0
Roots and tubers	19	18	33	38	53	55	108	0	0	0	5	0	5	5
Vegetables (upcountry)	38	17	83	31	121	48	169	0	0	0	6	0	6	6
Vegetables (low country)	17	18	20	18	36	36	72	0	0	0	2	0	2	2
Green chili	4	4	21	12	25	16	41	0	0	0	3	0	3	3
Total (seasonal crops)	1007	309	3336	1586	4343	1895	6238	68	26	1122	866	1188	892	2081
Annual/permanent crops														
Ginger + turmeric		36		32			67		0		2			2
Spices		105		98			204		0		0			0
Fruits		187		253			440		2		13			15
Sugar		2		76			79		1		26			27
Betel + tobacco		3		3			6		0		0			0

(continued)

Table 2.2 (continued)

Crops	Total CWU (Mm^3)							Irrigation CWU (Mm^3)						
	Wet zone		Dry zone		Sri Lanka			Wet zone		Dry zone		Sri Lanka		
	Maha	Yala	Maha	Yala	Maha	Yala	Total	Maha	Yala	Maha	Yala	Maha	Yala	Total
Tea		2747		600			3347		0		0			0
Rubber		1580		201			1781		0		0			0
Coconut		1669		4406			6075		0		0			0
Total (all crops)		7645		10,592			18,236		97		2029			2125

Source Authors estimation

- Seasonal crops, primarily paddy, account for 97% of the irrigation CWU, with 59% of this occurring during the Maha season.
- The dry zone accounts for 94% of the total irrigation CWU, emphasizing the importance of irrigation in this region.

However, the above estimates of irrigation CWU are likely to be significantly underestimated. It is well known that areas with coconut and tea plantations often pump groundwater from agro-wells or directly from streams to irrigate their land during periods of water stress. Many rainfed crops, including paddy in the dry zone, also rely on irrigation water pumped from small storage ponds or agro-wells to alleviate water stress during critical stages of crop growth. These sources of irrigation are not reflected in official statistics and, as a result, are not included in the irrigation CWU estimates.

2.4.2 Process Depletion

Irrigation CWU constitutes a significant portion of the total process depletion resulting from water withdrawals in Sri Lanka (Table 2.3). Process depletion from water withdrawals includes the irrigation CWU in the agricultural sector, as well as the CWU from water supplies to the domestic and industrial sectors. The domestic and industrial sectors accounted for only 7.7% of the total process depletion from 2020 to 2022.

Irrigation and rainfall CWU of all crops (except tea, rubber, and coconut) show vast spatial variation (Fig. 2.3). The wet zone districts account for 42% of the total CWU, but only 5% of the irrigation CWU (Annex Table 2.4). Seven districts account for 76% of the irrigation CWU (Fig. 2.4). These include large rainfall-dependent districts with substantial minor tank-irrigated areas, such as Anuradhapura, Kurunegala, Monaragala, Batticaloa, and Trincomalee.

The domestic sector CWU, which is assumed to be 25% of the sector's water supply, is only 8% of the total CWU. Approximately 66% of the domestic sector's water supply comes from piped water, and only 38% of the population has access to piped water (including filtered water). At the national level, households with a piped water supply use an average of 147 L per person per day, with usage ranging

Table 2.3 Process depletion from different sectors in 2020–2022 (in Mm^3)

	Irrigation			Domestic	Industrial	Total
	Maha season	Yala season	Total			
Dry zone	1,141	888	2,029	60	0.2	2,089
Wet zone	69	27	96	113	5.1	214
Total	1,210	915	2,125	173	5.3	2,303

Source Author's estimates

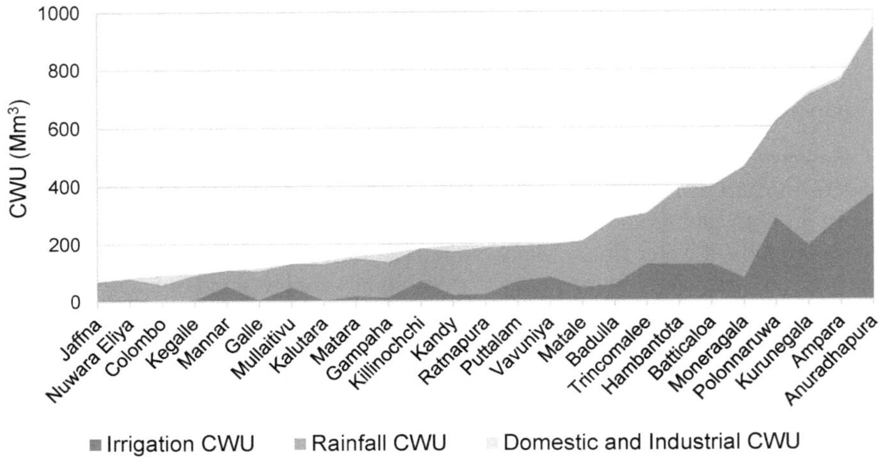

Fig. 2.3 Total rainfall and irrigation CWU across districts. *Notes* Irrigation and rainfall CWU all crops without tea, rubber, and coconut

from 58 to 222 L per person per day across the Jaffna to Kalutara Districts. The domestic sector CWU is a substantial part of the total CWU in the wet zone districts, comprising 97%, 71%, 69%, and 47% of the total CWU in Colombo, Gampaha, Kalutara, and Galle districts, respectively (see Annex Table 2.5).

The CWU in the industrial sector, which comprises construction and small to medium-sized industries, makes a very limited contribution to the total CWU. The irrigation sector contributes 92% of the total CWU, varying from 96% in the dry zone to 69% in the wet zone.

2.5 Water Stress

We use the water depletion ratio—the ratio of total (process and non-process) water depletion to utilizable water resources—to assess water stress in river basins. Process depletion refers to the CWU from water supplies to the three sectors (see Table 2.3), of which 90% currently occurs in the dry zone.

Non-process depletion includes evaporation from water bodies and fallow lands. The estimated evaporation, even at 10–15% of gross storage, indicates that non-process depletion from surface storage is approximately 0.845–1.268 billion cubic meters, which is significantly higher relative to total depletion across the three sectors. Given the spatial distribution of water bodies (Fig. 2.1), most non-process depletion also occurs in the dry zone. We estimate the spatial variation of reservoir evaporation using a relationship developed for the water surface area and storage data of reservoirs and tanks managed by the Irrigation Department (Irrigation Department, 2022). (See Annex 4 for further details on this estimation.)

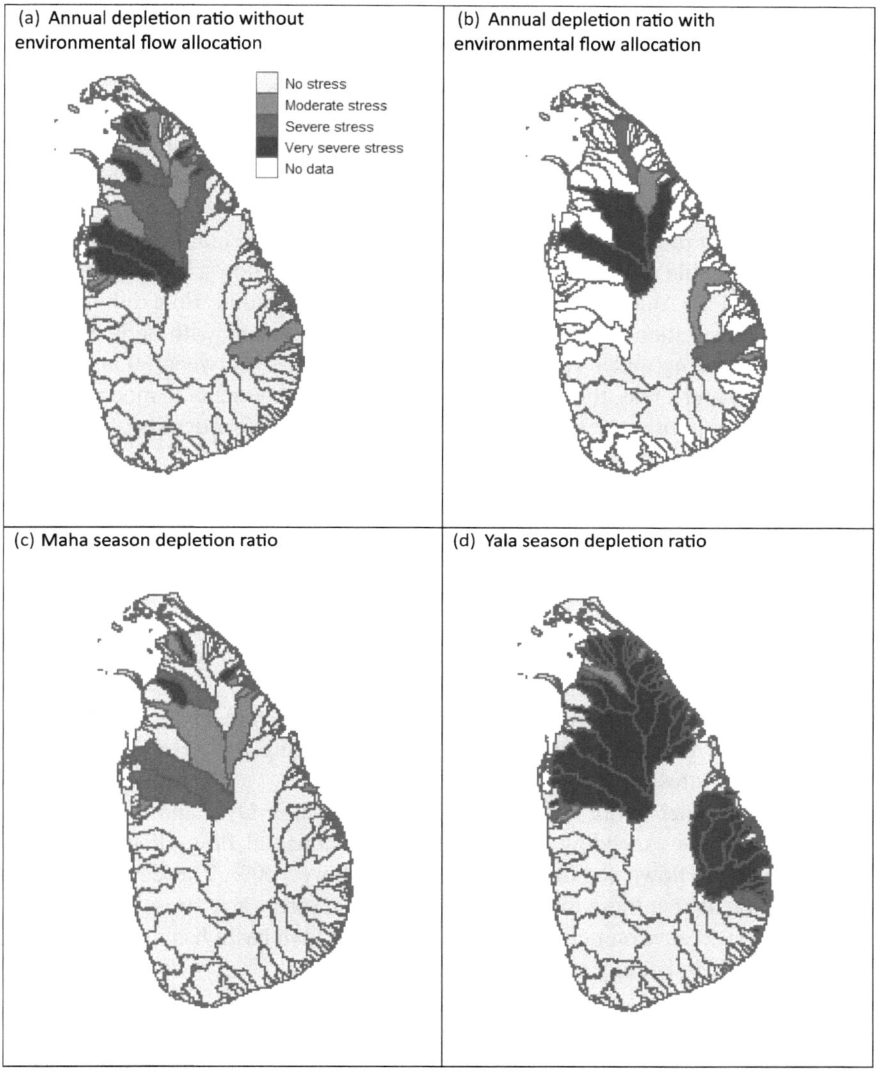

Fig. 2.4 **a** Maha season, **b** Yala season, and **c** Annual depletion ratio without allocating Eflows, and **d** annual depletion ratio after allocating Eflows

For utilizable runoff, this chapter considers two scenarios: one based on the 75% exceedance probability runoff (dependable, or P75, runoff) and the other based on the P75 runoff minus the environmental flow requirements of rivers (Annex Table 2.5). The estimated utilizable runoff for all basins, with and without environmental flow allocation, is 35 billion and 41 billion cubic meters, respectively. The environmental flows considered here are limited to the major rivers, with estimates based on gauging stations downstream where continuous flow data are available for a reasonable number of years (Eriyagama et al., 2015). Currently, there are no specific runoff allocations for environmental flows; however, many rivers in the dry zone have little to no flow to the sea during dry months. These two scenarios (Fig. 2.4a, b), which highlight the current state of stress, underscore the precarious situation in which the pressure to allocate environmental flows increases with development, while climate change diminishes surface runoff.

The depletion ratio for all river basins, without environmental flow allocation, is 9% of the utilizable runoff. With reserved environmental flows, the depletion ratio increases to 11% (Annex Table 2.5). However, the depletion ratio exhibits considerable spatial variation (Fig. 2.4). We categorize river basins with depletion ratios above 80% as being under very severe water stress, those between 60 and 80% as under severe water stress, those between 40 and 60% as experiencing moderate water stress, and those below 40% as having no water stress. A few details of individual basins include:

- River basins draining into the sea from the wet zone have very low depletion ratios. The depletion ratio for the Kirindi Oya to Yan Oya river basins, which drain into the sea from the dry zone, is also low.
- The annual depletion ratio for the Mahaweli basin is 9% without environmental flows and 11% with environmental flows. However, these figures may not fully reflect the situation due to substantial inter-basin water transfers from the Mahaweli basin.
- The depletion ratio for the Yan Oya basin increases from 71 to 110% with the inclusion of reserved environmental flows, indicating both current and future severe water stress.
- Among the other major basins, Kala Oya, Malwathu Oya, and Yan Oya have depletion ratios exceeding 70% without environmental flow allocations. With environmental flows, the depletion ratios rise above 100%. These high depletion ratios indicate that these three basins are under severe water stress.
- With reserved environmental flows, even the Gal Oya basin has a depletion ratio of over 67%, indicating severe water stress.

In fact, most river basins in the dry zone's northern regions—from Yan Oya to Kala Oya—exhibit high depletion ratios. These basins primarily support rainfed agriculture (including minor tank irrigation) and are particularly vulnerable to natural disasters such as droughts and extreme temperatures.

Many river basins experience significant seasonal water stress, particularly during the Yala season (Fig. 2.4c, d). The river basins from Karanda Oya to Yan Oya, except for the Mahaweli Ganga and the smaller basins between the Mahaweli and Maduru Oya, have high depletion ratios during the Yala season. A depletion ratio exceeding one during the Yala season indicates that not only the carry-over surface storage but also water reuse supports agriculture during this period. These river basins receive a substantial portion of their rainfall from the Second Inter-monsoon (SIM) and Northeast Monsoon (NEM), which support both Maha season agriculture and the surface storage that sustains Yala season agriculture.

In this analysis, water stress in river basins is defined as the ratio of the process CWU (depletion through evaporation and transpiration) to the utilizable supply. Another common indicator of water stress is the ratio of water withdrawals to utilizable water supply. Due to low application efficiencies, a significantly higher water supply is required to meet the evapotranspiration demand. Additionally, large paddy-irrigated areas require more water to satisfy percolation needs. Without sufficient infrastructure to capture and reuse return flows, the excess water supply, beyond what is needed for evapotranspiration, becomes part of the depleted water. Therefore, if water withdrawals (rather than CWU) were used as the numerator, the depletion ratios in many basins would be much higher.

2.6 Adaptation Challenges

The variability of surface runoff and significantly higher water consumption, especially for paddy irrigation, are the primary causes of seasonal water supply and demand gaps in river basins. While only a few river basins experience year-round water stress, the supply–demand gap during the Yala season is particularly pronounced in many river basins in the dry zone. This gap is expected to widen for several reasons.

With rising temperatures, evapotranspiration is expected to increase in the irrigation sector, while demand for drinking water and cooling will rise in the domestic sector. Additionally, changing demographic patterns, lifestyle shifts, and economic growth are expected to lead to increased water usage in the agricultural, domestic, and industrial sectors. On the other hand, the surface runoff in many river basins in the dry zone is likely to decrease due to reduced rainfall. Given the large share of consumptive water use, adaptation challenges will be most pronounced in the dry zone's agricultural sector.

There is an urgent need to implement demand management strategies, including increasing rainfall use efficiency, altering cropping patterns, promoting high-yielding and drought-tolerant varieties, improving irrigation water use efficiency, and reusing return flows. Better management of existing storage and augmenting surface and groundwater storage wherever possible are key supply-side adaptation strategies.

There is significant potential to increase rainfall efficiency in Maha season rice cultivation. In most months, even in the dry zone, effective rainfall exceeds potential evapotranspiration. Therefore, every effort should be made to optimize the use of Maha season rainfall CWU and conserve storage in water bodies for the Yala season. A slight adjustment to the crop calendar to align with the onset of rainfall could reduce irrigation requirements for land preparation in the Maha season.

Changes to the cropping calendar require improved area-specific weather forecasting systems and enhanced institutional coordination, including better water management at the irrigation system level and within canal water user associations.

Changing cropping patterns can help reduce non-process evaporation from paddy fields. However, the biggest barrier to crop diversification is the lack of access to seeds, rural service providers, and markets. These areas must be strengthened to enable farmers to diversify into high-value, low-water-consuming crops. Additionally, outdated policies and regulations that hinder crop diversification in paddy-irrigated areas should be revised.

Promoting high-yielding, drought-tolerant paddy varieties can improve land, water, and labor productivity, while also reducing drought risks. The adoption of efficient irrigation technologies, such as wet and dry irrigation for paddy fields, as well as sprinkler, drip, and underground pipe irrigation for other crops, can increase surface irrigation water use efficiency. These measures will help make efficient use of scarce water resources in areas already experiencing or facing emerging water stress.

Annex

See Tables 2.4, 2.5 and 2.6.

Annex 2.1. Estimating Gross Storage of Tanks and Reservoirs

The Department of Irrigation publishes the surface water spread area and gross storage of reservoirs and tanks that it manages. Many of the other small tanks and reservoirs in the country have no reliable estimates of gross storage. However, the surface water spread area of all water bodies can now be estimated with RS/GIS (Fig. 2.1). We use these estimates to assess the gross storage of all water bodies. First, we estimate a relationship between water spread area and gross water storage of reservoirs and tanks at the river basin level. These are the aggregated data of reservoir storage and surface area collected by the Department of Irrigation (Fig. 2.5). Next, we use this relationship to estimate the gross storage of water bodies in river basins, which are not under the management of the Irrigation Department.

Table 2.4 River basin characteristics

No	River basin	Drainage area	Seasonal runoff				Annual runoff		2022 Population	Per capita water resource in 2022	Surface area of water bodies	Gross storage estimate
	Name		P75		P50		P75	P50				
		km^2	Bm^3	Bm^3	Bm^3	Bm^3	Bm^3	Bm^3	1000 s	m^3/pc/year	Km^2	Mm^3
1	Kelani Ganga	2278	2.33	2.97	2.41	3.22	5.30	5.63	2775	2071	13.87	64.43
2	Bolgoda Lake	374	0.41	0.50	0.43	0.56	0.92	0.99	1421	720	11.04	51.07
3	Kaluganga	2688	3.17	3.70	3.35	4.23	6.87	7.58	1579	4976	0.87	3.07
4	Bentota Ganga	760	0.73	0.85	0.78	0.97	1.58	1.75	458	3950	0.11	0.31
5	Madu Ganga	59	0.07	0.08	0.07	0.09	0.15	0.17	72	2400	7.06	30.69
6	Madampe Lake	90	0.10	0.12	0.11	0.14	0.23	0.25	56	4628	1.58	5.94
7	Telwatte Ganga	51	0.06	0.07	0.07	0.08	0.13	0.14	46	3214	1.95	7.49
8	Ratgama Lake	10	0.01	0.01	0.01	0.01	0.02	0.03	17	1685	2.33	9.10
9	Gin Ganga	922	1.04	1.26	1.18	1.37	2.30	2.56	506	5181	1.05	3.90
10	Koggala Lake	64	0.06	0.07	0.07	0.08	0.14	0.16	159	1000	5.92	25.29
11	Polwatta Ganga	233	0.21	0.20	0.24	0.24	0.41	0.48	202	2424	0.21	0.65
12	Nilwala Ganga	960	0.73	0.61	0.88	0.76	1.34	1.63	633	2614	2.88	14.94
13	Sinimodara Oya	38	0.03	0.02	0.03	0.03	0.05	0.06	63	976	0.02	0.06
14	Kirama Oya	223	0.14	0.11	0.17	0.14	0.26	0.32	154	2081	1.49	9.78
15	Rekawa Oya	92	0.04	0.03	0.05	0.04	0.08	0.10	43	2255	4.86	20.36
16	Uruhokke Oya	348	0.18	0.14	0.22	0.17	0.32	0.39	126	3197	6.44	60.26
17	Kachigala Ara	220	0.10	0.07	0.12	0.09	0.18	0.22	79	2853	3.16	12.72

(continued)

Table 2.4 (continued)

No	River basin	Drainage area	Seasonal runoff				Annual runoff		2022 Population	Per capita water resource in 2022	Surface area of water bodies	Gross storage estimate
	Name		P75		P50		P75	P50				
		km^2	Bm^3	Bm^3	Bm^3	Bm^3	Bm^3	Bm^3	1000 s	m^3/pc/year	Km^2	Mm^3
18	Walawe Ganga	2442	0.98	0.66	1.20	0.86	1.65	2.06	693	3088	78.84	440.11
19	Karagan Oya	58	0.02	0.01	0.02	0.02	0.03	0.04	18	2381	7.45	32.57
20	Malala Oya	399	0.11	0.08	0.15	0.10	0.19	0.24	69	3653	17.75	50.60
21	Embilikala Oya	59	0.01	0.01	0.02	0.01	0.02	0.03	16	1954	5.11	21.55
22	Kirindi Oya	1165	0.20	0.15	0.29	0.15	0.35	0.43	188	2448	41.30	289.80
23	Bambawe Ara	79	0.01	0.01	0.02	0.01	0.02	0.03	15	1877	1.54	5.78
24	Mahasilawa Oya	13	0.00	0.00	0.00	0.00	0.00	0.00	3	1308	0.22	0.67
25	Butawa Oya	38	0.01	0.00	0.01	0.00	0.01	0.01	7	1528	0.36	1.17
26	Menik Ganga	1272	0.17	0.08	0.22	0.10	0.25	0.32	142	2140	20.62	124.79
27	Katupila Aru	86	0.01	0.00	0.02	0.01	0.02	0.02	11	2228	0.25	0.79
28	Kuranda Ara	131	0.02	0.01	0.03	0.01	0.03	0.04	10	3988	1.14	4.17
29	Namadagas Ara	46	0.02	0.01	0.03	0.01	0.03	0.04	8	4314	2.16	8.37
30	Kambe Ara	46	0.01	0.00	0.01	0.00	0.01	0.02	5	3132	2.64	11.83
31	Kumbukkan Oya	1218	0.30	0.05	0.39	0.08	0.36	0.47	146	3310	0.87	3.08

(continued)

Table 2.4 (continued)

No	River basin	Drainage area	Seasonal runoff				Annual runoff		2022 Population	Per capita water resource in 2022	Surface area of water bodies	Gross storage estimate
	Name		P75		P50		P75	P50				
		km^2	Bm^3	Bm^3	Bm^3	Bm^3	Bm^3	Bm^3	1000 s	m^3/pc/year	Km^2	Mm^3
32	Bangura Oya	92	0.02	0.00	0.03	0.01	0.03	0.04	1	63,656	0.02	0.05
33	Girikula Oya	15	0.00	0.00	0.01	0.00	0.01	0.01	0.00	NA	0.87	3.07
34	Heawa Ara	51	0.02	0.00	0.02	0.00	0.02	0.02	0.00	NA	6.30	21.88
35	Wila Ara	484	0.15	0.02	0.20	0.03	0.17	0.22	16	15,253	9.85	43.17
36	Heda Oya	604	0.23	0.02	0.31	0.03	0.25	0.34	24	14,623	9.65	37.97
37	Karanda Oya	422	0.17	0.01	0.22	0.02	0.18	0.24	7	34,909	0.46	2.22
38	Simena Ara	51	0.02	0.00	0.03	0.00	0.02	0.03	0.0	NA	3.38	13.67
39	Tandiadi Aru	22	0.01	0.00	0.01	0.00	0.01	0.01	0.0	NA	1.58	5.94
40	Kangikadichi Ara	56	0.02	0.00	0.03	0.00	0.03	0.03	0.0	NA	4.42	18.37
41	Rufus Kulam	35	0.01	0.00	0.02	0.00	0.02	0.02	0.0	NA	3.10	12.42
42	Pannel Oya	184	0.08	0.00	0.10	0.01	0.09	0.11	1	203,213	8.19	51.34
43	Ambalama Oya	115	0.05	0.00	0.07	0.00	0.05	0.07	0	NA	8.36	49.32
44	Gal Oya	1792	0.82	0.03	1.05	0.06	0.86	1.11	96	13,092	108.54	1235.49
45	Andella Oya	522	0.24	0.01	0.31	0.02	0.25	0.33	42	8788	11.95	68.79
46	Thumpankeni Tank	9	0.00	0.00	0.01	0.00	0.00	0.01	11	608	0.21	0.66

(continued)

Table 2.4 (continued)

No	River basin	Drainage area	Seasonal runoff				Annual runoff		2022 Population	Per capita water resource in 2022	Surface area of water bodies	Gross storage estimate
	Name		P75		P50		P75	P50				
		km^2	Bm^3	Bm^3	Bm^3	Bm^3	Bm^3	Bm^3	1000 s	m^3/pc/year	Km^2	Mm^3
47	Namakada Aru	12	0.01	0.00	0.01	0.00	0.01	0.01	209	42	0.41	1.36
48	Mandipattu Aru	100	0.05	0.00	0.06	0.00	0.05	0.07	12	6469	5.24	22.12
49	Pattanthe Dephue Aru	100	0.01	0.00	0.02	0.00	0.01	0.02	55	357	1.55	5.81
50	Magalatavan Aru	346	0.18	0.00	0.22	0.01	0.18	0.23	28	9595	14.48	82.04
51	Vett Aru	26	0.01	0.00	0.02	0.00	0.01	0.02	4	5838	0.02	0.06
52	Mundeni Aru	1280	0.58	0.02	0.74	0.03	0.60	0.77	100	8768	32.02	138.45
53	Miyangollal Ela	225	0.09	0.00	0.12	0.01	0.09	0.12	17	7988	8.94	39.80
54	Maduru Oya	1541	0.50	0.04	0.69	0.06	0.54	0.75	149	5071	98.21	527.39
55	Pulliyanpotha Aru	52	0.02	0.00	0.02	0.00	0.02	0.03	4	8118	14.17	65.93
56	Kirimechi Odai	77	0.03	0.01	0.04	0.01	0.04	0.05	4	12,022	4.81	24.35
57	Bodigoda Aru	164	0.07	0.02	0.09	0.03	0.09	0.11	10	11,642	3.32	13.43
58	Mandan Aru	13	0.01	0.00	0.01	0.00	0.01	0.01	1	14,210	0.07	0.18

(continued)

Table 2.4 (continued)

No	River basin	Drainage area	Seasonal runoff				Annual runoff		2022 Population	Per capita water resource in 2022	Surface area of water bodies	Gross storage estimate
	Name		P75		P50		P75	P50				
		km^2	Bm^3	Bm^3	Bm^3	Bm^3	Bm^3	Bm^3	1000 s	m^3/pc/year	Km^2	Mm^3
59	Makarachchi Aru	37	0.02	0.01	0.02	0.01	0.03	0.03	3	9597	3.54	7.63
60	Mahaweli Ganga	10,327	5.37	2.71	6.16	2.93	8.08	9.09	3716	2616	302.73	1835.42
61	Kantalai Basin Per Ara	445	0.13	0.01	0.18	0.02	0.14	0.20	52	3901	31.40	155.34
62	Panna Oya	69	0.02	0.00	0.03	0.00	0.02	0.03	126	231	1.57	5.90
63	Palampotta Aru	143	0.03	0.00	0.05	0.00	0.04	0.05	14	3971	15.92	74.95
64	Pankulam Ara	382	0.08	0.01	0.12	0.01	0.09	0.13	23	5794	20.45	107.10
65	Kanchikamban Aru	205	0.04	0.00	0.06	0.01	0.04	0.06	17	3977	10.70	48.47
66	Palakutti Aru	20	0.00	0.00	0.00	0.00	0.00	0.01	2	3228	0.77	2.69
67	Yan Oya	1520	0.20	0.02	0.32	0.04	0.22	0.36	135	2853	91.22	492.85
68	Mee Oya	90	0.01	0.00	0.02	0.00	0.01	0.02	8	2826	4.35	18.04
69	Ma Oya	1024	0.13	0.01	0.21	0.03	0.15	0.24	49	5211	47.64	242.08
70	Churian Aru	74	0.01	0.00	0.01	0.00	0.01	0.02	3	6263	8.23	36.33
71	Chavar Aru	31	0.00	0.00	0.01	0.00	0.00	0.01	2	3565	2.16	8.36

(continued)

Table 2.4 (continued)

No	River basin	Drainage area	Seasonal runoff				Annual runoff		2022 Population	Per capita water resource in 2022	Surface area of water bodies	Gross storage estimate
	Name		P75		P50		P75	P50				
		km^2	Bm^3	Bm^3	Bm^3	Bm^3	Bm^3	Bm^3	1000 s	m^3/pc/year	Km^2	Mm^3
72	Palladi Aru	61	0.01	0.00	0.01	0.00	0.01	0.01	8	1959	0.98	3.50
73	Nay ARa	187	0.02	0.00	0.04	0.01	0.03	0.04	88	533	29.19	145.78
74	Kodalikallu Aru	74	0.01	0.00	0.01	0.00	0.01	0.02	7	2560	4.70	19.64
75	Per Ara	374	0.05	0.01	0.07	0.01	0.05	0.08	19	4906	12.15	55.72
76	Pali Aru	84	0.01	0.00	0.02	0.00	0.01	0.02	6	3529	0.00	0.00
77	Muruthapilly Aru	41	0.00	0.00	0.01	0.00	0.01	0.01	3	3293	1.56	5.84
78	Thoravil Aru	90	0.01	0.00	0.02	0.00	0.01	0.02	6	3636	3.24	13.06
79	Piramenthal Aru	82	0.01	0.00	0.01	0.00	0.01	0.02	4	4562	3.50	14.22
80	Nethali Aru	120	0.01	0.00	0.02	0.00	0.02	0.02	5	6160	4.13	17.05
81	Kanakarayan Aru	896	0.10	0.02	0.15	0.03	0.12	0.18	58	3770	35.61	181.31
82	Kalawalappu Aru	56	0.01	0.00	0.01	0.00	0.01	0.01	10	1367	4.99	20.96
83	Akkarayan Aru	192	0.02	0.00	0.03	0.01	0.02	0.04	33	1414	32.84	165.88

(continued)

Table 2.4 (continued)

No	River basin	Drainage area	Seasonal runoff				Annual runoff		2022 Population	Per capita water resource in 2022	Surface area of water bodies	Gross storage estimate
	Name		P75		P50		P75	P50				
		km²	Bm³	Bm³	Bm³	Bm³	Bm³	Bm³	1000 s	m³/pc/year	Km²	Mm³
84	Mendekal Aru	297	0.03	0.01	0.05	0.01	0.04	0.06	33	2209	31.51	158.56
85	Pallarayan Kadu	159	0.02	0.00	0.03	0.01	0.02	0.03	14	2732	8.87	39.42
86	Pali Aru	451	0.05	0.01	0.07	0.01	0.06	0.08	11	10,050	19.59	94.10
87	Chappi Aru	66	0.01	0.00	0.01	0.00	0.01	0.01	2	7390	2.37	9.26
88	Parangi Aru	832	0.08	0.02	0.12	0.03	0.10	0.15	132	1516	29.81	149.20
89	Nay Aru	560	0.06	0.01	0.08	0.02	0.07	0.10	8	17,743	0.00	0.00
90	Malvatu Oya	3246	0.32	0.08	0.47	0.11	0.40	0.58	379	2041	234.77	1097.35
91	Kal Ara	210	0.02	0.01	0.03	0.01	0.02	0.04	16	3234	0.67	2.32
92	Moderagam Ara	932	0.08	0.03	0.13	0.04	0.11	0.17	91	2491	46.32	183.99
93	Kala Oya	2772	0.23	0.08	0.38	0.12	0.31	0.50	4	152,293	137.94	762.53
94	Moongil Aru	44	0.00	0.00	0.01	0.00	0.00	0.01	475	19	1.44	5.37
95	Mi Oya	1516	0.12	0.01	0.18	0.03	0.13	0.21	3	102,265	98.55	465.48
96	Madurankuli Aru	62	0.01	0.00	0.01	0.00	0.01	0.01	197	74	1.80	6.86

(continued)

Table 2.4 (continued)

No	River basin	Drainage area	Seasonal runoff				Annual runoff		2022 Population	Per capita water resource in 2022	Surface area of water bodies	Gross storage estimate
	Name		P75		P50		P75	P50				
		km^2	Bm^3	Bm^3	Bm^3	Bm^3	Bm^3	Bm^3	1000 s	m^3/pc/year	Km^2	Mm^3
97	Kalagamuwa Oya	151	0.02	0.01	0.03	0.01	0.03	0.04	72	631	10.60	47.98
98	Pantampola Oya	215	0.04	0.02	0.05	0.02	0.05	0.07	34	2358	11.31	51.51
99	Deduru Oya	2616	0.52	0.26	0.67	0.31	0.78	0.98	217	5224	65.71	324.19
100	Karambala Oya	589	0.17	0.12	0.20	0.14	0.29	0.35	972	381	9.53	43.40
101	Ratmal Oya	215	0.08	0.07	0.09	0.08	0.15	0.17	309	572	0.80	2.82
102	Maha Oya	1510	0.69	0.61	0.79	0.73	1.31	1.53	187	8229	2.23	8.66
103	Attanagalu Oya	727	0.54	0.62	0.58	0.69	1.16	1.27	1045	1231	0.83	2.93
104	Jaffna Peninsular	1025							1319		53.94	286.01

Table 2.5 River basin runoff, CWU, storage capacity, and depletion ratios

No	River basin[a]	Utilizable runoff (75% probability runoff) (Bm^3)		Process CWU (bm^3)				Storage (Bm^3)		Seasonal depletion ratios		Annual depletion ratios	
				Irrigation		Total (irrigation + Domestic + Industrial)		Capacity[b]	Surface evaporation[3]			With Eflows	Without Eflows
		Maha	Yala	Maha	Yala	Maha	Yala		Annual	Maha	Yala	Annual	Annual
1	Kelani Ganga	2.33	2.97	0.01	0.00	0.02	0.02	0.06	0.01	1	1	1	1
2	Bolgoda Lake	0.41	0.50	0.00	0.00	0.01	0.01	0.05	0.01	3	2	3	
3	Kalu Ganga	3.17	3.70	0.01	0.00	0.01	0.01	0.00	0.00	0	0	0	0
4	Bentota Ganga	0.73	0.85	0.00	0.00	0.00	0.00	0.00	0.00	0	0	0	0
5	Bentota-Nilwala	1.56	1.82	0.00	0.00	0.01	0.01	0.08	0.01	1	1	1	
6	Nilwala Ganga	0.73	0.61	0.01	0.00	0.01	0.01	0.01	0.00	2	1	2	3
7	Nilwala-Walawe	0.50	0.38	0.02	0.02	0.03	0.02	0.10	0.02	7	7	7	7
8	Walawe Ganga	0.98	0.66	0.03	0.02	0.04	0.03	0.44	0.07	7	9	8	12
9	Walawe-Krindi oya	0.15	0.10	0.02	0.01	0.02	0.01	0.10	0.02	17	21	19	
10	Kirindi Oya	0.20	0.15	0.01	0.01	0.01	0.01	0.29	0.04	19	23	20	23
11	Krindi oya- Manik Ganga	0.02	0.01	0.00	0.00	0.00	0.00	0.01	0.00	21	35	26	
12	Manik Ganga	0.17	0.08	0.01	0.01	0.01	0.01	0.12	0.02	13	27	17	24
13	Manik Ganga-Kumbumkan oya	0.07	0.02	0.00	0.00	0.00	0.01	0.03	0.00	10	38	16	
14	Kumbukkan Oya	0.30	0.05	0.01	0.01	0.01	0.01	0.00	0.00	3	12	4	5

(continued)

Table 2.5 (continued)

No	River basin[a]	Utilizable runoff (75% probability runoff) (Bm^3)		Process CWU (bm^3)				Storage (Bm^3)		Seasonal depletion ratios		Annual depletion ratios	
				Irrigation		Total (irrigation + Domestic + Industrial)		Capacity[b]	Surface evaporation[3]			With Eflows	Without Eflows
		Maha	Yala	Maha	Yala	Maha	Yala		Annual	Maha	Yala	Annual	Annual
15	Kumbukkan Oya-Karanda oya	0.43	0.04	0.01	0.01	0.01	0.01	0.11	0.02	3	36	6	7
16	Karanda oya-Gal Oya	0.37	0.02	0.02	0.02	0.02	0.02	0.15	0.02	9	167	18	20
17	Gal Oya	0.82	0.03	0.07	0.10	0.07	0.10	1.24	0.19	20	569	41	69
18	Gal Oya-Mundini Aru	0.51	0.02	0.05	0.07	0.05	0.08	0.18	0.03	12	557	29	32
19	Mundini aru + Miyangolla	0.67	0.03	0.01	0.02	0.01	0.02	0.18	0.03	4	124	8	12
20	Maduru Oya	0.50	0.04	0.05	0.05	0.05	0.05	0.53	0.08	17	219	32	48
21	Maduru Oya-Mahaweli Ganga	0.14	0.05	0.01	0.01	0.01	0.01	0.11	0.02	11	41	18	
22	Mahaweli Ganga	5.37	2.71	0.23	0.20	0.25	0.22	1.84	0.28	7	13	9	11
23	Mahaweli - Yan oya	0.31	0.03	0.03	0.02	0.03	0.03	0.39	0.06	20	215	35	
24	Yan Oya	0.20	0.02	0.05	0.03	0.05	0.03	0.49	0.07	45	358	71	110
25	Mee + Ma Oya	0.14	0.01	0.03	0.01	0.03	0.01	0.26	0.04	33	235	52	57
26	Ma oya-Kanakarayan aru	0.15	0.02	0.04	0.02	0.04	0.02	0.32	0.05	43	200	62	

(continued)

Table 2.5 (continued)

No	River basin[a]	Utilizable runoff (75% probability runoff) (Bm^3)		Process CWU (bm^3)				Storage (Bm^3)		Seasonal depletion ratios		Annual depletion ratios	
				Irrigation		Total (irrigation + Domestic + Industrial)		Capacity[b]	Surface evaporation[3]			With Eflows	Without Eflows
		Maha	Yala	Maha	Yala	Maha	Yala		Annual	Maha	Yala	Annual	Annual
27	Kanakarayan Aru	0.10	0.02	0.02	0.02	0.02	0.02	0.18	0.03	35	164	54	73
28	Kanakarayan Aru-Parangi Aru	0.13	0.03	0.03	0.03	0.03	0.03	0.49	0.07	54	241	86	
29	Parangi + Nay aru	0.14	0.03	0.04	0.01	0.04	0.01	0.15	0.02	38	58	41	339
30	Aruvi aru (Malwathu oya)	0.32	0.08	0.10	0.04	0.10	0.04	1.10	0.16	58	155	78	131
31	Kal Aru-Modaragam aru	0.10	0.03	0.02	0.01	0.02	0.01	0.19	0.03	38	90	50	
32	Kala oya + Wilpattu	0.23	0.08	0.09	0.05	0.10	0.06	0.76	0.11	67	144	87	180
33	Moongil oya + Rathambala oya	0.18	0.04	0.06	0.03	0.06	0.04	0.58	0.09	56	201	82	
34	Deduru Oya	0.52	0.26	0.07	0.04	0.07	0.04	0.32	0.05	18	27	21	35
35	Karabalan oya + Maha oya	0.25	0.18	0.02	0.01	0.02	0.01	0.05	0.01	8	9	8	
36	Maha oya	0.69	0.61	0.01	0.01	0.01	0.01	0.01	0.00	2	2	2	2
37	Attanagalu Oya	0.54	0.62	0.01	0.00	0.01	0.01	0.00	0.00	3	2	2	
38	Jaffa Peninsula	1.03	0.22	0.00	0.00	0.00	0.01	0.29	0.04	2	12	4	

Source Author's estimation

Notes [a] This includes 21 larger drainage basins and contiguous coastal area basins with smaller drainage areas; [b]See Annex 1 for storage estimation; [c]evaporation from storage surface area is estimated at 15% of the storage capacity

Table 2.6 2022 Paddy area and production in river basins

River Basin[a]	Major irrigation systems				Minor irrigation systems				Rainfed systems				Total	
	irrigated area ('000 Ha)		Production ('000 MT)		Irrigated area ('000 Ha)		Production ('000 MT)		Rainfed area ('000 Ha)		Production ('000 MT)		Area ('000 Ha)	Production ('000 MT)
	Maha	Yala	Maha	Yala	Maha	Yala	Maha	Yala	Maha	Yala	Maha	Yala		
Kelani Ganga	0.97	0.78	3.51	2.78	3.53	2.65	13.75	8.62	6.28	4.85	24.31	14.05	19.05	67.0
Bolgoda Lake	0.14	0.11	0.50	0.33	0.33	0.19	1.10	0.63	1.44	1.05	4.74	2.81	3.27	10.1
Kalu Ganga	2.10	2.06	6.55	6.26	3.99	3.35	9.49	6.92	7.90	6.12	23.76	16.39	25.53	69.4
Bentota Ganga	0.07	0.06	0.23	0.18	0.64	0.53	2.04	1.47	3.83	3.09	12.57	8.90	8.23	25.4
Bentota-Nilwala	0.81	0.73	3.06	2.41	1.96	1.56	6.26	4.13	7.75	6.28	25.25	17.34	19.09	58.4
Nilwala Ganga	2.90	2.67	11.66	9.65	2.25	2.17	7.81	6.98	4.72	4.44	15.71	13.94	19.16	65.8
Nilwala-Walawe	9.40	9.32	54.41	52.36	2.08	1.74	9.59	7.52	1.22	1.11	4.22	3.55	24.86	131.7
Walawe Ganga	6.35	5.85	30.65	28.67	4.75	3.30	14.87	9.31	2.54	0.86	7.20	1.39	23.66	92.1
Walawe-Krindi oya	5.35	5.26	31.29	30.20	1.25	0.88	5.71	4.12	0.49	0.17	1.89	0.54	13.39	73.7
Kirindi Oya	5.67	5.11	30.68	28.44	2.71	1.42	11.31	6.40	1.64	0.16	6.33	0.51	16.71	83.7
Krindi oya- Manik Ganga	2.83	2.82	16.76	16.23	0.52	0.42	2.59	1.97	0.11	0.08	0.42	0.28	6.80	38.2
Manik Ganga	2.98	2.34	14.65	13.12	2.64	1.20	9.70	5.36	2.29	0.12	9.02	0.38	11.58	52.2
Manik Ganga-Kumbumkan oya	3.57	3.50	20.86	20.13	0.83	0.59	3.81	2.74	0.32	0.11	1.26	0.36	8.93	49.2
Kumbukkan Oya	2.45	1.92	11.53	10.44	2.07	0.92	7.57	4.07	1.89	0.09	7.44	0.27	9.33	41.3
Kumbukkan Oya-Karanda oya	7.69	7.09	35.95	36.80	1.69	0.75	6.42	3.41	2.70	0.10	10.60	0.34	20.02	93.5

(continued)

Table 2.6 (continued)

River Basin[a]	Major irrigation systems				Minor irrigation systems				Rainfed systems				Total	
	irrigated area ('000 Ha)		Production ('000 MT)		Irrigated area ('000 Ha)		Production ('000 MT)		Rainfed area ('000 Ha)		Production ('000 MT)		Area ('000 Ha)	Production ('000 MT)
	Maha	Yala	Maha	Yala	Maha	Yala	Maha	Yala	Maha	Yala	Maha	Yala		
Karanda oya-Gal Oya	9.17	8.71	43.07	44.86	0.95	0.44	4.00	2.04	2.36	0.08	9.19	0.28	21.71	103.4
Gal Oya	10.86	10.00	50.77	51.91	2.49	1.10	9.40	5.01	3.90	0.15	15.29	0.49	28.49	132.9
Gal Oya-Mundini Aru	10.29	9.67	44.80	45.41	1.32	1.20	5.31	5.61	10.11	0.49	37.02	2.16	33.08	140.3
Mundini aru + Miyangolla	13.69	12.88	60.95	62.26	1.73	1.34	6.98	6.29	10.56	0.50	38.92	2.15	40.69	177.5
Maduru Oya	15.73	14.81	71.03	72.70	2.88	2.02	12.65	9.46	8.24	0.37	30.20	1.61	44.04	197.7
Maduru Oya-Mahaweli Ganga	4.17	3.98	18.34	18.94	0.58	0.56	2.52	2.69	3.24	0.16	11.77	0.71	12.69	55.0
Mahaweli Ganga	76.39	69.34	354.83	343.65	28.59	15.46	122.66	63.54	17.04	1.97	63.30	6.57	208.78	954.6
Mahaweli - Yan oya	12.68	10.63	56.14	52.55	5.74	2.63	23.34	11.11	6.08	0.17	19.42	0.75	37.93	163.3
Yan Oya	11.49	7.20	58.81	33.19	9.80	3.60	43.08	14.86	3.14	0.08	11.34	0.25	35.30	161.5
Mee + Ma Oya	6.58	3.98	32.61	18.57	6.98	2.08	29.47	8.94	2.17	0.04	8.01	0.13	21.84	97.7
Ma oya-Kanakarayan aru	10.91	3.75	47.41	18.12	4.71	0.84	19.69	3.81	4.17	0.00	15.41	0.00	24.38	104.4
Kanakarayan Aru	3.66	2.61	15.36	12.99	2.37	0.43	9.42	2.10	3.74	0.00	13.00	0.00	12.81	52.9

(continued)

Table 2.6 (continued)

River Basin[a]	Major irrigation systems				Minor irrigation systems				Rainfed systems				Total	
	irrigated area ('000 Ha)		Production ('000 MT)		Irrigated area ('000 Ha)		Production ('000 MT)		Rainfed area ('000 Ha)		Production ('000 MT)		Area ('000 Ha)	Production ('000 MT)
	Maha	Yala	Maha	Yala	Maha	Yala	Maha	Yala	Maha	Yala	Maha	Yala		
Kanakarayan Aru-Parangi Aru	7.48	5.28	31.75	26.27	2.16	0.43	8.85	2.00	7.52	0.00	25.59	0.00	22.87	94.5
Parangi + Nay aru	3.95	1.33	16.96	6.54	4.40	0.69	17.35	3.39	1.57	0.00	6.06	0.00	11.93	50.3
Aruvi aru	18.74	9.43	94.93	43.03	17.52	5.41	75.56	22.73	4.11	0.09	15.62	0.23	55.30	252.1
Kal Aru-Modaragam aru	11.39	4.78	56.42	21.74	7.94	2.53	34.53	10.36	1.77	0.05	6.66	0.12	28.46	129.8
Wilpattu + Kala oya	16.20	9.30	77.51	40.81	16.61	6.17	65.26	24.14	5.09	1.06	19.61	3.52	54.42	230.9
Moongil oya + Rathambala oya	7.24	2.88	15.67	11.34	11.35	4.33	24.51	15.05	4.17	2.66	15.85	9.04	32.63	91.5
Deduru Oya	9.47	6.73	42.72	28.39	18.38	11.12	69.76	41.84	12.68	8.17	49.12	28.06	66.55	259.9
Karabalan oya + Maha oya	3.27	2.11	13.15	8.86	6.17	3.52	21.67	13.13	3.93	2.55	15.18	8.75	21.55	80.7
Maha oya	3.59	2.55	16.05	10.72	7.32	4.50	27.74	16.58	6.01	4.12	23.58	13.63	28.08	108.3
Attanagalu Oya	0.99	0.74	3.23	2.53	2.03	1.40	7.05	4.36	3.41	2.47	12.45	7.02	11.04	36.6
Jaffa Peninsula	1.33	5.30	25.88	26.45	0.06	3.72	27.60	14.66	4.49	0.47	9.37	1.64	15.37	105.6
Sri Lanka	322.5	257.6	1520.7	1259.8	193.3	97.2	780.4	377.3	174.6	54.2	626.6	168.1	1100	4733

Source Author's estimation
Notes [a] This includes 21 larger drainage basins and contiguous coastal area basins with smaller drainage areas

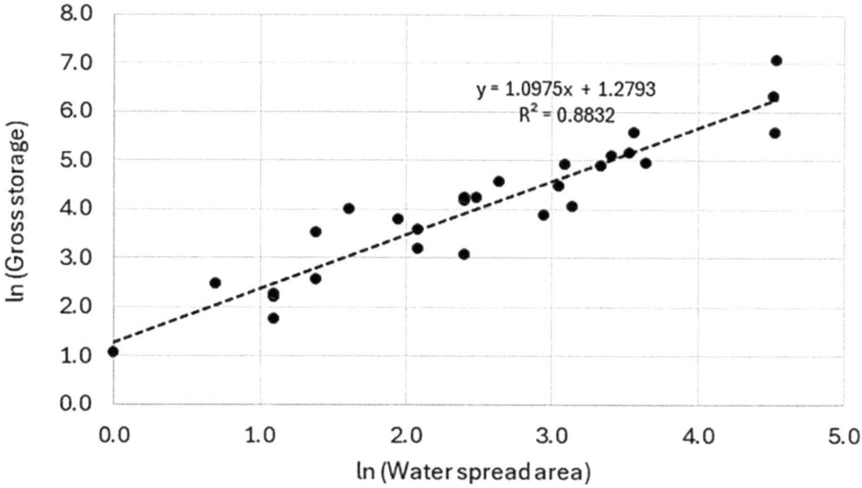

Fig. 2.5 Water spread area and gross storage capacity of DI-managed water bodies across river basins

References

Allen, R. G., Pereira, L. S., Raes, D., & Smith, M. (1998). *Crop evapotranspiration—Guidelines for computing crop water requirements.* FAO Irrigation and Drainage Paper No. 56. Food and Agriculture Organization of the United Nations (FAO), Rome, Italy.

Amarasinghe, U. A., Mutuwatte, L., & Sakthivadivel, R. (1999). *Water scarcity variations within a country: A case study of Sri Lanka.* International Water Management Institute (IWMI) (28 pp.). (IWMI Research Report 32). https://doi.org/10.3910/2009.038

Amarasinghe, U. A. (2010). Spatial variation of water supply and demand in Sri Lanka. In K. Jinapala, S. de Silva, & M. M. M. Aheeyar (Eds.), *Proceedings of the National Conference on Water, Food Security and Climate Change in Sri Lanka* (Vol. 3: Policies, institutions and data needs for water management, pp. 19–33). BMICH, Colombo, Sri Lanka: International Water Management Institute (IWMI).

Amarasinghe, U. A., Amarnath, G., Alahacoon, N., Aheeyar, M., Ghosh, S., Chandrasekharan, K., & Nakada, T. (2021). *Adaptation to climate variability in Sri Lanka: A case study of the Huruluwewa Irrigation System in the Dry Zone* (Vol. 200). International Water Management Institute (IWMI).

De Silva, C. S., Weatherhead, K., & Rushton, K. R. (1996). Sustainability of agro-well irrigation on hard rock aquifers of Sri Lanka. *Tropical Agricultural Research, 8*, 163–174.

Herath, G. (2006). Overview of groundwater management, the agro-well program, and the impact of the 2004 tsunami in Sri Lanka. *International Review for Environmental Strategies, 6*(2), 373–386.

DCS (Department of Census and Statistics). (2020). *Household income and expenditure survey 2019.* Department of Census and Statistics.

DCS (Department of Census and Statistics). (2024). *Agriculture.* Department of Census and Statistics. https://www.statistics.gov.lk/Agriculture/StatisticalInformation/new

Eriyagama, N., Smakhtin, V., & Jinapala, K. (2015). The Sri Lanka environmental flow calculator: A science-based tool to support sustainable national water management. *Water Policy, 17*(6), 1046–1064. https://doi.org/10.2166/wp.2015.158

Falkenmark, M., & Lundqvist, J. (1998). Towards water security: Political determination and human adaptation crucial. *Natural Resources Forum, 22*(1), 37–51. https://doi.org/10.1111/j.1477-8947.1998.tb00708.x
DoA (Department of Agriculture). (2024). *Crop Calendar*. National Agriculture Information and Communication Center, Department of Agriculture. https://doa.gov.lk/naicc-publications-crop-calender/
Harris, I., Osborn, T. J., Jones, P., et al. (2020). Version 4 of the CRU TS monthly high-resolution gridded multivariate climate dataset. *Scientific Data, 7*, 109. https://doi.org/10.1038/s41597-020-0453-3
IWMI (International Water Management Institute). (2003). *Planning groundwater use for sustainable rural development*. Water Policy Briefing Issue 14. International Water Management Institute, Colombo, Sri Lanka
Irrigation Department. (2022). *Scheme Register* (2nd ed.). Irrigation Department.
Kikuchi, M., Weligamage, P., Barker, R., Samad, M., Kono, H., & Somaratne, H. M. (2003). *Agro-well and pump diffusion in the Dry Zone of Sri Lanka: past trends, present status and future prospects*. IWMI Research Report No. 66. International Water Management Institute, Colombo, Sri Lanka.
MA (Mahaweli Authority). (2022). *Mahaweli statistical handbook 2022*. Planning and Monitoring Unit, Mahaweli Authority.
Panabokke, C. R. (2004). Small tank settlements in Sri Lanka. In M. M. M. Aheeyar (Ed.), *Small tank settlements in Sri Lanka: Proceedings of the symposium*. Hector Kobbekaduwa Agrarian Research and Training Institute, Colombo.
Panabokke, C. R., & Perera, A. P. G. R. L. (2005). *Groundwater resources of Sri Lanka*. Water Resources Board.
Shah, T., Samad, M., Ariyaratne, R., & Jinapala, K. (2013). Ancient small-tank irrigation in Sri Lanka: Continuity and change. *Economic and Political Weekly, 48*(11), 58–65.
Sood, A., Manthrithilake, H., Siddiqui, S., Rajah, A., & Pathmarajah, S. (2015). Managing shallow aquifers in the dry zone of Sri Lanka. *Environmental Monitoring and Assessment, 187*(7), 1–16.
Water Supply and Drainage Board. (2024). *Domestic and industrial sector piped water supply*. National Water Supply and Drainage Board.

Dr. Upali A. Amarasinghe Emeritus Scientist at the International Water Management Institute, has over 33 years of experience as a researcher and project leader in water sources management, as well as ten years of teaching experience as a university lecturer. He holds a Ph.D. degree in Statistics from the University of Illinois, Urbana-Champaign, USA. His research focused on water management issues at different spatial scales; which included estimating water productivity and footprints of crops to river basins; food and water supply and demand and projections of river basins to countries; benchmarking of performance of canal irrigation systems; poverty and vulnerability assessment; poverty mapping using small area estimation; water security assessment of industries; impact assessments of interventions in water resources management. He has extensive experience managing research projects in India and Sri Lanka, as well as short-term work experiences in other countries, including Pakistan, China, Malaysia, Thailand, and Vietnam. Dr. Amarasinghe has published over 130 research articles and edited four books.

Ms. Madhusha Perera is a Research Consultant at the International Water Management Institute, with a strong foundation in data analysis and advanced programming. She holds a bachelor's degree with a specialization in Statistics from the University of Sri Jayewardenepura. Her expertise lies in statistical analysis, data-driven research, and proficiency in programming languages.

Dr. Niranga Alahacoon has a strong background in remote sensing and disaster risk management. He holds a Ph.D. in Remote Sensing, awarded by the University of Colombo in 2021. He received the prestigious "Astronomer Royal Sir Arnold Wolfendale Memorial Gold Medal" for the Best Postgraduate Student in Physics at the University of Colombo. He completed an MSc in Geoinformation and Earth Observation (Geo-Hazard) in 2010 at the ITC in the Netherlands and obtained his B.Sc. (Hons) in Geology in 2007 at the University of Peradeniya. Currently, he works at the International Water Management Institute Headquarters as a Regional Researcher, contributes significantly with application of remote sensing data and image processing, geographic information systems (GIS), hydrological and hydrodynamic modelling, geology, and rapid emergency response mapping, and has a publication record of more than 40 scientific publications and 22 training manuals. Prior to joining IWMI, Niranga served as a lecturer at the Faculty of Fisheries and Marine Sciences and Technology (FMSC), University of Ruhuna, Sri Lanka.

Mr. Thilina Madushanka is a chartered civil engineer, graduated from the University of Moratuwa in 2013. He has dedicated research to the Water Resources Development sector in Sri Lanka, with a particular focus on Infrastructure development projects. Dr. Madushanka is a National researcher in IWMI, and works in several projects in Sri Lanka and the Southern African region (South Africa, Botswana, Zimbabwe, and Mozambique). From 2016 to 2023, he served as a Water Resources Engineer (Modelling) for the Mahaweli Water Security Investment Program. He has collaborated with various organizations such as ADB, IUCN, at UNDP. Dr. Madushanka is an expert in hydrological modeling using tools such as SWAT+, SWAT, and HEC-HMS, as well as Water allocation modeling utilizing Riverware, HEC-ResSim, and WEAP.

Chapter 3
Sri Lanka: Political, Economic, and Institutional Setting

Sachini Ukwattage, Kaushika Seelanatha, Madhusha Perera, and Upali A. Amarasinghe

Abstract Sri Lanka has experienced impressive socioeconomic development in recent decades. It boasts one of the highest literacy rates and per capita gross domestic product, as well as the lowest infant and maternal mortality rates and levels of child stunting in South Asia. However, significant non-climatic risks threaten to undermine the country's recent progress. This chapter examines the changes in the political, institutional, and administrative landscape, as well as demographic and economic developments, with a particular emphasis on the role of the agricultural sector in providing food, employment, and economic growth during the post-independence period since 1948. Factors such as slowing population growth, the declining contribution of agriculture to the gross domestic product, increasing urbanization, shifting lifestyles, and rising vulnerabilities in the dry zone are expected to influence future water demand patterns and investment in resource management.

Keywords Population growth · Poverty · Monsoon · Vulnerability · Gross domestic product

3.1 Introduction

Sri Lanka's socioeconomic and political landscape has undergone significant changes since its independence from the United Kingdom in 1948. Before independence, various colonial powers, including the Portuguese, Dutch, and British Empires, governed the country from 1505 to 1948 (Perera 1981). After independence, Sri

S. Ukwattage · K. Seelanatha · M. Perera · U. A. Amarasinghe (✉)
International Water Management Institute, Colombo, Sri Lanka
e-mail: u.amarasinghe@cgiar.org

K. Seelanatha
e-mail: k.seelanatha@cgiar.org

M. Perera
e-mail: m.perera@cgiar.org

U. A. Amarasinghe et al. (eds.), *Water Governance for Climate Change Adaptation in Sri Lanka*, Global Issues in Water Policy 13,
https://doi.org/10.1007/978-3-032-04950-6_3

Lanka established a multi-party-political system and became a democratic Socialist Republic with the constitution introduced in 1978. The president, who can serve a maximum of two five-year terms as per the 19th Amendment to the Constitution, is both the head of State and the Government (DeVotta, 2005).

The national government consists of 225 parliamentary members. Of these, 196 are elected through proportional representation from each district in a general election held every six years, while the remaining 29 members are nominated by political parties based on the proportion of votes received (GoSL, 2018). Elections are also held to select representatives for the nine provincial councils and 341 local government bodies, which include 29 municipal councils, 36 urban councils, and 276 Pradeshiya Sabhas (GoSL, 2025). Members of local government councils are elected through a mixed-member proportional representation system. A governor appointed by the President heads each provincial council, and a Chief Minister is chosen from among the elected members. The leader of the council is either the head of the party holding a clear majority or, in the absence of such a majority, elected through a vote among the council members.

Sri Lanka's population in 2023 is estimated at 21.9 million (DCS, 2025). The population growth rate, which peaked at 2.8% annually during the late 1940s to early 1950s, has gradually declined to 0.5% between 2012 and 2024. While detailed data from the 2023 Census is not yet available, the 2012 Census indicated that the major ethnic groups were Sinhalese (74%), Tamils (11%), Muslims (9%), and Burghers (6%). The major religions recorded were Buddhism (70.1%), Hinduism (12.6%), Islam (9.7%), Catholicism (6.2%), and other forms of Christianity (1.2%) (DCS, 2015a).

Poverty, measured by the percentage of the population living below the national poverty line—also known as the headcount ratio (HCR)—has declined significantly over the years. Based on the poverty line adjusted using the 2002 Colombo Consumer Price Index (CCPI), the HCR was 26% in 1990–91 (DCS, 2015a) and decreased to 3.4% by 2019, although it varied across districts, ranging from 0.3 to 10.8%. However, with the poverty line adjusted using the 2012/13 National Consumer Price Index (NCPI), the HCR for 2019 is higher, at 14.3%, with district-level variation ranging from 2.7 to 44.5%. Districts that were most severely affected by internal conflicts before 2010 generally have higher poverty rates.

Multidimensional poverty, which measures progress across several aspects of human development, was 0.067 in 2023. It varied from 0.014 to 0.193 across districts (DCS, 2019). Along with income poverty, MPI considers the health and educational levels of the population to estimate the overall as well as the severity of deprivations in the three dimensions. In Sri Lanka, inadequate access to cooking fuel, safe drinking water, healthcare, and other basic facilities contributes to the highest level of deprivation and accounts for most of the 2019 MPI variation (DCS, UNICEF, OPHI, 2019).

In 1960, the country's gross domestic product (GDP) was 5 billion, and agriculture contributed 25% (World Bank, 2024). The GDP has increased at an annual rate of 4.6% since 1960 and reached 87.8 billion USD by 2023. Services and industrial sectors are now contributing 90% of the total 2023 GDP, with the services sector holding the dominant share (60%). Per capita GDP has increased from 589 USD per person in the early 1960s to 4,122 USD per person in the early 2020s.

Sri Lanka has made remarkable progress in human development, despite its low-income level. The country has made significant strides in key health indicators, including increases in life expectancy among its citizens and the elimination of malaria, filariasis, polio, and neonatal tetanus. However, Sri Lanka's health system faces challenges due to an aging population and the burden of non-communicable diseases, which contribute to about 75% of deaths in the country.

Three internal civil disturbances impacted GDP growth in three distinct periods. The first internal disturbance since independence, which occurred in the form of a youth uprising by the Janatha Vimukthi Peramuna (JVP) against perceived discrimination, was between 1978 and 1979. The second major uprising was started in the northern province by a group of young Tamilians, called the Liberation Tigers of Tamil Eelam (LTTE). The LTTE fought for an independent Tamil state in the north-east of the island due to the perceived persecution and discriminatory policies of the Sinhalese-dominated governments of the country. The violence started on a small scale in the north in the 1980s, spread to the East, and then to the rest of the Island, causing negative GDP growth in the early 1990s. The economy, particularly in the north and east, was vastly affected by the LTTE uprising. The other major civil disturbance, which was started again by the JVP, occurred during the 1989–1990 period.

Although Sri Lanka won its independence from Great Britain in 1948, the right to free education was enshrined in the constitution only 30 years later. Nevertheless, it has the highest reported youth literacy rate in South Asia at 98.8%, compared to 89.7% in India and 83.2% in Bangladesh. The United Nations recognizes Sri Lanka, along with the Maldives, as one of only two countries in South Asia to have achieved "high human development." (Justine D'Souza, Credential Analyst, WES, and Thomas D. Moore, Credential Examiner, WES).

This chapter provides details of the physical and socio-economic landscapes of Sri Lanka, with a particular focus on regions where climate-related impacts are strongly felt and are expected to influence water governance in the future. The following section examines the variation in climate, demographic, and poverty trends, as well as the contributions of different sectors to the economy, and land use and cropping patterns that account for a significant portion of water withdrawals and contribute to seasonal water scarcities in river basins. We conclude the chapter by identifying the implications of spatial and temporal trends of key parameters for water resources management and governance.

3.2 Physical and Socio-economic Landscape

3.2.1 Administrative Divisions and Demography

Sri Lanka, with a total land area of 65,610 square kilometers, is divided into nine provinces: Northern, Eastern, Western, Southern, Central, North Central, North-western, Sabaragamuwa, and Uva. These provinces are further divided into 25 districts, which are then split into 331 Divisional Secretariat (DS) divisions (see Fig. 3.1a). At the local level, administration is handled through Grama Niladhari (GN) divisions, totaling 14,021 units nationwide.

These administrative divisions stretch across 103 river basins (Fig. 3.1b). Of these, 21 major basins are significant because they cover approximately 76% of the country's land area and account for 78% of the average annual surface water runoff, which totals 52 billion cubic meters. These water resources are vital for the nation's agricultural activities (see Chap. 2 for details).

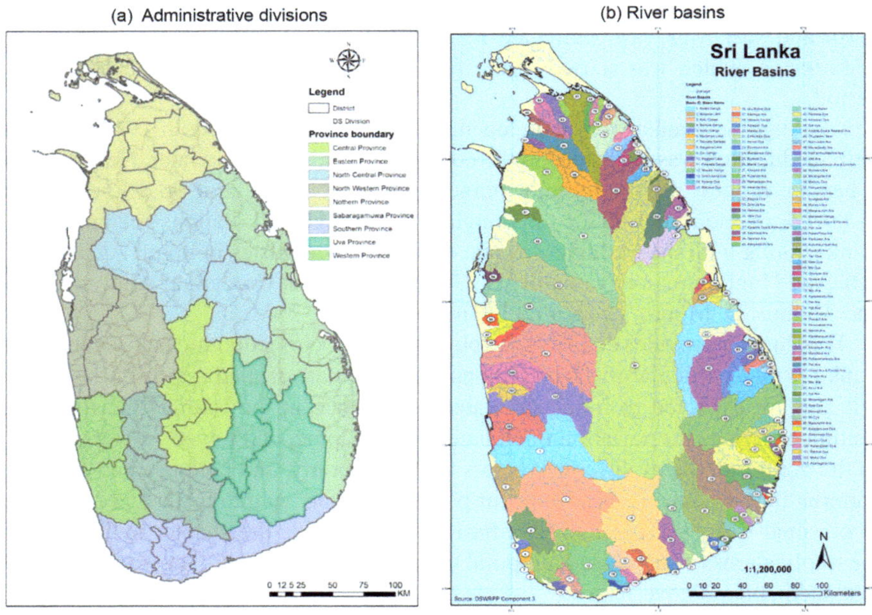

Fig. 3.1 Administrative divisions and river basins in Sri Lanka

3.2.2 *Agroecological Regions and Climate*

Sri Lanka is divided into 46 agroecological zones (AEZs), which fall under three main agro-climatic zones: the wet zone, the intermediate zone, and the dry zone (Fig. 3.2a). The wet zone includes nine districts, while the dry and intermediate zones cover the remaining 16 districts, though many of these districts span more than one zone.

The country receives an average annual rainfall of around 2,000 mm, but this varies significantly by region and season. In the wet zone, rainfall typically exceeds 2,000 mm, while the dry zone receives less than 2,000 mm (Fig. 3.2b). The intermediate zone experiences rainfall between 1,250 and 2,500 mm.

Sri Lanka has four distinct monsoonal seasons:

- **First Inter-Monsoon (FIM)**: March to April
- **Southwest Monsoon (SWM)**: May to September
- **Second Inter-Monsoon (SIM)**: October to November
- **Northeast Monsoon (NEM)**: December to February.

Rainfall from the FIM and SWM supports the **Yala** cultivation season (April to September), while rainfall from the SIM and NEM contributes to the **Maha** season (October to March) (Table 3.1). The wet zone, which comprises approximately 23% of Sri Lanka's land area, receives around 38% of the country's total annual rainfall, equivalent to approximately 120 billion cubic meters (Bm^3), based on the average from 1999 to 2022. The **Maha** season accounts for about 62% of the country's total rainfall, although this varies by region, ranging from 47% in the wet zone to 69% in

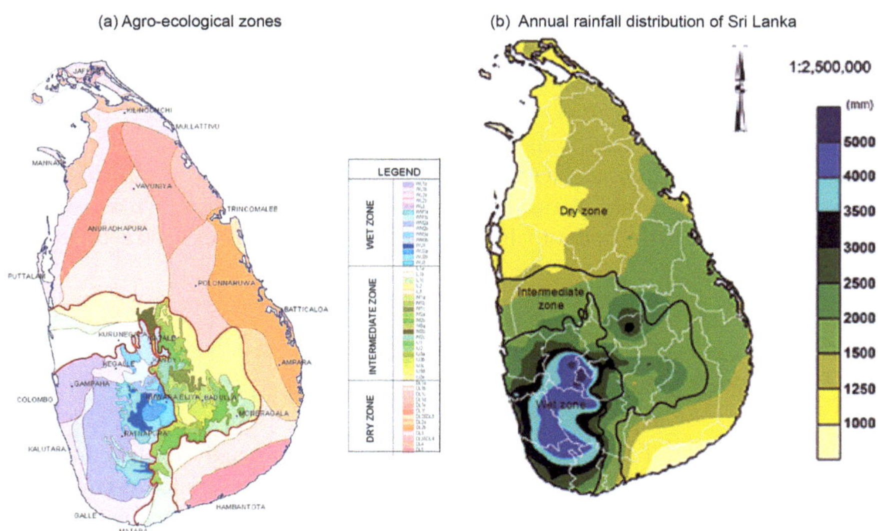

Fig. 3.2 Agro-ecological zones and annual rainfall distribution

Table 3.1 Rainfall distribution across wet- and dry-zones in Yalan and Maha seasons

Climatic zone	Parameter	Yala season rainfall (mm)			Maha season rainfall (mm)			Total (mm)
		FIM	SWM	Total	SIM	NEM	Total	
Wet zone (65,610 km^2)	Average	598	850	1449	709	586	1295	2744
	CV	(0.27)	(0.29)		(0.28)	(0.33)		
Dry zone (15,260 km^2)	Average	224	207	430	524	537	1061	1492
	CV	(0.35)	(0.43)		(0.27)	(0.44)		
Sri Lanka	Average	350	396	746	610	613	1224	1970

Source Authors' estimates based on data collected from the Meteorology Department, Sri Lanka

the dry zone. Rainfall during the **Yala** season is significantly more intense in the wet zone, where it is over three times higher than in the dry zone. The low intensity of Yala season rainfall, particularly in the dry zone, poses a major challenge to sustaining year-round agriculture in these areas.

Sri Lanka's dry zone is especially vulnerable to high year-to-year variability in rainfall, with significant fluctuations during the First Inter-Monsoon (FIM), the Southwest Monsoon (SWM), and the Northeast Monsoon (NEM). This variability, projected to worsen due to climate change, poses a significant challenge to agricultural productivity, particularly in rice cultivation, which is crucial to the country's food security. The dry zone contributes a substantial share of Sri Lanka's total rice output. Rainfall variability is especially problematic in districts that rely heavily on rain-fed agriculture, including Anuradhapura, Batticaloa, Kurunegala, and Monaragala. In these areas, more than 50% of agricultural activity depends on seasonal rainfall. Together, these districts are home to a large segment of the population and account for approximately 40% of the nation's rice production.

Climate change projections indicate that the wet zone will become wetter, while the dry zone will become drier, intensifying existing rainfall variability (Chap. 5). These changes are expected to have a significant impact on economic activities in both regions. In the wet zone, increased rainfall and flooding could disrupt industrial and service sectors, while worsening dryness in the dry zone will likely reduce agricultural productivity, both of which are critical to Sri Lanka's food security and economic growth.

3.2.3 Demographic Trends

An accurate picture of demographic trends is crucial, especially in the context of increasing climate variability and the associated risks of disasters, such as landslides, floods, and droughts. Accurate population trends are crucial for the effective planning and management of water resources, as well as for adaptation and development planning.

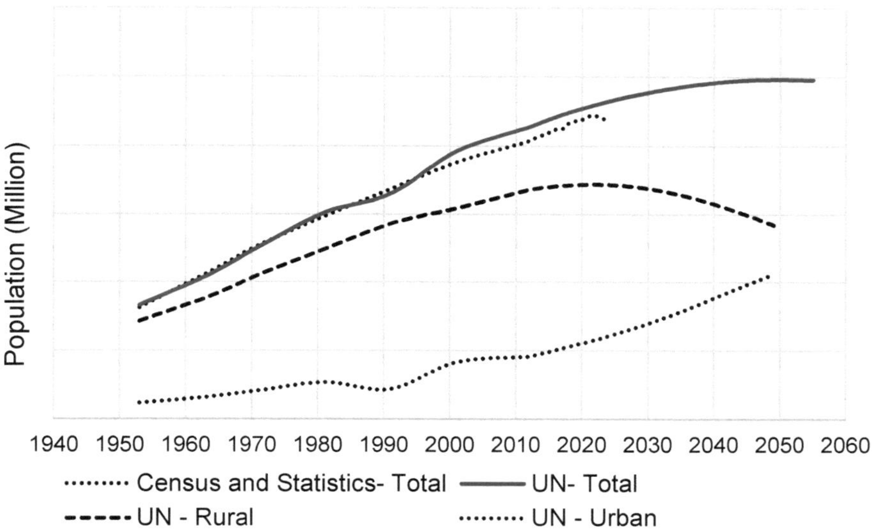

Fig. 3.3 Population growth. *Source* Department of Census and Statistics, Sri Lanka, and the United Nations Population projections

Sri Lanka's population has grown significantly since the first national census in 1871, rising from 2.4 million to 20.4 million by 2011, the year of the 14th census (Fig. 3.3). Based on mid-year estimates from the Department of Census and Statistics, the population is expected to increase by about 1.49 million by 2024 (DCS, 2025). However, the growth rate has steadily declined over the decades, from 2.8% in 1953 to just 0.7% in 2011 (DCS, 2011). The preliminary statistics of the 2023/24 census data suggest a further decline (0.5%) in overall population growth between 2012 and 2024, primarily due to lower birth rates and increased migration (DCS, 2025).

However, the United Nations' medium population projections suggest that Sri Lanka's population will continue to grow, reaching a peak of around 24.14 million by 2051 (UN, 2024). However, the 15th Population Census (2024–2025), just completed (DCS, 2025), confirms the recent trends, and it is not likely to reach UN projections by 2050. The population changes in recent years are being driven mainly by low birth rates and increased migration. These demographic shifts are common in most districts except the Monaragala district (Annex Table 3.5). The changing demographic shifts have major implications for investments in water resources management. Most major water resources development in the past had a long-term perspective on population growth and food demand patterns. However, with reduced population growth, investments in water resource management may have entered an era of managing existing rather than developing new infrastructure.

As of 2023, Sri Lanka's population density had increased to approximately 336 people per square kilometer, although this varies significantly by region (Table 3.2). For example, Colombo District in the Western Province has a very high population density of 3,438 people per square kilometer, while Mullaitivu District in the Northern Province has just 38 people per square kilometer (Annex Table 3.5). It's also worth noting that the Western Province, which makes up only 6% of the country's land area, is home to 22% of the total population. According to the 2011 census, approximately 77.4% of the population lives in rural areas, many of whom heavily depend on agriculture for their livelihoods.

Sri Lanka's urbanization has progressed at a relatively slower pace and on a smaller scale compared to other countries in the region. As of 2011, only 18.2% of the Sri Lankan population lived in urban areas, in contrast to 32% in India, 39% in Pakistan, and 32% in Bangladesh (UN, DESAPD 2024). Over the past 30 years, the urban population growth rate has remained below 3% per year (DCS, 2015a).

The Western Province is home to 61% of Sri Lanka's urban population, with Colombo District alone accounting for 48% of the total (Annex Table 3.5). Other significant urban centers include the districts of Kandy, Galle, Jaffna, Batticaloa, and Ampara. Meanwhile, the estate sector—largely dependent on plantation livelihoods—is concentrated in the Central, Sabaragamuwa, and Uva provinces, representing 4.4% of the population.

However, these low urbanization figures may not accurately reflect the widespread development occurring in Sri Lanka's peri-urban areas (Akalanka et al., 2024; UN-Habitat, 2018). According to the UN's *State of Sri Lankan Cities 2018* report (UN-Habitat, 2018), the actual urban population is likely much higher than what official census data suggests. In fact, the nine provincial capitals—Colombo (Western), Kandy (Central), Galle (Southern), Kurunegala (Northwestern), Anuradhapura (North Central), Jaffna (Northern), Trincomalee (Eastern), Badulla (Uva),

Table 3.2 Geographic and demographic details of Sri Lanka

Factors	2001	2011	2022	Wet zone	Dry zone
Area (km^2)	65,610	65,610	65,610	15,715	49,885
Total population (1,000 s)	18,797	20,359	22,037	11,744	8,615
Population density (persons/km^2)	286	325	336	782	181
Rural population (% of total)	80.0	77.4	77.4	69.9	87.6
Urban population (% of total)	14.6	18.2	18.2	24.1	10.2
Estate sector population (% of total)	5.4	4.4	4.4	6.0	2.2
Poverty at the national poverty line (% of total)	22.2	6.7	3.2	5.0	9.1
Rural poverty at the national poverty line (% of total)	24.7	7.6	3.3	–	–
Multidimensional poverty (% of total)	–	–	16	–	–

Sources All data except for GDP data are from the Department of Census and Statistics. GDP estimates are from the World Bank
Notes The wet zone comprises nine districts (Colombo, Gampaha, Kalutara, Galle, Matara, Ratnapura, Kandy, Nuwara Eliya, and Kegalle), while the intermediate and dry zones encompass the remaining 16 districts

and Ratnapura (Sabaragamuwa)—together may host over 7.9 million urban residents. The challenge for water resources managers will be supplying a clean and adequate drinking water supply for the growing urban population. Moreover, with the expansion of services and industrial sectors in the economy (see the next section), there will be a significant increase in water demand from these sectors.

3.2.4 *Gross Domestic Product*

Sri Lanka's industrial and service sectors have made significant contributions to the country's Gross Domestic Product (GDP). From 1975 to 2018, the GDP of the industrial and service sectors grew at annual rates of 5.45% and 5.60%, respectively (Fig. 3.4a). In contrast, the agriculture sector's GDP grew at a slower rate of 2.78%. The share of industrial and service sectors in total GDP increased from 78% in the 1980s to 92% by 2018. This GDP growth is despite notable declines in some periods, including a sharp dip in 2000–2001 due to Tamil insurgent attacks on the Colombo airport and port, and another decline after 2018, primarily due to the COVID-19 pandemic and the economic crisis of 2022 (Fig. 3.4b). The financial crisis in 2022 was triggered by poor macroeconomic management, unsustainable debt, and a depletion of foreign reserves, ultimately leading to a debt default.

Agriculture is highly sensitive to weather conditions, with significant declines in growth occurring in 1987, 1992, 1996, 2001, and 2016–17, primarily due to severe droughts (Fig. 3.4b). Despite these setbacks, agricultural GDP has grown at an average rate of 2.38% annually from 1975 to 2018. However, the sector's

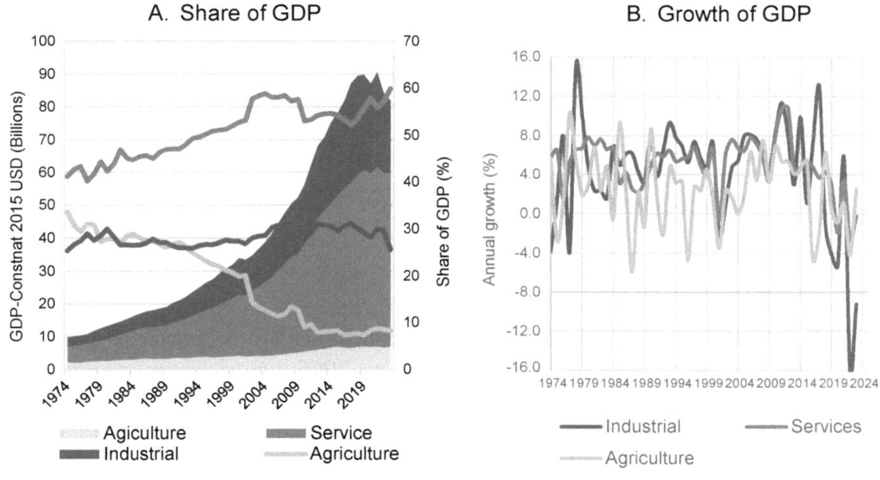

Fig. 3.4 Industrial, service, and agricultural sector contributions to GDP. *Source* World Development Indicators, World Bank (2024)

contribution to the total GDP has decreased over recent decades, falling from 27% in 1991 to approximately 8.2% in 2023 (Fig. 3.3a).

3.2.5 Poverty

Poverty, as measured by the headcount ratio (HCR), the percentage of people living below the national poverty line, has decreased significantly over the past three decades. It fell from 29% in the mid-1990s to approximately 3.2% in 2019 (Table 3.3). This substantial reduction in poverty has been observed in both urban and rural sectors. However, the head count ratio in the estate sector, which has remained notably higher, has not shown any significant decline since 2010.

At the sub-national level, the districts of Nuwara Eliya, Badulla, and Monaragala, which have a large population in the estate sector, had some of the highest poverty rates in 2019 (Annex Table 3.6). Districts in the Northern, Sabaragamuwa, Uva, and Eastern provinces also experience significantly higher poverty levels than the rest of the country. Further disaggregation of poverty using small-area estimates reveals that many Divisional Secretariat (DS) divisions have substantially higher headcount ratios than the national or provincial averages (DCS, 2015c).

Historical trends indicate that agricultural GDP growth is closely linked to reductions in rural poverty (Fig. 3.5). A 1% increase in agricultural GDP is associated with a 4.1% reduction in rural HCR, emphasizing the strong connection between agricultural development and poverty reduction in Sri Lanka. Higher agricultural wages, which have grown annually by an average of 5.7% since 2006, contributed to a rapid decline in rural poverty. However, the 2021–22 financial crisis, along with a sudden shift in agricultural policy towards 'organic fertilizer,' has led to a sharp increase in poverty in Sri Lanka, reaching 43% (FAO and WFP, 2023). The GINI coefficient of 0.46 indicates a significant disparity among household incomes and poverty levels (CBSL, 2021).

3.2.6 Contribution to Agricultural GDP Growth

In absolute terms, agricultural output increased by 29% between 1990 and 2022 (FAO, 2024). The civil unrest that prevailed in the country significantly affected agricultural output growth until 2005. Although unrest in the northeast persisted until 2009, growth accelerated considerably after 2005 due to sector-specific policies aimed at ensuring food security and enhancing farmers' profits. These policies included government support for prices and fertilizers. However, the COVID pandemic, followed by a sudden shift in policy towards organic fertilizer use and an economic crisis, contributed to the abrupt decline in agricultural GDP in the early 2020s.

Table 3.3 Headcount ratio and multidimensional poverty (MPI) indicators (% of total population)

Sector	HCR								MPI
	1990–1991	1995–1996	2002	2006–2007	2009–2010	2012–2013	2016	2019	2019
National level	26.1	28.8	22.7	13.3	8.9	6.7	4.1	3.2	16
Urban	16.3	14.0	7.9	6.7	5.3	2.1	1.9	1.1	4.4
Rural	29.5	30.9	24.7	15.7	9.4	7.6	4.3	3.3	16.6
Estate	20.5	38.4	30.0	32.0	11.4	10.9	8.8	10.4	51.3

Source Department of Census and Statistics (DCS, 2008, 2011, 2017, 2019)

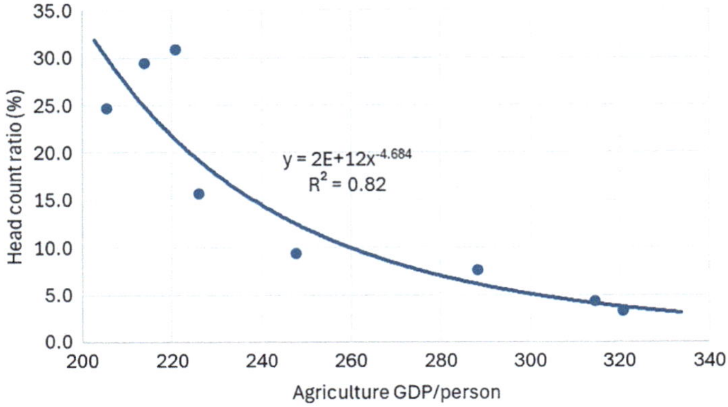

Fig. 3.5 Rural HCR versus per capita agricultural GDP. *Source* World Bank (2024), DCS (2024)

The agriculture sector's contribution to the overall economy is expected to decrease further; however, there is potential for growth by developing the commercial and export-oriented agriculture. The economic policies of recent governments aimed to support the development of commercial agriculture in response to increasing demands in local and export markets. Major and minor export crops (tea, rubber, coconut, and spices) account for the largest share (37%) of agricultural GDP (Fig. 3.6), followed by rice at 18%. The outputs of fruits, vegetables, and livestock products, including milk and meat, which are significant components in changing food consumption patterns, now account for 18% of agricultural output.

Rice has the largest output value by far. Rice production has nearly doubled over the last three decades (Fig. 3.7). Spice production, the second largest, increased by more than 180%, while tea and vegetable production increased by 210% and 108%, respectively. Animal products also show significant output growth, with meat increasing by 690%, milk by 120%, and eggs by 119%, respectively. These figures indicate the diversification trends in agricultural production patterns and highlight the sector's robustness, with exports and high-value crops making significant contributions to recent growth in agricultural GDP. However, despite these positive trends, the output values for tea, coconut, and rubber have declined since 2015.

3.2.7 Land Use Patterns

Arable land and cropped areas have increased by 52% and 25%, respectively, since the early 1990s (Fig. 3.8). However, this growth has been accompanied by a gradual decline in cropping intensity, from over 200% in 1991 to 173% by 2022.

The primary reasons for this decline include the expansion of rainfed crop areas, such as annual crops, tea, and spices, as well as the limited availability of irrigation facilities. In the mid-1990s, approximately 64% of arable land was equipped for

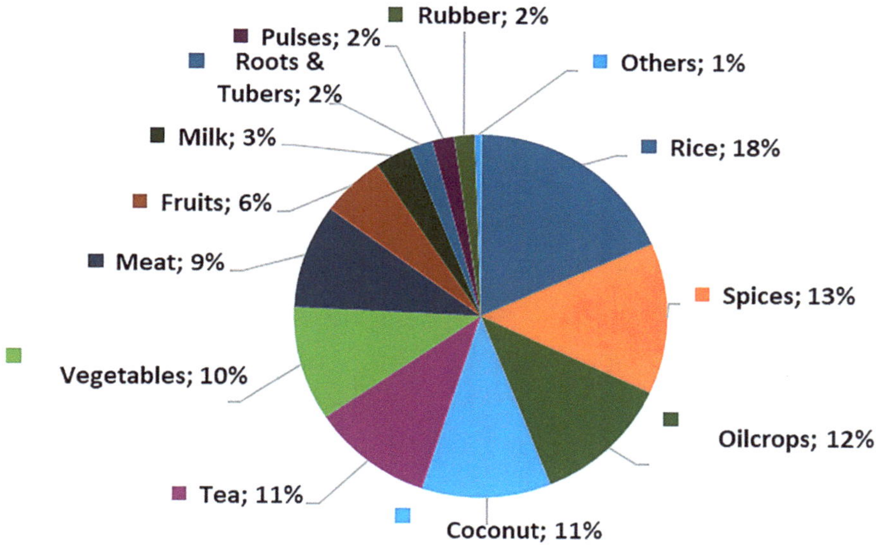

Fig. 3.6 Share of the value of the output of commodities in 2020–2022. *Source* FAOSTAT data (FAO, 2024)

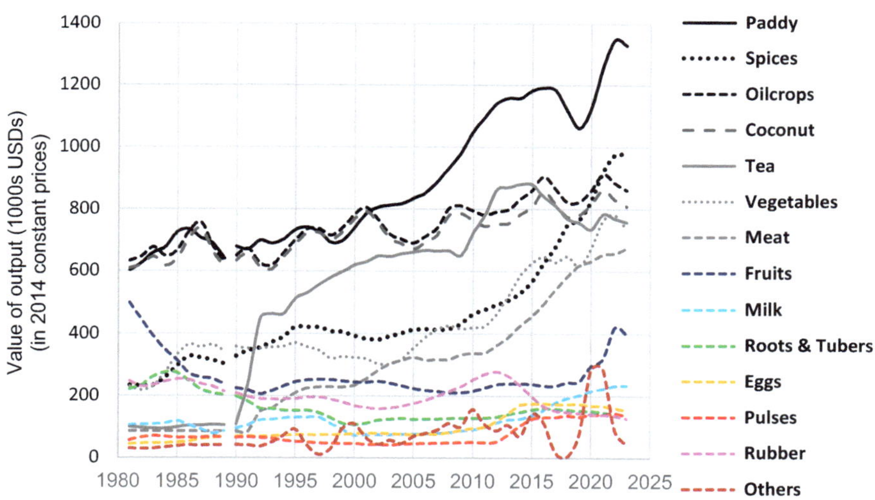

Fig. 3.7 Growth of the value of the output of commodities. *Source* FAOSTAT data (FAO, 2024)

irrigation, but by the early 2020s, this figure had decreased to 46%. With most of the potential areas for irrigation already developed, increasing the irrigated area during the Yala season presents a significant challenge under climate change. Since seasonal crops are susceptible to weather vagaries, water-related climate risks to agricultural

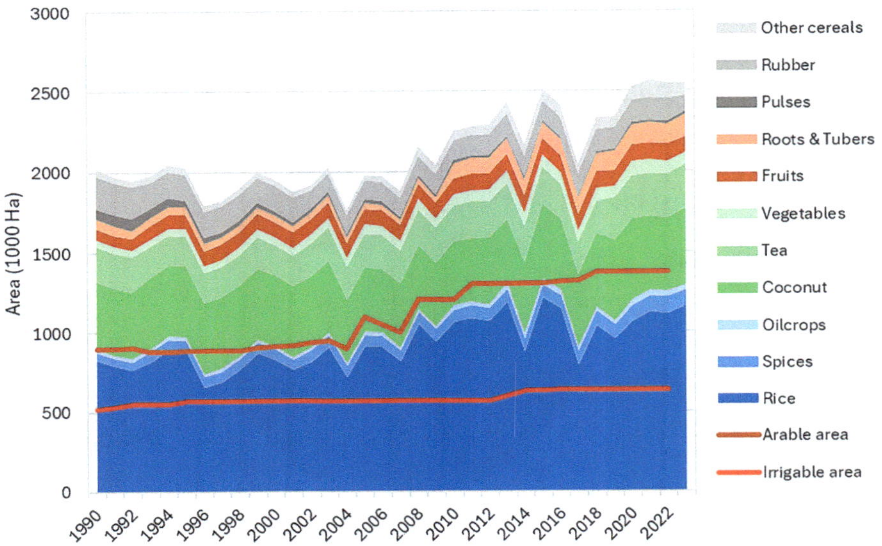

Fig. 3.8 Land use and cropping patterns. *Source* FAOSTAT DATA (FAO, 2024)

production, especially those outside the major irrigated areas, are prominent in the dry zone. Among these, floods and droughts are the most notable. A prolonged period of high temperatures can also affect agriculture in both the dry and wet zones.

Between 1990 and 2022, cropping patterns indicate that seasonal crops account for 60% of the total cropped area (Table 3.4), with 87% of this area located in the dry-zone districts. Paddy alone accounts for 80% of the total seasonal cropped area, with 88% of the paddy area located in the dry zone. Vegetables and other cereals, such as maize, account for 12% of the seasonal cropped area. While plantation crops like tea, rubber, and coconut dominate the non-paddy crop area in the wet zone, coconut (primarily in the intermediate zone), fruits, and spices are the main non-paddy crops in the dry zone. The country's total area equipped for irrigation is approximately 570,000 hectares (FAO, 2024), with 97% cultivated with paddy during the Maha season and 68% during the Yala season, reflecting a paddy irrigation intensity of 165%.

Additionally, certain areas dedicated to other field crops—such as maize, vegetables, chili, ginger, and sugar—receive supplemental irrigation, particularly during the Yala season. According to AQUASTAT (FAO, 2024), the irrigated area for other field crops comprises approximately 700 hectares of maize, 100 hectares of other cereals, 800 hectares of legumes, 4,300 hectares of roots and tubers, 12,000 hectares of vegetables, 17,400 hectares of sugarcane, and 4,100 hectares of various crops. However, sub-national statistics on the irrigated areas for these crops are unavailable, so it is assumed that only a portion of the Yala season non-paddy cropped area in the dry-zone districts receives irrigation. The total irrigated area for other field crops is approximately 29,000 hectares.

Table 3.4 Cropped and irrigated area in 2019–2022

Crop	Cropped area (1000 Ha)							Irrigated area (1000 Ha)						
	Wet zone		Dry zone		Sri Lanka			Wet zone		Dry zone		Sri Lanka		
	Maha	Yala	Maha	Yala	Maha	Yala	Total	Maha	Yala	Maha	Yala	Maha	Yala	Total
Paddy	82.3	66.5	667.6	376.3	750.0	442.8	1192.8	41.1	33.4	509.7	354.3	550.8	387.7	938.6
– Major irrigation	15.3	13.9	335.5	271.5	350.9	285.4	636.3	15.3	13.9	335.5	271.5	350.9	285.4	636.3
– Minor irrigation	25.8	19.5	174.2	82.8	200.0	102.3	302.3	25.8	19.5	174.2	82.8	200.0	102.3	302.3
– Rainfed	41.2	33.1	157.9	22.0	199.1	55.0	254.2	0.0	0.0	0.0	0.0	0.0	0.0	0.0
Maize	0.7	0.8	71.2	7.8	72.0	8.6	80.6	0.0	0.0	0.0	0.7	0.0	0.7	0.7
Other cereals	0.4	0.3	4.0	0.6	4.4	0.9	5.3	0.0	0.0	0.0	0.0	0.0	0.0	0.0
Pulses	0.4	0.7	16.5	11.9	16.9	12.6	29.5	0.0	0.0	0.0	0.8	0.0	0.8	0.8
Oilseeds	0.3	0.3	10.4	14.4	10.7	14.6	25.3	0.0	0.0	0.0	0.0	0.0	0.0	0.0
Roots and tubers	6.1	5.6	10.5	11.6	16.6	17.3	33.8	0.0	0.0	0.0	2.8	0.0	2.8	2.8
Vegetables (upcountry)	11.2	4.7	23.4	8.2	34.7	12.9	47.6	0.0	0.0	0.0	3.4	0.0	3.4	3.4
Vegetables (low country)	4.8	5.3	5.4	5.0	10.2	10.2	20.4	0.0	0.0	0.0	1.2	0.0	1.2	1.2
Green chili	1.3	1.1	5.7	4.0	7.0	5.1	12.1	0.0	0.0	0.0	2.9	0.0	2.9	2.9
Total (seasonal crops)	107.6	85.3	814.7	439.6	922.3	524.9	1447.2	41.1	33.4	509.7	366.1	550.8	399.5	950.4
Ginger + turmeric		5.9		5.0			10.9		0.0		1.1			1.1
Spices		25.9		23.5			49.4		0.0		0.0			0.0

(continued)

Table 3.4 (continued)

Crop	Cropped area (1000 Ha)							Irrigated area (1000 Ha)						
	Wet zone		Dry zone		Sri Lanka			Wet zone		Dry zone		Sri Lanka		
	Maha	Yala	Maha	Yala	Maha	Yala	Total	Maha	Yala	Maha	Yala	Maha	Yala	Total
Fruits		20.7		27.2			47.9		0.7		3.4			4.1
Sugar		0.4		12.0			12.4		0.4		12.0			12.4
Betal + tobacco		0.6		0.7			1.3		0.0		0.0			0.0
Tea		219.0		47.0			266.0		0.0		0.0			0.0
Rubber		122.5		15.4			137.9		0.0		0.0			0.0
Coconut		130.5		343.4			473.9		0.0		0.0			0.0
Total (all crops)		718.5		1728.5			2447.0		75.6		892.4			968.0

Source Authors' estimates from the Department of Census and Statistics and FAOSTAT data

Notes

1. Other cereals (kurakkan, sorghum, and meneri) show 2011–2013 average
2. Uses include black gram, green gram, and cowpea
3. Oilseeds include gingelly, groundnut, and mustard
4. Roots and tubers include manioc, sweet potatoes, potatoes, red and big onions
5. Vegetables include brinjals, bitter and snake gourds, tomatoes, cucumber, carrot, nokhol, radish, beetroot, beans, leeks, pumpkin, ash plantain, and capsicum
6. Fruits include oranges, limes, mangoes, plantain, papaw, pineapple, and passion fruit
7. Spices include cinnamon, pepper, nutmeg, and cloves

3.2.8 *Vulnerability to Water and Agricultural Risks*

Given the water stress, agricultural-based livelihoods, climate variability, exposure to disasters, and socio-economic sensitivity and coping capacity, many rural areas in Sri Lanka are vulnerable to climate change risks. These regions are mostly concentrated in the north and east (Fig. 3.9). These vulnerability maps are created using 41 indicators, of which 13, 14, and 14 indicators, respectively, assessed the exposure (Fig. 3.9a), sensitivity (Fig. 3.9b), adaptive capacity (Fig. 3.9c), and overall vulnerability (Fig. 3.9d) to hazards and risks (Amarasinghe et al., 2024).

Figure 3.9 shows that 27 river basins have very high vulnerability, 49 river basins have high vulnerability, and 23 river basins have moderate vulnerability to water and agricultural risks. Only six river basins have little or no vulnerability. The vulnerable basins are primarily water-stressed river basins that were identified in Chap. 2. These basins span multiple districts in the dry zone, where agricultural-related activities are the dominant livelihood forms, and climate extremes have the most significant impact. Thus, stress on water, food, and livelihood systems is interrelated and requires well-targeted adaptation interventions.

3.3 Conclusion

The socio-economic trends indicate that Sri Lanka needs to reassess its investments in addressing climate change-related disasters and risks. These investments require better targeting, which necessitates further refinement in identifying vulnerable locations. Socio-economic trends and climate change impacts need to be studied at the divisional secretariat (DS) level, with the grama niladari (GN) level being the ideal situation.

Sri Lanka should pay particular attention to demographic trends in planning long-term investments. Decelerating population growth, which differs from past projections, and a rapidly changing urban–rural structure with internal and external migration, could require a completely different focus on water resources development to meet future needs. Effective management of both large and small storage facilities is a key to addressing short-term water governance challenges. A related issue is the structure of the cropping patterns, which is currently dominated by paddy cultivation.

With a strong focus on improving land and water productivity in high-performing rice irrigated areas, crop diversification to high-value crops should also be a priority to increase income in rural areas. These require investments, technical interventions, and policy interventions. These include investments in tank and reservoir rehabilitation and management, improving soil health, implementing water conservation and management practices, enhancing water use efficiency and productivity, and developing infrastructure, such as roads and regional markets, and policy and technical investments to facilitate better targeting of these financial investments.

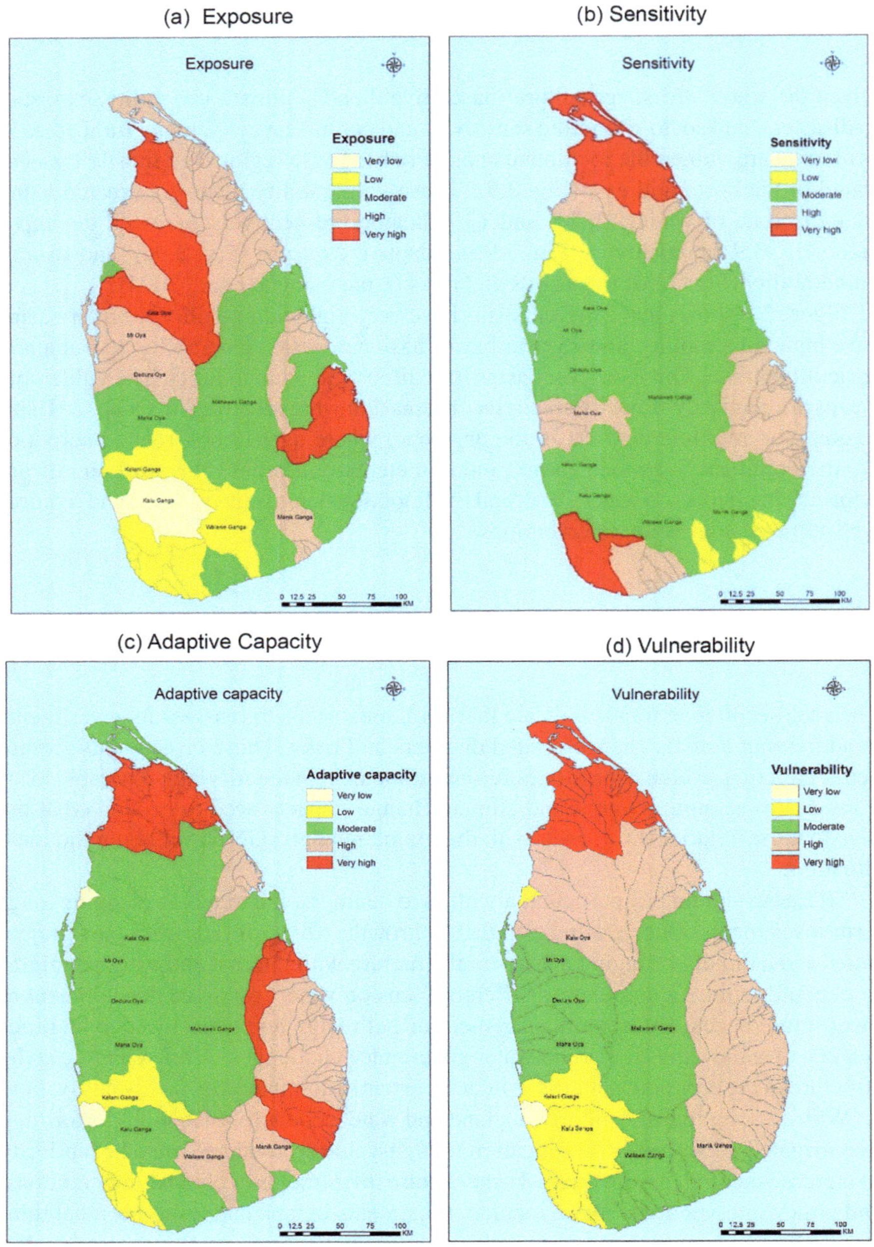

Fig. 3.9 Exposure, sensitivity, adaptive capacity, and vulnerability to water and agricultural risks. *Source* Amarasinghe et al. (2024)

Annex

See Tables 3.5 and 3.6.

Table 3.5 District-level demographic trends

District/region	2011 population (millions)	Sector population—% of 2011 total			2023 total (millions)	Annual growth (%)	Population density 2024 (persons/km^2)
		Rural	Urban	Estate		2011–2023	
Sri Lanka	20.359	78.3	17.1	4.6	21.763	0.61	329
Urban	3.65	–	100	–	Na	Na	–
Rural	16.65	100	–	–	Na	Na	–
Estate	0.96	–	–	100	Na	Na	–
Colombo	2.324	22	78	0	2.374	0.19	3,549
Gampaha	2.305	84	16	0	2.434	0.50	1,774
Kalutara	1.222	88	9	3	1.306	0.60	804
Kandy	1.375	81	12	6	1.461	0.55	779
Matale	0.485	84	12	4	0.527	0.76	267
Nuwara Eliya	0.712	41	6	53	0.725	0.17	427
Galle	1.063	86	13	2	1.097	0.28	694
Matara	0.814	85	12	3	0.838	0.26	648
Hambantota	0.600	95	5	0	0.671	1.02	272
Jaffna	0.584	80	20	0	0.594	0.16	630
Mannar	0.100	100	0	0	0.124	1.99	66
Vavuniya	0.172	75	25	0	0.172	0.01	96
Mullaitivu	0.092	100	0	0	0.123	2.62	50
Kilinochchi	0.114	80	20	0	0.136	1.69	115
Batticaloa	0.527	76	24	0	0.595	1.12	251
Ampara	0.649	71	29	0	0.744	1.25	178
Trincomalee	0.380	78	22	0	0.442	1.40	180
Kurunegala	1.618	98	2	0	1.761	0.77	379
Puttalam	0.762	91	9	0	0.818	0.64	278
Anuradhapura	0.861	94	6	0	0.960	0.99	147
Polonnaruwa	0.406	100	0	0	0.447	0.88	137
Badulla	0.815	73	9	19	0.872	0.61	314
Moneragala	0.451	98	0	2	0.527	1.43	96
Ratnapura	1.088	91	2	7	1.145	0.47	355
Kegalle	0.841	82	9	9	0.870	0.31	529

Source DCS (2015a, 2025)

Table 3.6 Poverty trends and access to services

District/ region	Head count ratio (HCR)—% population below the national poverty line (%)				MPI in 2019	Female–headed households—% of total in 2019	Education up to Grade 11 (G.C.O) level	Chronic illness/ disability (% of total population)	Population with moderate to severe food insecurity—% of total
	1990–1991	2002–2003	2012–2013	2019					
Sri Lanka	16.2	6.4	1.4	0.3	0.067	25.3	68.2	17.8	7.31
Urban						24.9	57.6	20.5	9.68
Rural						25.4	69.5	17.7	26.52
Estate						25.8	87.8	9.9	6.79
Colombo	16.2	6.4	1.4	0.3	0.014	22.6	54.2	22.3	5.86
Gampaha	14.7	10.7	2.1	1.0	0.019	23.9	57.9	19.2	7.7
Kalutara	32.3	20	3.1	2.7	0.046	22.2	61.3	19.4	9.07
Kandy	35.9	24.9	6.2	3.0	0.096	30.4	68.1	18	8.74
Matale	28.7	29.6	7.8	3.9	0.081	27.4	71.8	19.3	23.04
Nuwara Eliya	20.1	22.6	6.6	8.5	0.193	28.3	77.5	8.4	7.21
Galle	29.7	25.8	9.9	3.0	0.041	25.3	68.4	20.4	5.88
Matara	29.2	27.5	7.1	2.3	0.074	27.3	65.6	17.4	8.36
Hambantota	32.4	32.2	4.9	1.1	0.081	25.6	73.4	17.9	12.58
Jaffna			8.3	9.1	0.039	23.0	72.9	17.8	18.05
Mannar			20.1	1.3	0.112	18.1	64.2	17.4	25.87
Vavuniya			3.4	5.5	0.11	28.3	81.1	19.9	19.46
Mullaitivu			28.8	14.3	0.054	22.2	79.9	12.3	8.8
Kilinochchi			12.7	7.0	0.06	22.7	80.5	13.3	11.38
Batticaloa			19.4	5.0	0.085	32.3	78.1	11.2	15.57
Ampara			5.4	2.8	0.058	24.2	70.5	14.4	16.14

(continued)

Table 3.6 (continued)

District/ region	Head count ratio (HCR)—% population below the national poverty line (%)				MPI in 2019	Female–headed households—% of total in 2019	Education up to Grade 11 (G.C.O) level	Chronic illness/ disability (% of total population)	Population with moderate to severe food insecurity—% of total
	1990–1991	2002–2003	2012–2013	2019					
Trincomalee			9	3.6	0.059	24.6	77.2	10.1	13.27
Kurunegala	27.2	25.4	6.5	2.0	0.048	25.2	70.3	19.6	9.75
Puttalam	22.3	31.3	5.1	1.9	0.044	24.8	79.9	18.2	8.99
Anuradhapura	24.4	20.4	7.6	0.9	0.071	29.2	73.5	18.7	10.27
Polonnaruwa	24.9	23.7	6.7	4.6	0.085	23.8	75.9	21.6	11.23
Badulla	31	37.3	12.3	7.1	0.153	27.9	76.4	12.7	16.13
Monaragala	33.7	37.2	20.8	5.6	0.141	24.2	80.6	18.5	8.47
Ratnapura	30.8	34.4	10.4	6.9	0.116	22.9	72.1	15.2	10.5
Kegalle	31.2	32.5	6.7	5.3	0.075	23.5	69.8	17.7	7.31

Sector/district	Drinking water supply				Access to infrastructure			Natural disaster		
	From a safe Source	From a pipe water supply	Inadequate	On the premises	Gas cooking fuel	Toilet within premises	Mobile phones	Total	Floods	Droughts
Sri Lanka	88.5	41.5	10.2	82.9	39.9	92.8	92.4	**11.8**	**3.7**	**5.0**
Urban	99.1	79.0	2.7	94.3	76.9	93.4	95.3	5.7	3.3	0.5
Rural	89.2	34.8	11.1	80.3	33.2	93.3	92.2	13.6	3.9	6.2
Estate	33.2	20.4	22.1	85.9	18.3	79.3	84.9	2.8	0.4	1.0
Colombo	99.1	80.6	1.8	97.9	80.0	94.0	95.8	5.5	4.2	
Gampaha	98.8	38.9	3.1	96.3	62.3	93.5	95.2	4.5	3.7	0.4
Kalutara	92.9	32.6	9.5	92.3	58.7	94.1	94.3	7.7	4.3	0.7

(continued)

Table 3.6 (continued)

Sector/district	Drinking water supply				Access to infrastructure			Natural disaster		
	From a safe Source	From a pipe water supply	Inadequate	On the premises	Gas cooking fuel	Toilet within premises	Mobile phones	Total	Floods	Droughts
Kandy	87.2	69.6	8.5	89.4	37.1	92.0	93.6	12.8	0.4	2.2
Matale	89.9	44.4	14.4	71.1	19.5	92.5	87.3	14.2	0.3	2.2
Nuwara Eliya	32.7	28.5	15.1	93.5	22.4	88.4	92.1	4.1	0.6	0.1
Galle	91.1	35.1	10.6	91.9	40.8	93.6	91.1	8.3	4.7	0.3
Matara	80.7	38.3	10.9	92.6	38.3	94.3	91.1	12.4	4.6	0.9
Hambantota	96.1	72.0	3.7	89.0	22.4	93.9	91.8	10.1	0.7	1.3
Jaffna	95.0	12.0	4.0	64.1	27.5	92.7	92.7	12.8	3.2	11.9
Mannar	94.4	29.4	9.6	59.8	31.6	92.3	93.8	21.2	5.3	15.9
Vavuniya	98.9	0.2	21.2	25.5	43.7	90.7	95.2	8.5	4.3	7.9
Mullaitivu	94.4	3.3	19.3	78.5	10.5	95.9	90.2	40.1	35.6	16.5
Kilinochchi	90.1	2.1	14.8	59.5	22.8	91.8	95.2	22.5	18.5	4.0
Batticaloa	97.1	32.5	8.1	80.4	47.1	90.6	88.2	15.8	14.7	3.3
Ampara	98.6	69.5	6.4	89.5	51.0	95.0	91.3	18.0	14.0	13.4
Trincomalee	93.6	66.8	27.6	81.7	41.2	89.7	92.4	10.1	0.5	4.3
Kurunegala	94.6	14.4	21.2	67.7	23.1	92.7	93.1	19.2	1.8	14.0
Puttalam	92.7	13.6	8.4	61.1	42.6	90.3	93.6	20.4	5.2	14.3
Anuradhapura	89.3	26.6	10.6	45.8	21.3	91.8	91.1	26.3	0.3	21.6
Polonnaruwa	96.0	37.7	9.1	56.7	19.5	91.3	92.1	10.6	2.5	4.7
Badulla	61.4	40.1	15.0	84.8	12.1	92.3	83.9	8.1	0.8	2.2

(continued)

Table 3.6 (continued)

Sector/district	Drinking water supply				Access to infrastructure			Natural disaster		
	From a safe Source	From a pipe water supply	Inadequate	On the premises	Gas cooking fuel	Toilet within premises	Mobile phones	Total	Floods	Droughts
Monaragala	93.4	51.0	22.6	72.5	10.7	91.8	90.5	33.5	0.4	24.8
Ratnapura	71.5	32.7	9.9	86.7	20.2	91.6	91.7	10.0	5.3	0.1
Kegalle	75.8	28.6	16.4	88.0	24.9	96.0	90.2	5.1	0.5	0.6

Source DCS (2015b, 2019), DCS, UNICEF, OPHI (2019)

References

Akalanka, N., Kankanamge, N., Munasinghe, J., & Yigitcanlar, T. (2024). Urban big data analytics: A novel approach for tracking urbanization trends in Sri Lanka. *Land, 13*(6), 888.

Amarasinghe, U. A., Ekanayake, C., Amarnath, G., & Alahacoon, N. (2024). *Improving water governance in climate adaptation for Sri Lanka's tank cascade systems.* Policy Brief. International Water Management Institute.

DeVotta, N. (2005). Sri Lanka in 2004: Enduring political decay and a failing peace process. *Asian Survey, 45*(1), 98–104.

DCS (Department of Census and Statistics). (2008). *Poverty Indicators. Household income and expenditure survey 2006/07.* Department of Census and Statistics. https://www.statistics.gov.lk/Poverty/StatisticalInformation/PovertyIndicators2006-07

DCS. (2011). *Poverty Indicators. Household income and expenditure survey 2009/10.* Department of Census and Statistics. https://www.statistics.gov.lk/Poverty/StatisticalInformation/PovertyIndicators-HIES2009-10

DCS. (2015a). *Census of population and housing 2012.* Department of Census and Statistics. http://203.94.94.83:8041/index.php?fileName=FinalReportE&gp=Activities&tpl=3

DCS. (2015b). *Poverty indicators. Household income and expenditure survey 2012/13.* Department of Census and Statistics. https://www.statistics.gov.lk/Poverty/StatisticalInformation/PovertyIndicators-HIES-2012-13

DCS. (2015c). *The spatial distribution of poverty in Sri Lanka.* Department of Census and Statistics. https://www.statistics.gov.lk/Poverty/StaticalInformation/SpatialDistributionPovertySriLanka-2012-13

DCS. (2017). *Poverty indicators. Department of Census and Statistics Household Income and Expenditure Survey – 2016.* Colombo, Sri Lanka: Department of Census and Statistics. Available at https://www.statistics.gov.lk/Poverty/StaticalInformation/PovertyIndicators-2016

DCS. (2019). *Poverty indicators-2019.* Department of Census and Statistics. https://www.statistics.gov.lk/Poverty/StatisticalInformation/PovertyIndicators-2019

DCS. (2024). *Agriculture.* https://www.statistics.gov.lk/Agriculture/StaticalInformation/new

DCS. (2025). *Population by districts in Sri Lanka.* Provisional Report. Department of Census and Statistics. Available at EM-DAT 2018. EM-DAT (2018) The CRED/OFDA international disaster database. Retrieved Feb 2018, from http://www.emdat.be

DCS, UNICEF, OPHI. (2019). *Multidimensional Poverty in Sri Lanka.* Department of Census and Statistics. https://www.ophi.org.uk/sites/default/files/2023-12/mpi_in_sri_lanka_briefing_2021.pdf

FAO (Food and Agriculture Organization). (2024). *FAOSTAT database.* https://www.fao.org/faostat/en/#data

FAO and WFP (World Food Programme). (2023). Special report. FAO/WFP crop and Food Security Assessment Mission (CFSAM) to the Democratic Socialist Republic of Sri Lanka. CFSAMs Special Reports. 25 May 2023. Rome. https://doi.org/10.4060/cc6202en

GoSL. (2018). *The local government system in Sri Lanka.* Country Profile-2017–18. https://www.clgf.org.uk/default/assets/File/Country_profiles/Sri_Lanka.pdf

GoSL. (2025). Election Commission of Sri Lanka: List of Local Authorities (PDF). https://www.elections.gov.lk

Perera, K. C. (1981). *The development of Sri Lanka's policy of nonalignment, 1947–1977.* The Australian National University (Australia).

UN, DESAPD (The United Nation, Department of Economic and Social Affairs, Population Division). (2024). *World population prospects.* Online Edition. https://population.un.org/wpp/downloads?folder=Standard%20Projections&group=Most%20used

UN-Habitat. (2018). *State of the Sri Lankan Cities.* https://unhabitat.org/sites/default/files/2020/09/soslc_report_final_low-r_0.pdf

World Bank. (2024). *World development indicators.* Retrieved 2024, from http://databank.worldbank.org/data/home.aspx

Ms. Sachini Ukwattage was a project consultant at IWMI. She completed her Bachelor of Industrial Studies—Agriculture and Plantation Engineering with first-class honors and top of the Batch in the Department of Agriculture and Plantation, Faculty of Engineering, The Open University of Sri Lanka. She received the Best Academic Performance Gold medal and the Most Outstanding Student in 2017 and 2018. She has experience as a demonstrator and research assistant in The Open University, ITI, and the University of Sri Jayawardhanapura field of organic agriculture and industrial collaborative projects. She has been nominated to represent Sri Lanka in SAARC Countries' first Batch JENESYS 2019—Agricultural and Environmental Program in Japan. She has publications on 'OURS' and ITI research symposiums. Agriculture, Hydrology, and Soil and Water Management are her main areas of competence. She is a Member of the National Science Foundation in Sri Lanka and the JICE SAARC Membership Association "JENESYS" in Japan.

Ms. Kaushika Seelanatha is a consultant at the International Water Management Institute, Sri Lanka. She completed her Bachelor of Science (Honors) in Agriculture degree program from the Faculty of Agriculture and Plantation Management, Wayamba University of Sri Lanka, specializing in Agricultural Economics and Business. She has consecutively won the Faculty Award for outstanding academic and extracurricular achievements for three years. She has experience as a Temporary Demonstrator, as an editorial assistant (The Journal of Applied Economics and Business) at her alma mater, as a Lesson writer (Agriculture) at Natural Enterprise and Industrial Training Authority, and as an In-plant trainee at the Natural Resources and Management Center. She is skilled in modeling and mapping Eco-systems and statistical data analysis. She has publications/conference proceedings related to agricultural economics and business. She is completing her Post Graduate Diploma in Marketing Management and preparing for M.Phil. in Environmental Economics at Wayamba University of Sri Lanka.

Ms. Madhusha Perera is a Research Consultant at the International Water Management Institute, with a strong foundation in data analysis and advanced programming. She holds a bachelor's degree with a specialization in Statistics from the University of Sri Jayewardenepura. Her expertise lies in statistical analysis, data-driven research, and proficiency in programming languages.

Dr. Upali A. Amarasinghe Emeritus Scientist at the International Water Management Institute, has over 33 years of experience as a researcher and project leader in water sources management, as well as ten years of teaching experience as a university lecturer. He holds a Ph.D. degree in Statistics from the University of Illinois, Urbana-Champaign, USA. His research focused on water management issues at different spatial scales; which included estimating water productivity and footprints of crops to river basins; food and water supply and demand and projections of river basins to countries; benchmarking of performance of canal irrigation systems; poverty and vulnerability assessment; poverty mapping using small area estimation; water security assessment of industries; impact assessments of interventions in water resources management. He has extensive experience managing research projects in India and Sri Lanka, as well as short-term work experiences in other countries, including Pakistan, China, Malaysia, Thailand, and Vietnam. Dr. Amarasinghe has published over 130 research articles and edited four books.

Chapter 4
Climate-Induced Vulnerability and Risk Mapping in Sri Lanka Using Historical Hazard Events

Niranga Alahacoon, Amarnath Giriraj, and Piriyanka Sivananthan

Abstract This chapter traces the climate variability-induced vulnerabilities and risks of disasters in Sri Lanka in the past and assesses the implications of increasing risks from the water sector perspective. Between 1977 and 2024, more than 7,800 floods and over 8,700 strong wind incidents occurred across the districts. Droughts are the next biggest hazard. The multi-hazards, some of which have geographic clustering, inflict major damage to people and households. A greater understanding of the nature and impacts of disasters, as well as the links to vulnerability and risks and their implications from the perspective of the water sector, is critical for preparing and implementing adaptation strategies.

Keywords Multi-hazards · Floods · Droughts · Strong winds · Damage · Vulnerability

4.1 Introduction

Climate-induced vulnerability and risk have become pressing challenges worldwide, particularly for developing nations with high exposure to natural hazards and limited adaptive capacities (Perera et al., 2020). Positioned within the tropical belt, Sri Lanka is subject to complex climatic influences, including seasonal monsoons, cyclones,

N. Alahacoon (✉) · A. Giriraj · P. Sivananthan
International Water Management Institute, Colombo, Sri Lanka
e-mail: n.alahacoon@cgiar.org

A. Giriraj
e-mail: a.giriraj@cgiar.org

P. Sivananthan
e-mail: apspiriyanka1996@gmail.com

U. A. Amarasinghe et al. (eds.), *Water Governance for Climate Change Adaptation in Sri Lanka*, Global Issues in Water Policy 13,
https://doi.org/10.1007/978-3-032-04950-6_4

droughts, and varying rainfall patterns (Abeysingha & Rajapaksha, 2020; Alahacoon & Edirisinghe, 2021). In recent decades, the frequency, intensity, and unpredictability of climate-related hazards in the country have increased, driven by global climate change (Weerasekara et al., 2021).

Sri Lanka's susceptibility to climate risks stems from physical, environmental, and socio-economic factors. The country's varied topography, which includes central highlands and low-lying coastal plains, interacts with climate variability to create a wide range of hazards (Samaraweera et al., 2024). Seasonal rainfall patterns (Alahacoon et al., 2021) and cyclonic rainfall can lead to floods and landslides, while extended dry spells often result in severe droughts, particularly in the north-central and eastern regions (Alahacoon & Amarnath, 2022; Alahacoon et al., 2018). Coastal districts are vulnerable to cyclones and storm surges, and areas with poor infrastructure are at increased risk from strong winds and lightning (Edirisinghe & Maduranga, 2021).

Beyond geography and weather, Sri Lanka's socio-economic conditions also amplify climate risks. A significant portion of the population relies on agriculture and natural resources for livelihoods, making them particularly sensitive to climatic disruptions (Marambe et al., 2015). Rural communities often lack access to climate-resilient infrastructure and early warning systems, while urban areas face challenges related to poor drainage, informal settlements, and unregulated expansion into hazard-prone zones (Hussainzad & Gou, 2024). Furthermore, systemic issues such as poverty, weak governance, and limited disaster preparedness compound the exposure and vulnerability of various population groups, particularly women, children, the elderly, and marginalized communities (Mendis et al., 2023).

Disaster records from the DesInventar disaster management information database reveal the vast scale of these challenges (https://www.desinventar.net/). According to this database, Sri Lanka experienced over 23,000 climate-related hazard events between 1974 and 2022 (Alahacoon et al., 2024), including floods, droughts, cyclones, landslides, lightning, and strong winds. The impacts of these hazards include loss of life, injuries, displacement, and damage to property and infrastructure. They also disrupt education, health, and economic systems. In many cases, the same districts experience repeated exposure to multiple hazards, creating cycles of vulnerability that are difficult to break without intervention and result in long-term impacts.

This chapter explores climate-induced hazards, vulnerability, and risk in Sri Lanka. By analyzing historical disaster data and considering both spatial and socio-economic impacts, the chapter aims to clarify the complex interactions between natural hazards and human systems at the Divisional Secretariat Division (DSD) administrative level. The analysis serves as a basis for creating targeted, inclusive, and data-driven strategies for disaster risk reduction and climate adaptation. By doing so, it supports ongoing efforts to develop a more resilient Sri Lanka capable of effectively handling the increasing challenges brought about by climate change. An in-depth analysis of nearly five decades of disaster records (1974–2022), primarily sourced from the DesInventar database, explores the spatial and temporal patterns of hazards and assesses their socio-economic implications. The goal is to inform integrated approaches to disaster risk reduction and climate adaptation tailored to Sri Lanka's unique context.

4.2 Climate Context and Hazard Landscape of Sri Lanka

Sri Lanka's climate is primarily influenced by its tropical location and two major monsoonal systems: the Southwest Monsoon (*Yala* season) and the Northeast Monsoon (*Maha* season). These seasonal weather patterns, along with inter-monsoonal periods, determine how rainfall is distributed and its intensity across the island. The country's diverse topography further contributes to spatial variation in rainfall, temperature, and wind patterns. Together, these factors create a complex and dynamic climate system that significantly impacts local vulnerability to hazards (Alahacoon & Edirisinghe, 2021).

The country experiences four distinct climate seasons:

- **First Inter-monsoon Season (March–April)**—is characterized by thunderstorm-type rainfall in most parts of the country.
- **Southwest Monsoon (May–September)**—brings heavy rainfall mainly to the southwestern quarter of the island, including the districts of Galle, Matara, Ratnapura, and Colombo.
- **Second Inter-monsoon Season (October–November)**—features widely distributed rain across the island, often causing flash floods and landslides.
- **Northeast Monsoon (December–February)**—primarily affects the northern and eastern parts of the island with rainfall.

This seasonal climatic pattern, however, has become increasingly erratic due to global climate change, leading to more frequent and severe climate-related hazards. Historical data illustrate an alarming rise in extreme weather events, including more prolonged droughts, heavier rainfall leading to floods, and heightened risks of landslides in the central highlands. Additionally, the incidence of tropical cyclones forming in the Bay of Bengal and affecting Sri Lanka's eastern and northern coasts has increased, particularly during the *Maha* season.

Sri Lanka's topography further amplifies the climate hazard landscape. The central highlands, with steep slopes and intense rainfall, are especially prone to landslides and flash floods. Low-lying coastal areas are vulnerable to storm surges, sea-level rise, and cyclonic winds. The north-central and southeastern dry zones frequently experience drought conditions due to low annual rainfall and limited water storage infrastructure.

The main climate-induced hazards observed in Sri Lanka include:

- **Floods**: Caused by monsoonal rains, overflowing rivers, and inadequate urban drainage.
- **Droughts**: Result from prolonged dry spells and reduced monsoonal rainfall, particularly in agricultural districts.

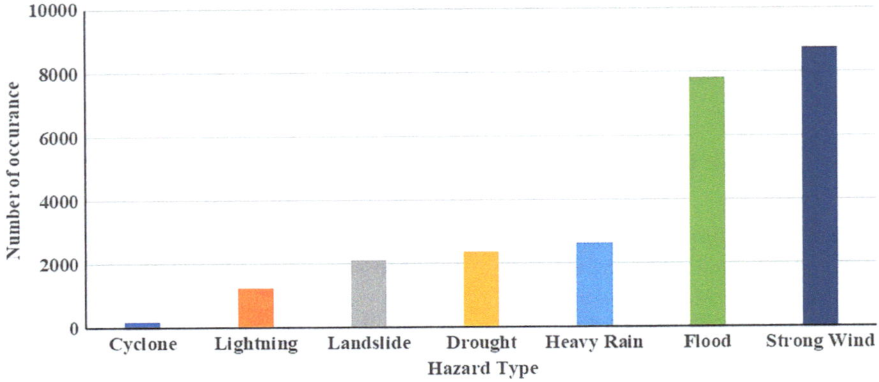

Fig. 4.1 Number of hazard occurrences from 1972 to 2022. *Source* DesInventar disaster database

- **Landslides**: Triggered by intense rainfall in the hill country, exacerbated by deforestation and improper land use.
- **Cyclones**: Relatively infrequent but extremely damaging, especially in coastal areas like Batticaloa, Jaffna, and Ampara.
- **Strong Winds**: Occur year-round but intensify during transitional periods between monsoons.
- **Lightning and Heavy Rain**: Often underreported yet significant. These events cause damage and fatalities at the local level.

According to the DesInventar disaster database, strong winds and floods are the most frequently reported hazards. At the same time, cyclones constitute only about 1% of total disaster occurrences, although their impacts on infrastructure, human life, and livelihoods are disproportionately high.

Understanding the climatic context and hazard landscape is essential for effective disaster risk reduction planning. It reveals the seasonal nature of risks and the long-term trends that demand a shift toward proactive, adaptive strategies. This context forms the foundation for deeper analyses of district-level vulnerabilities, hazard-specific impacts, and integrated response mechanisms, all of which are explored in subsequent sections of this chapter.

Strong winds and floods emerged as the most frequent hazards, while landslides were the deadliest (Fig. 4.1). The spatial distribution of hazards is uneven, shaped by geographic and topographic factors, with certain districts facing multiple concurrent threats.

4.2.1 District-Level Vulnerability and Hazard Hotspots

Understanding district-level vulnerability in Sri Lanka is crucial for formulating targeted disaster risk reduction strategies. Climate-induced hazards impact each of Sri

Lanka's 25 administrative districts due to variations in geography, topography, land use, population density, and socio-economic conditions. The DesInventar disaster database (1974–2022) offers detailed insights into how these vulnerabilities manifest across different hazard types and regions. Table 4.1 provides a breakdown of natural hazard occurrences across districts.

These data show that:

- **Strong winds** are the most frequently reported hazard, with 8,754 occurrences nationwide—notably high in Ratnapura (1,370), Kandy (960), and Kegalle (805) districts.

Table 4.1 Number of natural hazard occurrences across Sri Lankan districts (1974–2022)

District	Cyclone	Drought	Flood	Heavy rain	Landslide	Lightning	Strong wind
Ampara	18	97	250	34	0	28	219
Anuradhapura	11	285	134	25	0	31	176
Badulla	3	48	107	177	350	42	544
Batticaloa	8	156	557	6	0	37	159
Colombo	2	0	428	18	9	28	317
Galle	0	14	328	33	39	55	282
Gampaha	5	28	785	16	7	57	349
Hambantota	0	105	163	64	15	60	401
Jaffna	62	95	676	37	0	28	289
Kalutara	5	22	400	8	43	59	331
Kandy	0	66	197	966	750	56	960
Kegalle	1	53	131	21	287	240	805
Kilinochchi	4	80	217	1	0	14	65
Kurunegala	3	610	526	350	31	63	477
Mannar	6	30	178	0	0	13	49
Matale	2	171	111	148	37	16	194
Matara	0	14	452	24	48	29	176
Monaragala	0	55	122	192	5	46	401
Mullaitivu	7	38	91	1	0	20	113
Nuwara eliya	1	3	120	49	247	14	269
Polonnaruwa	6	70	182	27	1	38	131
Puttalam	6	134	442	39	0	77	356
Ratnapura	1	67	954	365	240	151	1370
Trincomalee	26	102	217	28	0	24	224
Vavuniya	5	27	64	4	0	17	97
Total	182	2370	7829	2633	2109	1243	8754

Source DesInventar disaster database (1974–2022)

- **Floods** are the second most common hazard (7,829 occurrences), with significant impacts in Ratnapura (954), Gampaha (785), and Batticaloa (557).
- **Droughts** account for 2,370 events, with Kurunegala (610) and Anuradhapura (285) being the most affected districts.
- **Landslides** (2,109 total) are most prevalent in Kandy (750), Badulla (350), and Kegalle (287)—areas known for hilly terrain and slope instability.
- **Lightning** incidents are comparatively low (1,243 in total), but Kegalle (240) and Ratnapura (151) report high frequencies.
- **Cyclones**, the least frequent hazard (182 events), are most concentrated in Jaffna (62) and Trincomalee (26)—coastal districts prone to tropical storms.

These data reveal geographic clustering of specific hazards. For instance, floods and strong winds tend to dominate low-lying coastal areas, while landslides and heavy rains are prevalent in the central highlands. Many districts experience multiple hazards at high frequencies, indicating a need for targeted disaster risk reduction strategies. For example, Ratnapura is the most flood-prone district, experiencing a significantly higher frequency of floods compared to heavy rain events. In contrast, the Kandy district experiences the highest number of heavy rain events, which are closely associated with landslides. Districts such as Badulla, Nuwara Eliya, and Kegalle—with substantial sloping topography—are more prone to landslides even during less intense heavy rainfall events. These observations suggest that such districts require localized early warning systems and targeted resilience-building initiatives to address both floods and landslides.

Kurunegala, Anuradhapura, Puttalam, Batticaloa, and Hambantota are among the most drought-prone districts, but they also experience a high incidence of floods, albeit for different reasons. Heavy rains primarily cause floods in Kurunegala, while floods in the other districts are mainly due to receding floodwaters from upstream river basins.

Mapping the spatial distribution of vulnerabilities at the district level highlights the unequal burden of climate hazards and provides a foundation for regionally tailored disaster risk management plans (Fig. 4.2). Targeted investments, infrastructure upgrades, and community preparedness programs must consider localized risk profiles to reduce vulnerability and enhance resilience throughout Sri Lanka.

A series of seven thematic maps of Sri Lanka illustrates the spatial distribution of different natural hazard events across districts. The hazard types visualized include cyclones, droughts, floods, heavy rain, landslides, lightning, and strong winds. Each map utilizes a color gradient to represent the frequency of hazard events, enabling easy comparison of risk intensity across regions. Certain districts, such as Ratnapura, Kandy, Kegalle, and Kurunegala, emerge as multi-hazard hotspots.

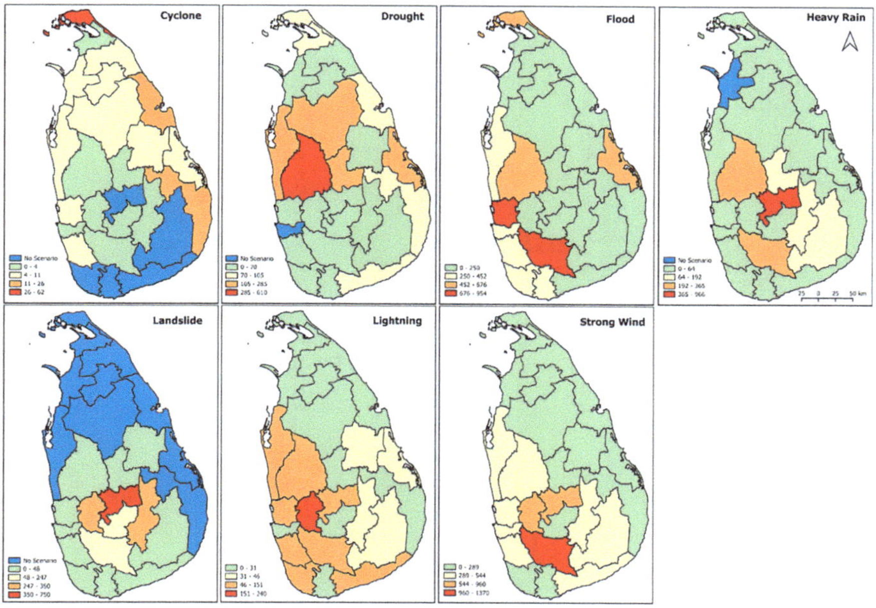

Fig. 4.2 District-level number of records for different hazards

4.2.2 DSD-Level Hazard Occurrence

While district-level hazard occurrence data offer a broad overview of disaster patterns, they often lack the spatial resolution needed to inform targeted risk reduction strategies. For more precise and actionable insights, mapping hazards at the Divisional Secretariat Division (DSD) level (or finer) becomes critical. This finer scale of analysis allows identification of localized vulnerability and exposure, which can vary significantly within a single district. Figure 4.3 presents the distribution of hazard occurrences at the DSD level for six key hazards: flood, drought, lightning, landslide, cyclone, and forest fire. Visualizing data at this granular level enables recognition of high-risk subregions, prioritization of areas for early warning systems, and more effective allocation of resources for disaster preparedness, mitigation, and response planning.

4.2.3 DSD-Level Hazard Impacts

An in-depth understanding of how each climate-induced hazard affects Sri Lanka's population, economy, and infrastructure is essential for risk-informed planning. This

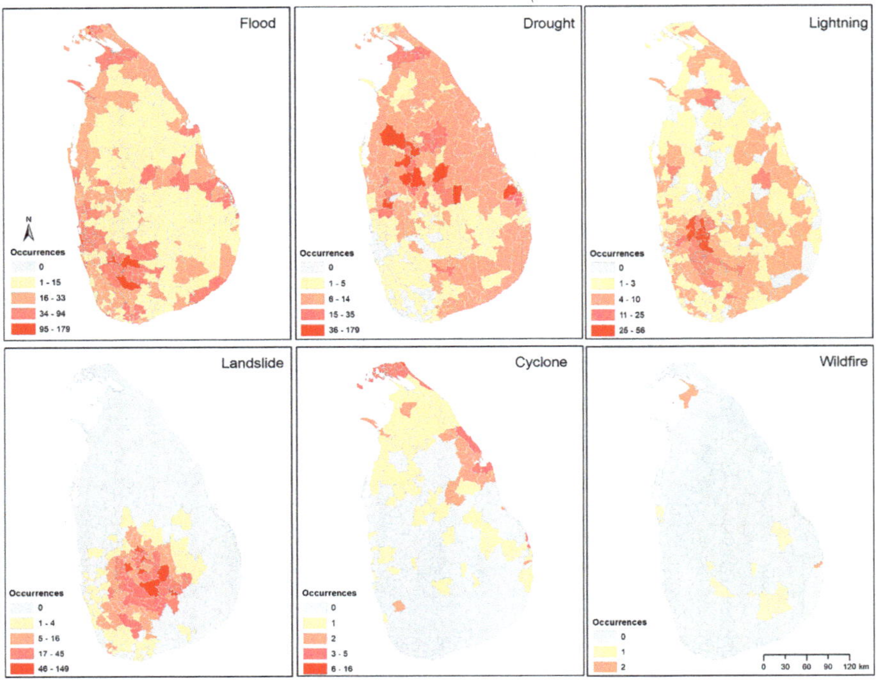

Fig. 4.3 Divisional secretariat (DS) divisional level hazard occurrences from 1972 to 2022

section provides a detailed examination of the impacts of the most frequently occurring and damaging hazards: droughts, floods, cyclones, and landslides. Data are primarily drawn from the DesInventar disaster database, covering the period from 1974 to 2022.

Figures 4.4, 4.5, and 4.6 provide a detailed representation of disaster impacts at the DSD level, focusing on three key indicators: the number of people affected, the number of deaths, and the number of people evacuated due to various hazard events. These visualizations are essential for understanding the human consequences of natural disasters at a more localized scale.

Figure 4.4 illustrates the number of people affected. Floods and droughts have the most widespread impact, affecting a significantly larger number of people across DSs. Figure 4.5, depicting the number of disaster-related deaths, reveals that landslides and floods are the deadliest hazards. This is particularly critical in regions with steep terrain and inadequate slope management, where landslides can occur suddenly and with little warning, often resulting in a higher mortality rate.

Figure 4.6 presents the number of individuals evacuated due to imminent or ongoing hazard events. Similar to the patterns observed in the affected population and mortality, floods and landslides are the leading causes of evacuation. This indicates

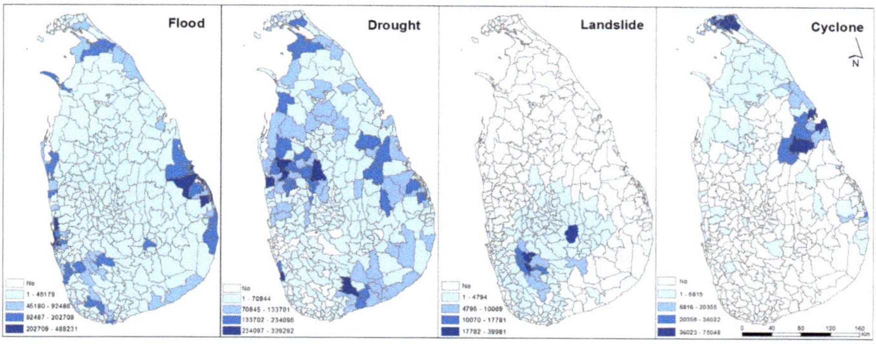

Fig. 4.4 Number of affected people from flood, drought, landslide, and cyclone hazards

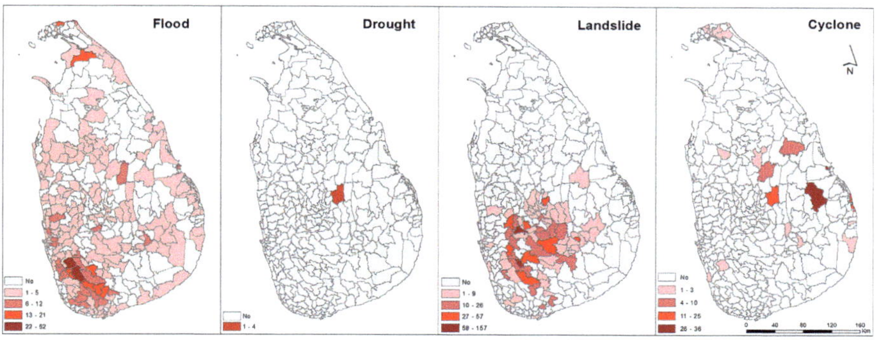

Fig. 4.5 Number of deaths from flood, drought, landslide, and cyclone hazards

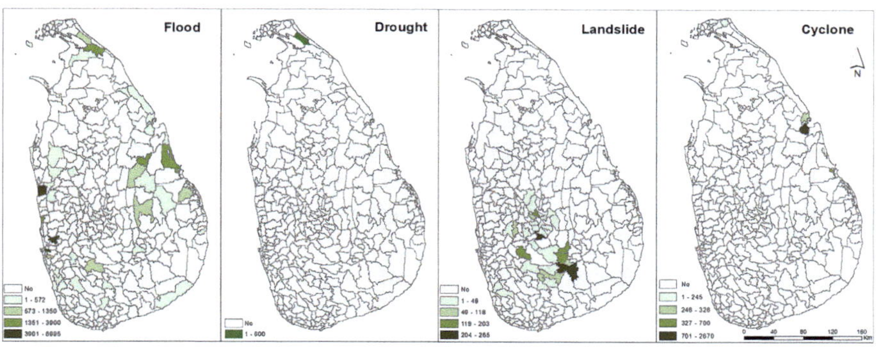

Fig. 4.6 Number of evacuees from flood, drought, landslide, and cyclone hazards

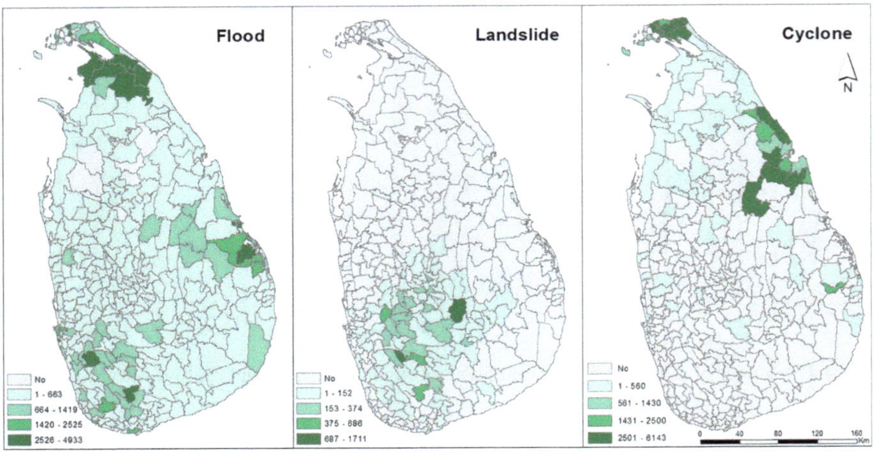

Fig. 4.7 Number of houses damaged due to flood, landslide, and cyclone hazards

both the severity of these hazards and the responsiveness of emergency management systems in relocating at-risk populations.

Together, Figs. 4.4, 4.5 and 4.6 emphasize the importance of localized disaster risk assessments. They highlight where physical impacts are most significant and where human vulnerability is most pronounced, thus informing more strategic and equitable disaster preparedness and response planning.

Unlike droughts (which primarily cause agricultural and economic losses rather than direct physical destruction), natural hazards such as floods, landslides, and cyclones tend to result in significant physical damage and human casualties. Figures 4.7, 4.8 and 4.9 illustrate the distribution of three critical impact indicators at the DSD level: the number of houses damaged, the number of homes destroyed, and the number of injuries reported due to these hazards.

Figure 4.7, which displays damage to houses, shows that cyclones have caused the highest structural damage compared to floods and landslides. However, while cyclones tend to result in concentrated and severe damage, floods are far more widespread across the country, making them the most common cause of housing damage overall.

A similar trend is observed in Fig. 4.8, which presents the number of houses destroyed. Although cyclones cause more destruction per event, the cumulative impact of floods is greater due to their frequency and geographic spread.

Figure 4.9 illustrates the number of injuries and follows the same pattern. Cyclones often result in higher injury counts in localized areas, while floods contribute to a broader distribution of injuries across many regions. Landslides, although less frequent, also contribute notably to both property destruction and human injuries, especially in the central highlands where terrain and rainfall intensity increase vulnerability.

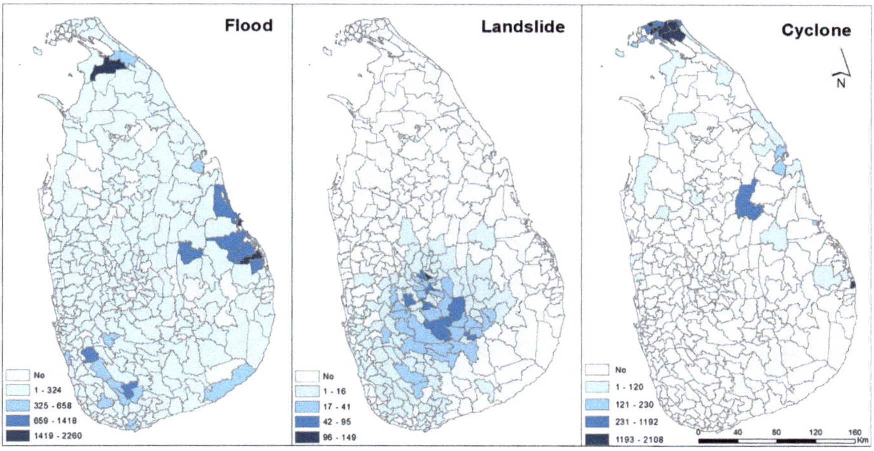

Fig. 4.8 Number of houses destroyed due to flood, landslide, and cyclone hazards

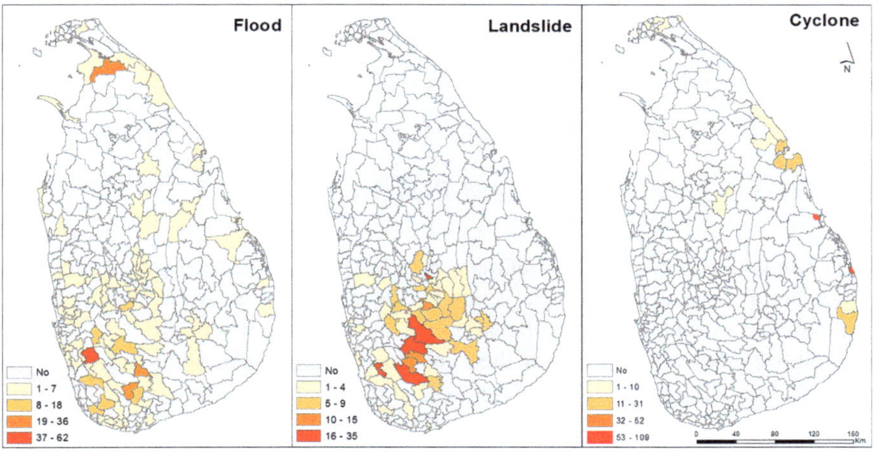

Fig. 4.9 Number of injuries due to flood, landslide, and cyclone hazards

Figures 4.7, 4.8 and 4.9 highlight the importance of hazard-specific response planning. While cyclone impacts may require targeted structural resilience measures, the pervasive nature of flood-related damage highlights the urgency of widespread flood mitigation and preparedness efforts across Sri Lanka.

As illustrated in Fig. 4.10, the highest number of missing persons is reported in association with landslides, followed by floods and cyclones. The inherent characteristics of these hazards largely influence this pattern. Landslides, in particular, tend to occur suddenly and with little or no warning, often burying homes and individuals under debris and making rescue and recovery operations extremely challenging.

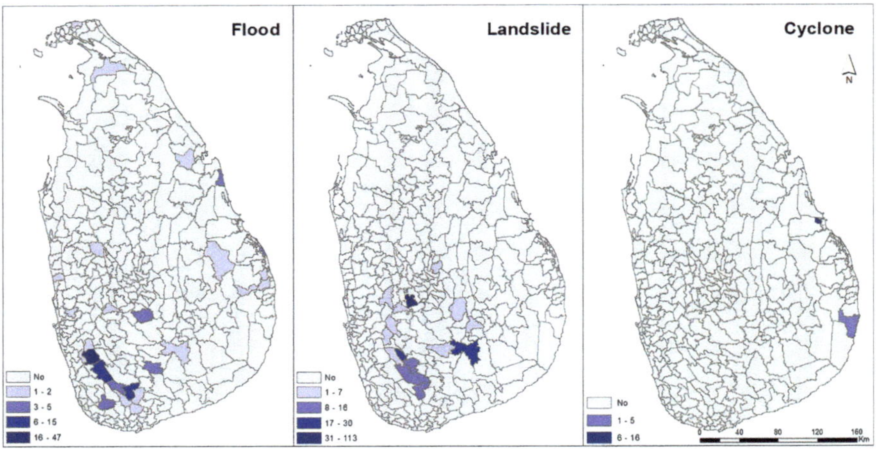

Fig. 4.10 Number of missing persons from flood, landslide, and cyclone hazards

Similarly, floods and cyclones can lead to people being swept away or displaced, resulting in many missing persons, especially in areas with poor communication or limited access during disasters.

4.2.4 DSD-Level Multi-hazard Risk Mapping

The complexity of Sri Lanka's climate risk landscape necessitates an integrated approach to disaster assessment beyond isolated hazard analyses. Multi-hazard risk mapping provides a comprehensive understanding of where different climate-induced hazards overlap and how their combined impacts can amplify community vulnerabilities. This spatial synthesis is critical for effective planning, risk reduction, and resource allocation.

Individual hazard data for floods, droughts, landslides, and cyclones were carefully compiled and analyzed for each DSD across the study area to develop a multi-hazard map. By aggregating the frequency of these four major hazards at the divisional level, a comprehensive map was created to highlight the DSDs that are most vulnerable to multiple natural hazards. The final output, offering a clearer understanding of the spatial distribution of multi-hazard exposure, is presented in Fig. 4.11.

When examining the distribution of multi-hazard occurrences across DSDs, it becomes evident that the most common hazard combination is **flood + drought**, affecting 99 DSDs. This indicates a significant overlap in geographical areas vulnerable to water-related extremes of excess and scarcity. The second most frequent combination is **flood-drought-landslide**, reported in 97 DSDs, highlighting regions highly exposed to multiple compounding hazards. Interestingly, only 11 DSDs have

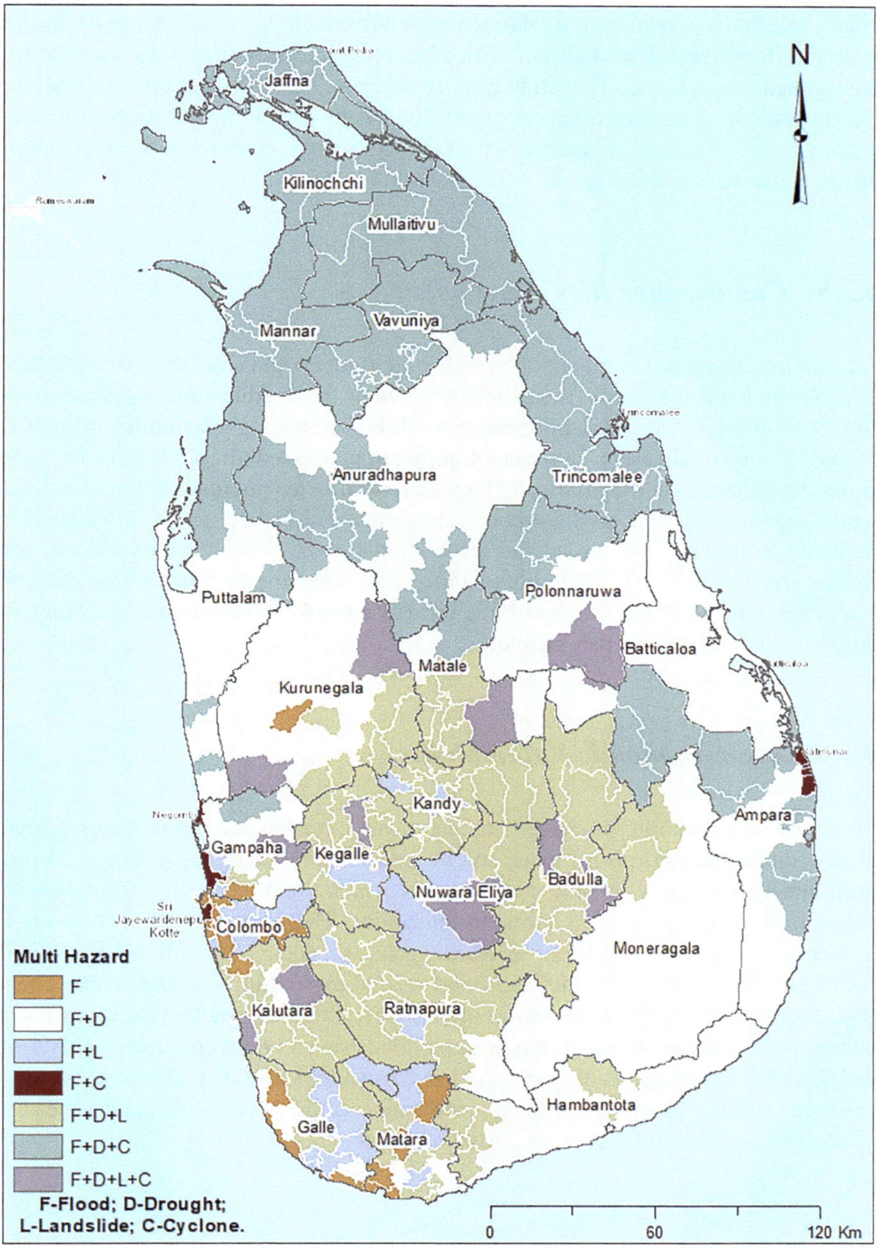

Fig. 4.11 Divisional secretariat division (DSD) level multi-hazard risk map

been identified as experiencing **all four** major hazards (flood, drought, landslide, and cyclone), highlighting a small but critically vulnerable subset of areas that require comprehensive risk reduction strategies. In contrast, 17 DSDs are affected **only by floods**, indicating localized but recurrent flood risks that do not overlap with other hazard types. This analysis is crucial for targeted disaster preparedness and planning efforts at the sub-national level.

4.2.5 Comparative Risk Assessment

A cross-hazard impact comparison reveals important insights into the varying effects of different natural hazards. Landslides emerge as the deadliest hazard, accounting for ~30% of total fatalities, while strong winds lead to the highest number of injuries (about 27% of total cases). In terms of population affected, droughts have the most significant impact, encompassing 53% of the total affected population. Floods, meanwhile, consistently rank as the second most impactful hazard across all categories, demonstrating their widespread and recurring nature. Although cyclones occur less frequently, they tend to result in high fatality rates when they do strike. This comparative analysis is valuable for identifying priority hazards and designing interventions tailored to each hazard's risk profile.

4.2.6 Socio-economic Vulnerability

Sri Lanka's vulnerability is exacerbated by poverty, inadequate infrastructure, urban expansion into hazard-prone areas, and limited adaptive capacity. Disasters disproportionately affect low-income communities, which often reside in informal settlements in floodplains or areas prone to landslides. The financial toll of disasters is immense: drought-related relief in Kurunegala exceeded 143 million LKR, while Jaffna and Galle led in cyclone- and flood-related expenditures, respectively. These costs highlight the need for preventive investment rather than reactive spending. Agriculture is particularly at risk. Droughts and floods severely disrupt paddy cultivation and seasonal cropping, leading to food insecurity, debt, and displacement. Women, children, and the elderly face heightened vulnerabilities due to limited access to healthcare, shelter, and resources during disasters.

4.3 Governance and Institutional Frameworks and Policy Recommendations

Sri Lanka's disaster risk governance has improved with the establishment of the Disaster Management Centre (DMC) and the development of a National Disaster Management Plan. The DesInventar system is a critical tool for tracking disaster events and informing risk reduction policies. However, gaps persist in early warning coverage, decentralized response capacity, and inter-agency coordination. The integration of climate adaptation into urban planning, land use, and agriculture remains inconsistent.

To effectively build resilience against natural disasters, a comprehensive and integrated approach is essential. One key priority is developing multilingual, community-based early warning systems, ensuring that alerts are accessible to all language groups and tailored to the specific hazards of each locality. Strengthening resilient infrastructure is equally critical, especially in high-risk districts, where investments should focus on upgrading housing, transport networks, and water systems to withstand extreme weather events. In the agricultural sector, promoting climate-smart practices, such as drought-resistant crops, agroforestry techniques, and efficient irrigation systems, can greatly enhance food security and livelihood resilience in vulnerable communities.

Moreover, enhancing community preparedness by forming and training local disaster management committees and integrating disaster education in schools empowers individuals to respond effectively during emergencies. Advanced technologies (e.g., satellite imagery, Geographic Information Systems (GIS), and real-time data collection tools) should be expanded to support accurate forecasting, risk modelling, and timely decision-making. Finally, sustainable land-use planning must be enforced by implementing zoning regulations and restricting development in areas prone to floods, landslides, or other hazards. Together, these strategies form a holistic framework for reducing disaster risk and fostering long-term resilience.

4.3.1 Enhancing Digital Interventions Through Fine-Scale Hazard Mapping

Digital technological interventions, such as the **AWARE** platform, play a crucial role in facilitating anticipatory action. By leveraging geospatial analytics and analyzing past event records, AWARE (a digital platform enabling Early Warning, Early Action, and Early Finance) identifies climate hazard hotspots and vulnerable communities at a fine scale (e.g., within a DSD). The interconnectivity between AWARE and DSD-level hazard hotspots facilitates timely alerts and decision-making, helping communities prepare for recurring threats such as floods, droughts, and landslides. By combining historical hazard patterns with near-real-time data, it is possible to provide more accurate and location-specific early warnings. This approach is vital for reducing disaster losses and improving local resilience.

Complementing this, the **Climate-Smart Governance (CSG) Dashboard** provides a strategic planning tool that links vulnerability data with socio-economic and environmental indicators to support climate adaptation interventions across multiple hazards. Using outputs from hazard and impact assessments, the CSG Dashboard helps identify where improvements in governance, infrastructure, and community preparedness are most needed. It enables planners and policymakers to prioritize interventions in high-risk DSDs, allocate resources efficiently, and develop cross-sectoral strategies that address both short-term risks and long-term climate impacts. Together, platforms like AWARE and CSG contribute to a robust digital ecosystem for proactive, evidence-based climate risk management in Sri Lanka.

4.4 Conclusion

Sri Lanka stands at a critical juncture in addressing climate-induced vulnerability and disaster risk. The country's geographic diversity and exposure to multiple climate-related hazards—ranging from floods and droughts to landslides and cyclones—present complex challenges that are further exacerbated by socio-economic and infrastructural vulnerabilities. Analysis of nearly five decades of disaster data reveals the frequency and intensity of these events, as well as their disproportionate impacts on specific districts, DSDs, and communities (particularly those in hazard-prone and economically marginalized areas).

These findings underscore the need to shift from reactive disaster response to proactive, data-informed, and locally tailored risk reduction and climate adaptation strategies. The multi-hazard risk mapping and district-level vulnerability assessments presented in this study provide a compelling foundation for targeted interventions that can break cycles of vulnerability and build long-term resilience. As climate change accelerates, Sri Lanka must harness this knowledge to inform policy, enhance institutional coordination, and empower communities to mitigate and withstand the growing threats posed by climate-related disasters.

Recommendations

- **Tailor** disaster preparedness plans using hazard-specific, district-level data, especially in high-risk areas.
- **Include** spatial risk mapping, effective early warning systems, and robust community engagement in all disaster management plans.
- **Strengthen** early warning capabilities at the DSD level, **integrate** climate risk considerations into development planning, and **avoid** unsafe land use (e.g., building in floodplains or unstable slopes).
- **Ensure** disaster risk governance is transparent and inclusive, with particular focus on protecting vulnerable groups.

- **Promote** resilient agriculture and climate-smart practices and pursue long-term investments (supported by public–private partnerships and climate finance) to achieve sustainable disaster risk reduction.

References

Abeysingha, N. S., & Rajapaksha, U. R. L. N. (2020). SPI-based spatiotemporal drought over Sri Lanka. *Advances in Meteorology, 2020*(1), 9753279.

Alahacoon, N., & Amarnath, G. (2022). Agricultural drought monitoring in Sri Lanka using multisource satellite data. *Advances in Space Research, 69*(11), 4078–4097.

Alahacoon, N., & Edirisinghe, M. (2021). Spatial variability of rainfall trends in Sri Lanka from 1989 to 2019 as an indication of climate change. *ISPRS International Journal of Geo-Information, 10*(2), 84.

Alahacoon, N., Edirisinghe, M., & Ranagalage, M. (2021). Satellite-based meteorological and agricultural drought monitoring for agricultural sustainability in Sri Lanka. *Sustainability, 13*(6), 3427.

Alahacoon, N., Matheswaran, K., Pani, P., & Amarnath, G. (2018). A decadal historical satellite data and rainfall trend analysis (2001–2016) for flood hazard mapping in Sri Lanka. *Remote Sensing, 10*(3), 448.

Alahacoon, N., Thathsarani, G., Sivananthan, P., & Amarnath, G. (2024). Evaluating climate hazard hotspots and hazard impacts in Sri Lanka: An analysis of historically reported data. *Under Review/Publisher info pending.*

Edirisinghe, M., & Maduranga, U. G. D. (2021). Distribution of lightning accidents in Sri Lanka from 1974 to 2019 using the DesInventar database. *ISPRS International Journal of Geo-Information, 10*(3), 117.

Hussainzad, E. A., & Gou, Z. (2024). Climate risk and vulnerability assessment in informal settlements of the Global South: A critical review. *Land, 13*(9), 1357.

Marambe, B., Punyawardena, R., Silva, P., Premalal, S., Rathnabharathie, V., Kekulandala, B., & Howden, M. (2015). Climate, climate risk, and food security in Sri Lanka: The need for strengthening adaptation strategies. In *Handbook of climate change adaptation* (pp. 1759–1789).

Mendis, K., Thayaparan, M., Kaluarachchi, Y., & Pathirage, C. (2023). Challenges faced by marginalized communities in a post-disaster context: A systematic review of the literature. *Sustainability, 15*(14), 10754.

Perera, A. T., Nik, V. M., Chen, D., Scartezzini, J. L., & Hong, T. (2020). Quantifying the impacts of climate change and extreme climate events on energy systems. *Nature Energy, 5*(2), 150–159.

Samaraweera, W. G. R. L., Dharmadasa, R. A. P. I. S., Kumara, P. H. T., & Bandara, A. S. G. S. (2024). Evidence of climate change impacts in Sri Lanka—a review of literature. *Sri Lanka Journal of Economic Research, 11*(2), 69–94.

Weerasekara, S., Wilson, C., Lee, B., Hoang, V. N., Managi, S., & Rajapaksa, D. (2021). The impacts of climate-induced disasters on the economy: Winners and losers in Sri Lanka. *Ecological Economics, 185*, Article 107043.

Dr. Niranga Alahacoon has a strong background in remote sensing and disaster risk management. He holds a Ph.D. in Remote Sensing, awarded by the University of Colombo in 2021. He received the prestigious "Astronomer Royal Sir Arnold Wolfendale Memorial Gold Medal" for the Best Postgraduate Student in Physics at the University of Colombo. He completed an M.Sc. in Geoinformation and Earth Observation (Geo-Hazard) in 2010 at the ITC in the Netherlands and obtained his BSc (Hons) in Geology in 2007 at the University of Peradeniya. Currently, he works at the International Water Management Institute Headquarters as a Regional Researcher, contributes significantly with application of remote sensing data and image processing, geographic information systems (GIS), hydrological and hydrodynamic modelling, geology, and rapid emergency response mapping, and has a publication record of more than 40 scientific publications and 22 training manuals. Prior to joining IWMI, Niranga served as a lecturer at the Faculty of Fisheries and Marine Sciences and Technology (FMSC), University of Ruhuna, Sri Lanka.

Dr. Amarnath Giriraj has a Ph.D. in Remote Sensing. He is the Research Group Leader for Water Risks and Disaster at IWMI, Sri Lanka, a Steering Committee member of Sentinel Asia by JAXA, and Chair of the Working Group on Water-related Disasters at the Asia Pacific Space Agency Forum. Dr. Giriraj's primary expertise includes disaster risk assessment, hydrological process research, agricultural disaster risk management, climate services, and food security, with a focus on developing solutions for pressing climate risk challenges. Many of his research and development projects have demonstrated the impact of climate risks, utilizing advanced science and technology in various hydro-climatic regions worldwide. He is keen on translating science-based knowledge into effective policies and strategies contributing to environmental, economic, and societal sustainability. He established the Water Risks and Disasters mission at IWMI and developed regional and national water-risk data systems and reports to inform policy and decision-makers on key disaster, agricultural, and water challenges and solutions.

Mrs Piriyanka Sivananthan holds a Bachelor of Science in Surveying Science, specialising in Geographic Information Systems (GIS), from the Faculty of Geomatics at Sabaragamuwa University of Sri Lanka. Her academic and professional experiences have focused on utilizing geospatial tools to tackle real-world challenges, including infrastructure planning and environmental management. Driven by her innovative spirit, Piriyanka became proficient in Google Earth Engine (GEE), which she employed in crop mapping initiatives to enhance agricultural risk management for stakeholders. This technology has facilitated the rapid processing of large datasets, significantly improving the efficiency and impact of the International Water Management Institute's (IWMI) environmental research. From engaging directly with vulnerable communities to conducting advanced geospatial analyses, Piriyanka has consistently deepened her passion for geospatial science, anticipatory action, and sustainable development. She strives to connect cutting-edge research with solutions that enhance resilience and improve livelihoods.

Chapter 5
Impact of Climate Change Projections on Water Resources in Sri Lanka

Piriyanka Sivananthan and Emmanuel Attoh

Abstract Climate projections for Sri Lanka indicate significant warming, erratic precipitation patterns, and an increase in extreme weather events, leading to floods and droughts. These changes are expected to have a severe impact on the country's water resources, exacerbating water stress. This chapter examines the climate change projections and their implications for Sri Lanka's water resources. Results indicate that rising temperatures will have the greatest impact on the Northern, North Central, and Eastern provinces. Precipitation patterns are changing across the country. The Western, Sabaragamuwa, Northwestern, and Southern provinces are projected to experience higher cumulative precipitation, thereby increasing flood risks. Increased rainfall from May to November could benefit agriculture but also elevate flood risks in these regions.

Keywords Climate change · Projections · Risks · Water resources

5.1 Introduction

Climate change impacts global water resources by altering precipitation patterns, leading to more frequent droughts and severe flooding. Melting glaciers disrupt water supply, while rising sea levels cause saltwater intrusion, affecting water availability, quality, agriculture, ecosystems, and biodiversity (Chaturvedi et al., 2021; IPCC, 2023). Sri Lanka, despite its minimal contribution to global climate change, faces severe climate impacts due to its tropical climate. Rising temperatures, changing precipitation patterns, sea level rise, and extreme weather events such as droughts and floods are intensifying (Jayawardena et al., 2024; Ratnayake et al., 2023). These changes pose a threat to agriculture, water resources, biodiversity, and human

P. Sivananthan (✉) · E. Attoh
International Water Management Institute, Colombo, Sri Lanka
e-mail: apspiriyanka1996@gmail.com

E. Attoh
e-mail: e.attoh@cgiar.org; enyadzi@yahoo.com

U. A. Amarasinghe et al. (eds.), *Water Governance for Climate Change Adaptation in Sri Lanka*, Global Issues in Water Policy 13,
https://doi.org/10.1007/978-3-032-04950-6_5

health (MoE, 2016). Prolonged droughts and rising sea levels also increase saltwater intrusion into coastal aquifers, depleting freshwater availability (Nianthi & Shaw, 2015).

Climate change projections are crucial for understanding the global environmental shifts that lie ahead. Country-level projections are useful for policymakers, businesses, and communities to plan mitigation and adaptation strategies (World Bank Group, 2021). Given the technical challenges in downscaling global projections, only a few studies have explored climate change projections for Sri Lanka, especially regarding water resources (Eriyagama et al., 2010; Ratnayake et al., 2023). This chapter examines the climate change projections and their implications for Sri Lanka's water resources.

5.2 Description of the Study Area

Sri Lanka is a tropical island nation in the Indian Ocean. It has a diverse topography, ranging from low plains to central mountains exceeding 2,500 m, which significantly influence its climate. The island experiences average temperatures of 27–28 °C, with cooler temperatures at higher elevations. It has four distinct monsoon seasons, divided into wet, dry, and intermediate climatic zones. Abundant freshwater resources are supplied by rainfall, river systems, and groundwater (Fig. 5.1).

5.3 Methodology

5.3.1 Data Collection and Analysis

Data collection for the research followed three main steps:

Step 1: A traditional synthesis review of 32 documents was conducted. Initially, 10 papers on climate change projections were selected, leading to the identification of an additional 8 research articles, 11 reports 12 reports, and 2 policy documents.

Step 2: Current and future climate conditions were analyzed using data from the World Bank Climate Change Knowledge Portal (World Bank Group, 2021) and the Climatic Research Unit (CRU) for historical temperature and precipitation data (1991–2020). The analysis, performed at a $0.5° \times 0.5°$ resolution, aimed to understand current climate conditions and inform future projections. For future climate scenarios, data included mean temperature, precipitation, and extreme events at a $0.25° \times 0.25°$ resolution. These data were analyzed from 2015 up to the end of the twenty-first century using a reference period of 1995–2014, under the SSP1-2.6 (sustainable future with low emissions) and SSP5-8.5 (fossil-fueled development with high emissions) scenarios from CMIP6 (Coupled Model Intercomparison

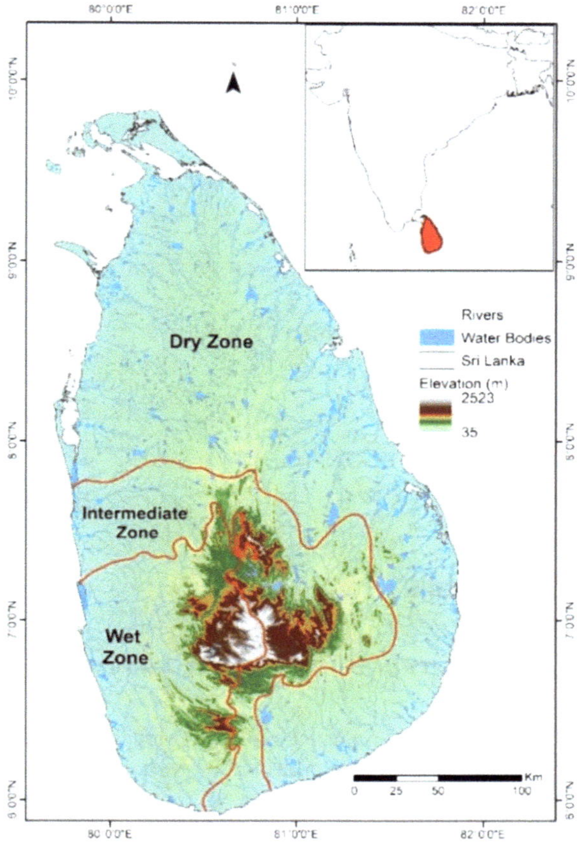

Fig. 5.1 Hydrological map of Sri Lanka. *Source* Somasundaram et al. (2020)

Project Phase 6). Extreme events such as droughts and floods were assessed using the Standardized Precipitation Evapotranspiration Index (SPEI), which accounts for both precipitation and evapotranspiration, and cumulative precipitation events, which represent the total amount of precipitation (rain, snow) accumulated over a specific period. Table 5.1 presents the details of the climate variables used for the analysis. Step 3 reviewed relevant literature on climate change and the water sector, noting a gap in studies at this intersection, and the findings were synthesized in the discussion.

Table 5.1 Selected variables for analysis

Current climate	Projected climate
Observed climatology of average mean surface air temperature 1991–2020	Temperature and precipitation projection for Sri Lanka from different sources in the literature
Annual rainfall climatology (1981–2010)	Province-wise CMIP6 annual projected temperature and precipitation anomaly and mean temperature for 2080–2099
Annual precipitation of Sri Lanka for 1951–2020	Monthly projected temperature and precipitation climatology for 2080–2099
Average monthly temperature and rainfall in Sri Lanka	Projected temperature and precipitation anomaly for the end of the 21st century for Sri Lanka
	Projected climatology of annual SPEI drought index for 2080–2099, Sri Lanka
	Largest 1-day precipitation and largest 5-day precipitation

5.4 Results

5.4.1 Current Climate of Sri Lanka

Results show that Sri Lanka has experienced a significant rise in temperatures over recent decades, with notable regional variations. Between 1896 and 1996, the average temperature increased by 0.003 °C annually, accelerating to 0.025 °C per year from 1987 to 1996 (Eriyagama et al., 2010). Specific regions, such as Nuwara Eliya, saw increases of 2.0 °C from 1961 to 1990 (USAID, 2015). The dry and intermediate zones experienced more consecutive dry days and rising temperatures (MoE, 2010). Surface air temperatures in Sri Lanka rose from 25.23 to 27.23 °C between 1951 and 2020 (Fig. 5.2a). Monthly trends indicate peak temperatures in May and the lowest temperatures in January.

Precipitation trends are less clear, with some reports indicating a decrease in average rainfall and others showing conflicting regional data (Fig. 5.2). Future projections indicate an increase in heavy precipitation events and changes in monsoon patterns, which could potentially impact water availability and flood risks (MoE, 2016). Notably, annual rainfall declined by approximately 7% relative to the 1931–1960 reference period, especially during the northeast monsoon and second inter-monsoon seasons (MoE, 2010). Monthly trends indicate that the highest rainfall occurs in November and the lowest in July (Fig. 5.3).

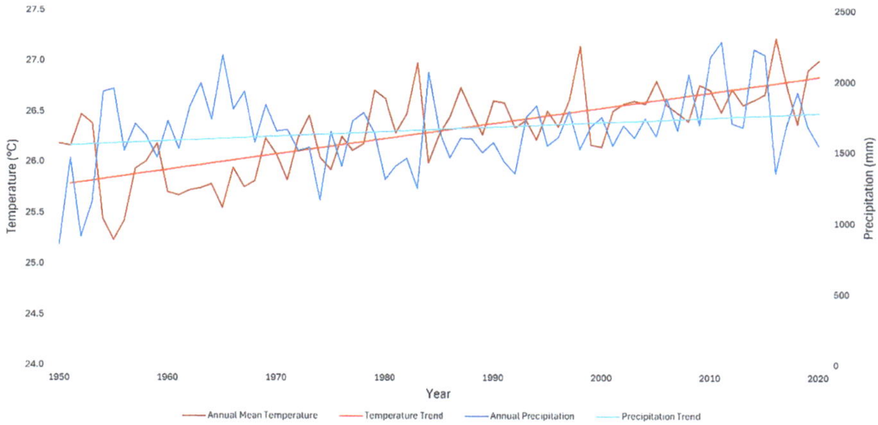

Fig. 5.2 Average surface temperature (**a**) and average precipitation (**b**) of Sri Lanka, 1951–2020. *Source* Authors

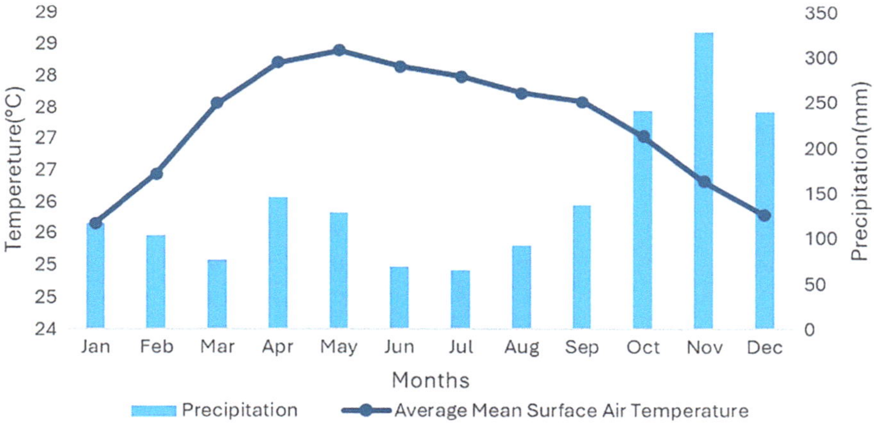

Fig. 5.3 Average monthly temperature and rainfall in Sri Lanka (1991–2020). *Source* Authors'

5.5 Climate Change Projections

5.5.1 Projected Changes in Temperature

The literature review shows that various climate models and studies indicate significant temperature increases in Sri Lanka due to climate change. By 2030, temperatures are expected to rise by 0.47–1.1 °C, with further increases of 0.77–2.04 °C by 2050, depending on the scenario (USAID, 2015). By 2080, temperatures could rise between 2.30 and 4.50 °C, with models showing even higher increases under more extreme scenarios (ADB, 2014; Punyawardena et al., 2012). From 2015 to 2100, predictions indicate a temperature increase ranging from 1.1 to 3.49 °C, depending on the Shared Socioeconomic Pathways (SSPs) and climate models used (World Bank, 2024).

Monthly analysis shows that temperature projections for the SSP1-2.6 and SSP5-8.5 scenarios indicate a rising trend for each month by the end of the 21st century, with the most significant increases expected in April, May, and June (Table 5.2). SSP1-2.6 shows a gradual rise in average temperature with a relatively narrow range of outcomes, whereas SSP5-8.5 indicates a more uncertain future with a wider range of potential temperatures (Fig. 5.4).

The temperature ranges for different periods under the SSP1-2.6 and SSP5-8.5 scenarios are as follows (Table 5.2). From 2020 to 2039, temperatures range from 27.16 to 27.84 °C under SSP1-2.6 and from 27.04 to 28.17 °C under SSP5-8.5. For the 2040–2059 period, the ranges are 27.54 to 28.08 °C (SSP1-2.6) and 27.62 to

Table 5.2 Province-wise CMIP6 projected annual temperature anomaly and mean (2080–2099) in Sri Lanka under SSP1-2.6 and SSP5-8.5 (reference period: 1995–2014), multi-model ensemble

Province	Climatology of temperature (°C, 1991–2020)	Mean (°C) SSP1-2.6	Mean (°C) SSP5-8.5	Anomaly (°C) SSP1-2.6	Anomaly (°C) SSP5-8.5
Northern	28.53	28.87	31.17	0.87	3.23
Northwestern	27.86	28.09	30.46	0.90	3.27
North Central	28.11	28.74	31.07	0.90	3.24
Eastern	28.11	28.91	31.22	0.91	3.23
Central	24.63	25.24	27.63	0.93	3.33
Western	27.10	27.58	29.97	0.92	3.33
Sabaragamuwa	25.69	25.65	28.03	0.93	3.34
Uva	26.15	27.35	29.72	0.93	3.31
Southern	26.98	27.66	30.01	0.92	3.27

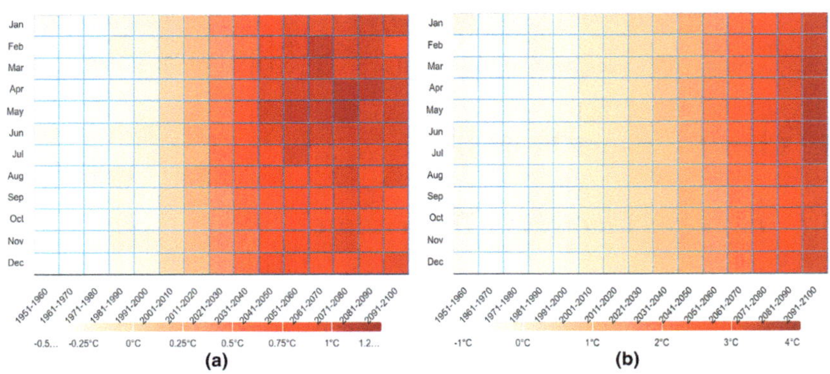

Fig. 5.4 Projected average mean surface air temperature anomaly in Sri Lanka; **a** SSP1-2.6 and **b** SSP5-8.5 scenarios (reference period 1995–2014, projected for the end of the century)

29.10 °C (SSP5-8.5). In the period 2060–2079, temperatures range from 27.76 to 28.08 °C (SSP1-2.6) and 28.40 to 30.13 °C (SSP5-8.5). Finally, from 2080 to 2099, the temperature ranges from 27.64 to 28.06 °C under SSP1-2.6 and from 29.44 to 30.96 °C under SSP5-8.5 (World Bank, 2024).

5.5.2 Projected Changes in Precipitation

The results of the literature review indicate that projected precipitation changes for Sri Lanka vary depending on the model, scenario, and time period. According to the Asian Development Bank (ADB, 2014), under a 2000 baseline regional climate model (CMIP3), Sri Lanka is expected to see an increase in precipitation of 3.6–11.0% by 2030, 15.8–25.0% by 2050, and 31.3–39.6% by 2080. Meanwhile, a study by Punyawardena et al. (2012) using the ECHAM4 general circulation model (GCM) through the PRECIS regional climate model (RCM) under the A2 scenario predicts a 17% decrease in precipitation by 2050 for the eastern half of Sri Lanka, while certain districts like Galle, Matara, Puttalam, and Kurunegala may experience a 50% increase in rainfall. The same study indicates that districts such as Anuradhapura, Polonnaruwa, and others will experience below-normal annual rainfall under the B2 scenario, with a 22% decrease in precipitation by 2080 across the Dry Zone, excluding Jaffna, Mannar, and Puttalam.

Other models offer different projections. The World Bank (2018) reports a 7% reduction in precipitation by 2050, based on the 1981–2010 Global Climate Model under RCP4.5 and RCP8.5 scenarios. The USAID (2015) multi-model ensemble predicts a 25% decrease in precipitation by 2030 (10th percentile) under RCP4.5 and a 31% increase (90th percentile) under RCP8.5. For 2050, the USAID ensemble projects a 22% decrease (10th percentile) and a 48% increase (90th percentile) under the same scenarios. Additionally, the World Bank (2021) predicts that by 2080, precipitation could increase by 5–30% under RCP6.0 and RCP8.5, based on GCMs for the 1986–2005 baseline period.

Projections for 2030, 2050, and 2080 show both increases and decreases in precipitation. For example, in 2050, the RCM projects a 15.8–25% increase (ADB, 2014), while the GCM indicates a 7% decrease (World Bank, 2018). Variations also occur within the same scenario; under the A2 scenario for 2050, precipitation is projected to decrease by 17% in the eastern half of the country and increase by 50% in the Galle, Matara, Puttalam, and Kurunegala districts (Punyawardena et al., 2012). Table 5.3 indicates that the highest projected precipitation occurs in the Western and Sabaragamuwa provinces, with reductions in the Northern, North Central, Eastern, and Uva provinces (Fig. 5.5).

Precipitation in January and February is projected to increase under the SSP1-2.6 scenario but decrease under the SSP5-8.5 scenario. Both scenarios show a decrease in precipitation in March and April (First Inter-monsoon period). For other months, precipitation is projected to increase, with significant rises in September and November. Results indicate that precipitation anomalies generally increase from

Table 5.3 Province-wise CMIP6 projected annual precipitation anomaly and mean (2080–2099) in Sri Lanka under SSP1-2.6 and SSP5-8.5 (reference period: 1995–2014), multi-model ensemble

Province	Climatology precipitation (mm, 1991–2020)	Mean (mm) SSP1-2.6	Mean (mm) SSP5-8.5	Anomaly (mm) SSP1-2.6	Anomaly (mm) SSP5-8.5
Northern	1237.89	1133.82	1246.49	74.41	184.02
Northwestern	1651.22	1689.78	1815.07	107.91	247.08
North Central	1506.78	1252.69	1355.35	76.73	191.82
Eastern	1689.76	1491.67	1586.39	83.85	230.15
Central	1865.88	2120.00	2230.16	112.33	277.87
Western	2502.26	3069.13	3254.58	203.34	438.79
Sabaragamuwa	2303.41	3282.68	3449.82	173.28	444.63
Uva	1759.45	1646.35	1757.39	93.98	248.66
Southern	2048.95	2095.45	2234.08	148.86	303.77

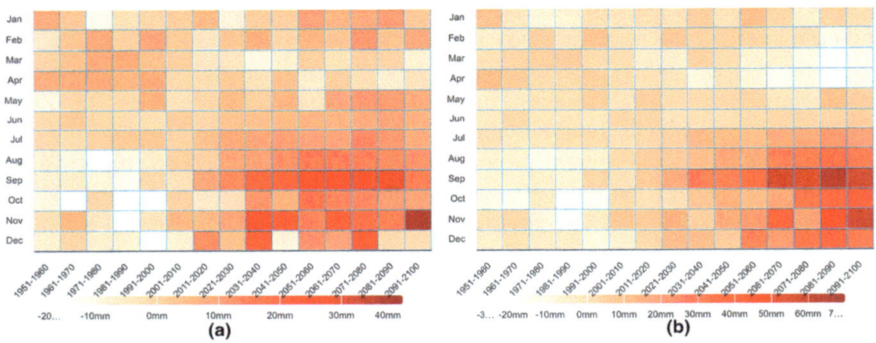

Fig. 5.5 Projected precipitation anomaly in Sri Lanka; **a** SSP1-2.6 and **b** SSP5-8.5 scenarios (reference period 1995–2014, projected for end of century)

May to November (Southwest Monsoon), particularly toward the end of the century. Conversely, March and April (part of the First Inter-monsoon period) show a notable decrease in precipitation anomalies. These trends highlight the variability in future precipitation patterns across different months and scenarios.

The projected change in precipitation distribution shows moderate increases with slightly shifted distribution peaks by the end of the century under SSP1-2.6. In contrast, SSP5-8.5 indicates larger increases, greater variability, and peaks shifting further to the right, highlighting more uncertainty under high-emission scenarios compared to low-emission ones.

5.5.3 *Predicted Changes in Drought and Flood*

The analysis shows that the projected Standardized Precipitation Evapotranspiration Index (SPEI) under both SSP1-2.6 and SSP5-8.5 reveals an increasing trend throughout the country for each month toward the end of the century. These trends suggest a future that is expected to be wetter than the reference period.

Under the SSP1-2.6 and SSP5-8.5 scenarios, the mean temperature increase, and the temperature anomaly vary across Sri Lanka's provinces. In the SSP1-2.6 scenario, mean temperature increases range from 0.16 °C in the Northern Province to 0.32 °C in the Western Province. Anomalies show a change of 0.35–0.63 °C in these regions. Under the SSP5-8.5 scenario, the mean temperature increase is higher, ranging from 0.49 °C in the Northern and North Central provinces to 0.78 °C in the Western Province, with anomalies ranging from 0.68 to 1.02 °C. This highlights the regional disparities in climate impacts.

Under the SSP1-2.6 scenario, the results indicate that the return periods for the largest 1-day and 5-day cumulative precipitation events are projected to decrease over time. From 2010 to 2039, major precipitation events are confined to specific provinces but are expected to spread island-wide by the end of the century. For 1-day precipitation, return periods decrease from 3.95 years to 3.38 years across provinces. For 5-day cumulative precipitation, return periods decrease from 3.81 years to 3.44 years. Under SSP5-8.5, return periods for 1-day precipitation drop from 4.18 years in 2010–2039 to 2.01 years in 2070–2085, with similar decreases for 5-day cumulative precipitation. By the end of the century, the Western, Sabaragamuwa, Southern, and Northwestern provinces are projected to experience the most frequent flooding.

5.6 Sri Lanka's Water Resources and Management

The water resources in Sri Lanka are crucial for domestic, agricultural, and industrial needs, while also supporting ecosystems. Despite this, access to safe drinking water remains limited (MoE, 2016). Groundwater is vital for both drinking and agriculture (Weerasinghe et al., 2023). The country features numerous surface water bodies, including over 100 rivers and approximately 16,000 reservoirs, which cover 260,000 hectares (NWSDB, 2015). Sri Lanka has diverse water bodies totaling 261,941 hectares: large reservoirs (70,850 ha), medium reservoirs (17,004 ha), minor reservoirs (39,271 ha), seasonal tanks (100,000 ha), flood lakes and villus (4,049 ha), upland reservoirs (8,097 ha), and Mahaweli reservoirs (22,670 ha). Sri Lanka also has numerous man-made water bodies, primarily used for irrigation (MoE, 2010). Agriculture is the largest water consumer, accounting for 87.34% of freshwater withdrawals, with domestic and industrial use at 6.22% and 6.42%, respectively (Rajakaruna et al., 2023).

Water management in Sri Lanka, particularly in the Dry Zone, has deep historical roots and cultural significance (Abeywardana et al., 2018). From ancient irrigation to modern water supply and sanitation projects, the country has continually evolved its practices to support its population and economy (NWSDB, 2024). Historically, the tank cascade system—with interconnected tanks reusing water—was a key innovation, especially in dry areas (Abeywardana et al., 2018). Today, challenges such as climate change, population growth, and agricultural expansion necessitate advanced techniques for ensuring future water security (World Bank Group, 2016). Rainwater harvesting, including runoff and rooftop methods, helps address shortages in places where groundwater is contaminated (Balasuriya & Arachchige, 2021). The National Water Supply and Drainage Board currently provides piped water to about 60% of the population and continues to expand clean water access and improve sanitation, especially in rural areas (NWSDB, 2024).

5.7 Implications of Climate Change Projections on Water-Dependent Sectors in Sri Lanka

Climate change intensifies water-related shocks, exacerbating challenges for public consumption and irrigated agriculture in Sri Lanka, which relies heavily on surface and groundwater resources (Ahmed & Suphachalasai, 2014; Eriyagama et al., 2010; Weerasinghe et al., 2023; World Bank Group, 2016). Climate change significantly impacts groundwater availability, quality, and quantity. Precipitation is vital for replenishing groundwater through soil percolation. However, heavy rainfall can cause runoff, reducing infiltration and recharge, and thus decreasing groundwater levels. Additionally, runoff from heavy rainfall can contaminate surface and groundwater by transporting pollutants from agricultural fields and urban areas. Conversely, limited rainfall increases groundwater salinity and mineralization (Abeysinghe et al., 2023).

For over 2,500 years, Sri Lanka's agrarian society has relied heavily on agriculture in the water-scarce Dry Zone (Ahmed & Suphachalasai, 2014). Climate change, with rising temperatures and shifting rainfall patterns, threatens agricultural productivity (Seenithamby & Nandalal, 2021). Water resources play a crucial role in the agricultural sector, where a substantial volume of water is necessary for a successful harvest, directly influencing the country's agricultural productivity. High temperatures and a heavy reliance on agriculture highlight the vulnerability of irrigation water availability, especially during dry periods when the volume and extent of reservoirs tend to drop.

Paddy cultivation dominates Sri Lanka's agricultural landscape, covering 58% of the Dry Zone, 22% of the Wet Zone, and 20% of the Intermediate Zone (MoE, 2022). Paddy requires significant irrigation water. However, higher temperatures and increased evaporation rates threaten yields. A 1 °C rise in growing-season minimum temperature during the dry season can reduce paddy yield by 10% (Peng et al., 2004). The Dry Zone, contributing over 70% of paddy production, faces negative rainfall anomalies from the northeast monsoon (Jayawardena et al., 2024). High-value crops like tea, rubber, coconuts, and upcountry vegetables are also affected by temperature anomalies and changing precipitation patterns (MoE, 2021). By 2080, Sri Lanka's agricultural productivity could decline by 0–15% due to climate change (Eriyagama et al., 2010).

Sri Lanka's economic stability is threatened by climate-induced changes in the water sector, affecting agriculture, livestock, hydropower, and health, while increasing forest fire risks (MoE, 2022). Table 5.4 shows how climate change affects water-dependent sectors in Sri Lanka.

5.8 Reflections and Conclusions

Climate projections for Sri Lanka predict significant warming, erratic precipitation patterns, and an increase in extreme weather events, leading to increased flooding and droughts. These changes are expected to have a severe impact on the country's water resources, exacerbating water stress. Rising temperatures will most affect the Northern, North Central, and Eastern provinces. Higher temperatures will elevate evaporation rates, reduce soil moisture, and decrease irrigation water availability, particularly in the Dry Zone, which already receives less rainfall during the dry season.

Changes in precipitation patterns are also expected to exacerbate water stress in Sri Lanka. While some areas may experience increased rainfall, others may see reductions, resulting in uneven water distribution. For example, the Western, Sabaragamuwa, Northwestern, and Southern provinces are projected to experience higher cumulative precipitation, increasing flood risks. Increased rainfall from May to November could benefit agriculture but also elevate flood risks in these regions. Extreme weather events, such as intense monsoon rains, will strain water infrastructure, potentially damaging supply systems and contaminating water sources. Conversely, prolonged droughts could deplete reservoirs, reduce groundwater recharge, and increase salinity, complicating water management.

A combination of projected rising temperatures and reduced precipitation in the Northern, North Central, Eastern, and Uva provinces will heighten drought risk. Specifically, decreased First Inter-monsoon rains will exacerbate drought conditions. Uva Province, crucial for regulating river flow from the highlands to coastal areas, faces increased vulnerability due to projected climate change. These changes threaten streamflow and overall water availability nationwide, impacting water resources in numerous regions. Sea level rise, projected to reach 0.7 m over the next 200 years, will intensify coastal flooding, inundating areas 3.5–15 m inland and causing saltwater intrusion into rivers and groundwater. This will adversely affect drinking water supplies and coastal agriculture (Palamakumbure et al., 2020; World Bank Group, 2021).

While the results of this study may be useful for future water planning and management, using SPEI and accumulated rainfall/temperature data as proxies for floods and droughts has limitations. This approach overlooks factors like land use changes and direct human interventions. Shared Socioeconomic Pathways (SSPs) capture future

Table 5.4 Climate change impact on water-dependent sectors in Sri Lanka

Sectors	Climate change factor					Sources
	Temperature	Rainfall		Extreme events		
	Increase	Increase	Decrease	Flood	Drought	
Agriculture	Reduced irrigation water availability leads to crop failure	Waterlogging and soil erosion causing crop damage	Increased irrigation demands	Crop damage and loss due to excessive water and increased pests-related diseases	Crop failure and soil degradation due to water scarcity	Eriyagama et al. (2010), MoE (2021, 2022), MoE (2016)
Energy supply	Decreased hydropower generation	Infrastructure damage	Reduce water availability for hydropower and cooling processes in thermal plants	Infrastructure damage	Reduce water availability for hydropower and cooling	Perera and Rathnayake (2019), MoE (2011, 2022), MoE (2016
Hydrology	Increased evaporation leads to lower water levels in lakes, reservoirs, and groundwater, and reduce groundwater recharge and River flow	Altered hydrological cycles impact water storage, polluted water and distribution systems	Decreased groundwater recharge and streamflow affect the water supply	Increase sedimentation and alter river courses, polluted water and distribution systems	Reduce groundwater levels and recharge, and river flow	MoE (2016, Eriyagama et al. (2010), MoE (2010, 2022), Ahmed and Suphachalasai (2014)
Health	Heat-related illnesses, Waterborne diseases, and malnutrition	Vector-borne, rodent-borne, and water-borne diseases	Kidney disease	Vector-borne, rodent-borne, and water-borne diseases	Kidney disease, reduced nutritional status and increased psychiatric illnesses	MoE (2016, MoE (2011, 2022)

(continued)

Table 5.4 (continued)

Sectors	Climate change factor					Sources
	Temperature	Rainfall		Extreme events		
	Increase	Increase	Decrease	Flood	Drought	
Livestock	Heat stress reduces livestock productivity, increases water and feed requirements	Increased risk of diseases and pests	Reduced water availability for drinking and fodder	Loss of livestock due to drowning and disease outbreaks	Reduced pasture quality and water availability, leading to lower livestock productivity	MoE (2022), FAO (2016)
Forest	Increase the risk of forest fires, Lower growth rates, and overall forest health	Alter forest growth patterns	Lower growth rates and overall forest health, causing shifts in species composition	Damage forest ecosystems and lead to increased soil erosion	Reduce forest cover and biodiversity, leading to increased forest degradation and loss of ecosystem services	FAO (2016), MoE (2011, 2022), Ahmed and Suphachalasai (2014)

socioeconomic scenarios, but do not fully address direct human impacts on water resources. We recommend using hydrological impact models that account for river runoff, evaporation, soil infiltration, and groundwater recharge to more accurately predict water availability under climate change.

Finally, rising temperatures and erratic rainfall pose major challenges for water-dependent sectors. They disrupt water availability, impact agriculture and hydropower, compromise watershed health, and alter river flow patterns and water supply systems. These changes highlight the urgent need for adaptive water management strategies.

References

Abeysinghe, A. M. N. P. B., & Rajapakse, R. R. G. R. (2023). *Impacts on groundwater resources by rainfall variations due to climate changes in Sri Lanka.* Water Resources Board. https://wrb.lk/wp-content/uploads/2023/06/Climate-Change-Impacts-on-Groundwater-Resources-in-Sri-Lanka-1.pdf

Abeywardana, N., Bebermeier, W., & Schütt, B. (2018). Ancient water management and governance in the dry zone of Sri Lanka until abandonment, and the influence of colonial politics during reclamation. *Water, 10*(12), 1746. https://doi.org/10.3390/w10121746

Asian Development Bank (ADB). (2014). *Climate change vulnerability in the mountain region of Sri Lanka* (Draft consultant's report, TA 8572-REG). [Unpublished manuscript].

Ahmed, M., & Suphachalasai, S. (2014). *Assessing the costs of climate change and adaptation in South Asia.* Asian Development Bank.

Balasuriya, B. M. C. M., & Arachchige, U. S. P. R. (2021). Rainwater harvesting for drinking purposes in Sri Lanka. *Journal of Research Technology and Engineering, 2*(3), 37–46. https://www.jrte.org/wp-content/uploads/2021/07/Rainwater-Harvesting-for-Drinking-Purposes-in-Sri-Lanka.pdf

Chaturvedi, A., Pandey, B., Yadav, A. K., & Saroj, S. (2021). An overview of the potential impacts of global climate change on water resources. In *Water conservation in the era of global climate change* (pp. 99–120). Elsevier.

Eriyagama, N., Smakhtin, V., Chandrapala, L., & Fernando, K. (2010). *Impacts of climate change on water resources and agriculture in Sri Lanka: A review and preliminary vulnerability mapping.* International Water Management Institute (Research Report 135). https://doi.org/10.5337/2010.211

Food and Agriculture Organization (FAO). (2016). *The state of food and agriculture: Climate change, agriculture and food security.* FAO. https://openknowledge.fao.org/ (accessed July 10, 2024).

Intergovernmental Panel on Climate Change (IPCC). (2023). *Climate change 2023: Synthesis Report. Contribution of Working Groups I, II and III to the Sixth Assessment Report of the IPCC* (Core Writing Team, H. Lee & J. Romero, Eds.). IPCC. https://doi.org/10.59327/IPCC/AR6-9789291691647

Jayawardena, I. S. P., Darshika, D. W. T. T., Herath, H. M. R. C., & Hapuarachchi, H. A. S. U. (2024). Climate variability, observed climate trends, and future climate projections for Sri Lanka. In *The role of tropics in climate change* (pp. 77–119). Elsevier.

Ministry of Environment (MoE). (2010). Sector vulnerability profile: Water. In *National climate change adaptation strategy for Sri Lanka 2011–2016*. Climate Change Secretariat, Ministry of Environment. https://www.climatechange.lk/adaptation/Files/Water_SVP_Nov-16-2010.pdf

Ministry of Environment (MoE). (2011). *Sri Lanka's second national communication on climate change.* Ministry of Environment. https://unfccc.int/sites/default/files/resource/SNC_final.pdf

Ministry of Mahaweli Development and Environment (MoE). (2016). *National adaptation plan for climate change impacts in Sri Lanka: 2016–2025*. Climate Change Secretariat, Ministry of Mahaweli Development and Environment.

Ministry of Environment (MoE). (2021). *Updated nationally determined contributions under the Paris agreement on climate change – Sri Lanka*. Climate Change Secretariat, Ministry of Environment. http://www.climatechange.lk/CCS2021/NDC%202021%20-%20English.pdf

Ministry of Environment (MoE). (2022). *Third national communication on climate change in Sri Lanka*. Climate Change Secretariat, Ministry of Environment. https://unfccc.int/sites/default/files/resource/Third%20National%20Communication%20of%20Sri%20Lanka.pdf

Nianthi, K. W. G. R., & Shaw, R. (2015). Climate change and its impact on the coastal economy of Sri Lanka. *The Global Challenge*, 1–21. [Publisher: Unknown].

National Water Supply and Drainage Board (NWSDB). (2015). *Caring for Water* (2nd ed.). NWSDB.

National Water Supply and Drainage Board (NWSDB). (2024). *National water safety policy*. NWSDB.

Palamakumbure, L., Ratnayake, A. S., Premasiri, H. M. R., Ratnayake, N. P., Katupotha, J., Dushyantha, N., Weththasinghe, S., & Weerakoon, W. A. P. (2020). Sea-level inundation and risk assessment along the south and southwest coasts of Sri Lanka. *Geoenvironmental Disasters, 7*(1), Article 54. https://doi.org/10.1186/s40677-020-00154-y

Perera, A., & Rathnayake, U. (2019). Impact of climate variability on hydropower generation in an ungauged catchment: Erathna run-of-the-river hydropower plant, Sri Lanka. *Applied Water Science, 9*(3), Article 92. https://doi.org/10.1007/s13201-019-0925-9

Punyawardena, B. V. R., Mehmood, S., Hettiarachchi, A. K., Iqbal, M., De Silva, A. S. H. S. A., & Goheer, A. (2012). Future climate of Sri Lanka: An approach through dynamic downscaling of ECHAM4 GCM. *Tropical Agriculturist, 161*, 35–50.

Peng, S., Huang, J., Sheehy, J. E., Laza, R. C., Visperas, R. M., Zhong, X., & Cassman, K. G. (2004). Rice yields decline with higher night temperature from global warming. *Proceedings of the National Academy of Sciences, 101*(27), 9971–9975.

Rajakaruna, R. M. A. S. D., Najim, M. M. M., & Sewwandi, B. G. N. (2023). Future of water consumption in Sri Lanka: A comparison with a water-scarce and a water-rich country. *[Unpublished manuscript]*. https://www.researchgate.net/publication/375552439_Future_of_Water_Consumption_in_Sri_Lanka_A_comparison_with_a_water-scarce_and_a_water-rich_country

Ratnayake, S. S., Reid, M., Larder, N., Kadupitiya, H. K., Hunter, D., Dharmasena, P. B., Kariyawasam, C. S., et al. (2023). Impact of climate change on paddy farming in the village tank cascade systems of Sri Lanka. *Sustainability, 15*(12), 9271.

Seenithamby, M., & Nandalal, K. D. W. (2021). Water resource development planning around village cascades: Piloting a scientific methodology in Yan Oya river basin of Sri Lanka. *Water Policy, 23*(4), 946–969. https://doi.org/10.2166/wp.2021.098

Somasundaram, D., Zhang, F., Ediriweera, S., Wang, S., Li, J., & Zhang, B. (2020). Spatial and temporal changes in surface water area of Sri Lanka over a 30-year period. *Remote Sensing, 12*(22), 3701. https://doi.org/10.3390/rs12223701

United States Agency for International Development (USAID). (2015). *Climate information fact sheet: Sri Lanka*. USAID. https://www.climatelinks.org/sites/default/files/asset/document/Sri%20Lanka%20Climate%20Info%20Fact%20Sheet_FINAL.pdf

Weerasinghe, W. A. R. H., Prasadani, H. A. M., Amarakoon, A. M. L. R., & Rankoth, R. P. I. T. (2023). *Importance of groundwater regulation in Sri Lanka*. Water Resources Board. https://wrb.lk/wp-content/uploads/2023/06/Arti-Importance-of-Groundwater-Regulation-in-Sri-Lanka.pdf

World Bank Group. (2016). *High and Dry: Climate change, water, and the economy*. World Bank. (License: CC BY 3.0 IGO). http://hdl.handle.net/10986/23665

World Bank Group. (2021). *Climate risk country profile: Sri Lanka*. World Bank. https://climateknowledgeportal.worldbank.org/sites/default/files/2021-05/15507-WB_Sri%20Lanka%20Country%20Profile-WEB.pdf

World Bank. (2018). *South Asia's Hotspots: The impact of temperature and precipitation changes on living standards*. World Bank.
World Bank. (2024). Climate change knowledge Portal: Sri Lanka. Retrieved March 4, 2024, from https://climateknowledgeportal.worldbank.org/

Piriyanka Sivananthan served as a Research Consultant at the International Water Management Institute (IWMI) from July 2023 to March 2025, where she contributed to the ClimBeR project focused on enhancing climate resilience. Her responsibilities included GIS analysis, development of data repositories, field data collection, and the preparation of technical reports to support evidence-based decision-making. Prior to this role, she completed an internship at Sanasa General Insurance, Sri Lanka, where she worked as a GIS Analyst for the Earth Observation for Agricultural Risk Management (EO4ARM) platform developed by IWMI. She holds a Bachelor of Science degree in Surveying Sciences with a specialization in Geographic Information Systems (GIS) from Sabaragamuwa University of Sri Lanka. Dr. Emmanuel M.N.A.N. Attoh is a climatologist with over a decade of experience across Africa, Europe, and Asia. His work focuses on co-developing data-driven solutions with diverse stakeholders to address climate risks, hydrological extremes, and water insecurity at local, national, and regional scales. A widely published researcher, he has made significant contributions to policymaking, mentoring, and higher education. Dr. Attoh holds a PhD in Climate and Environmental Science from Wageningen University and Research, Netherlands. His research interests include water and climate information services, hydro-climatology, climate adaptation, and ethnoclimatology. He currently works as a Programme Specialist- Natural Sciences at UNESCO.

Part II
Water Governance and Climate Change Adaptation: Status and Challenges

Chapter 6
Water Governance Challenges for Climate Change Adaptation in Sri Lanka: Analyzing Strategic Frameworks

Hasula Wickremasinghe

Abstract This chapter explores the challenges of water governance in Sri Lanka within the context of climate change adaptation. It critically examines the strategic frameworks outlined in the National Climate Change Strategy (2011–2016), Nationally Determined Contributions (NDC) 2021, the NDC Implementation Plan (2021–2030), the National Environment Policy (2022), the National Policy on Climate Change (2023), and the National Adaptation Plan (2016–2024). The analysis identifies key gaps and opportunities within existing governance structures, offering recommendations to strengthen adaptive capacity in the water sector. The article aims to advance the discourse on sustainable water management and climate resilience in Sri Lanka by providing a comprehensive review of these national policies and action plans.

Keywords Sri Lanka · Climate change adaptation · National Adaptation Plan · Nationally Determined Contribution · Water governance · Sectors

6.1 Introduction

Climate change affects various sectors, including agriculture, livestock, health, urban human settlements, industry, tourism and recreation, power generation, and waste management. Water plays a critical role across all these sectors. Recognizing this interconnection, international climate agreements have increasingly emphasized the integration of water considerations into all areas of planning and development. Despite this, water management in Sri Lanka has traditionally been limited to the drinking water and irrigation sectors. This narrow focus has contributed to siloed policies and sectoral planning, undermining coordinated climate action. Climate change

H. Wickremasinghe (✉)
Climate Change Secretariat, Ministry of Environment, Battaramulla, Sri Lanka
e-mail: hasulaw@yahoo.com

U. A. Amarasinghe et al. (eds.), *Water Governance for Climate Change Adaptation in Sri Lanka*, Global Issues in Water Policy 13,
https://doi.org/10.1007/978-3-032-04950-6_6

and water are now being addressed in several climate action plans across different sectors.

Sri Lanka became a Party to the United Nations Framework Convention on Climate Change (UNFCCC) in 1993. As a non-Annex I Party, it ratified the Kyoto Protocol in 2002 and the Paris Agreement in 2016. The Ministry of Environment, through the Climate Change Secretariat (CCS), serves as the National Focal Point for the UNFCCC and leads the implementation of Sri Lanka's international and national climate commitments.

Acknowledging the urgency of climate action, the Climate Change Secretariat has launched several key initiatives over the years. These include:

- The National Climate Change Adaptation Strategy (NCCAS) for Sri Lanka 2011–2016 (MoE 2010)
- The National Climate Change Policy (NCCP) introduced in 2012 and revised National Policy on Climate Change (NPCC) in 2023 (MoE 2023a)
- The National Adaptation Plan (NAP) for 2016–2025 (MoE 2016), currently under revision
- The Nationally Determined Contributions (NDCs) submitted in 2016 and 2021 (MoE 2023b).

The revised NAP now includes an expanded focus on provincial-level (sub-national) adaptation plans, enhancing the country's resilience to climate impacts in a more localized and inclusive manner.

This chapter examines Sri Lanka's water governance challenges, focusing on the context of the NDC Implementation Plan (2021–2030), the NPCC of 2023, and the NAP 2016–2024.

6.2 Climate Action Plans

6.2.1 National Climate Change Adaptation Strategy (2011–2016)

In 2011, Sri Lanka's National Climate Change Adaptation Strategy (NCCAS) identified five strategic thrust areas to guide the country's response to climate risks:

- Integrating climate change adaptation into national planning and development
- Fostering climate-resilient and healthy human settlements
- Minimizing climate change impacts on food security
- Enhancing the climate resilience of key economic sectors
- Protecting natural resources and biodiversity from the impacts of climate change.

The strategy underscored the importance of mainstreaming climate adaptation to ensure that Sri Lanka's economic development progresses without disruption. It aimed to safeguard critical investments in poverty reduction, food and water security, and public health from being compromised by climate-related impacts.

Under Strategic Thrust 3—*Minimizing Climate Change Impacts on Food Security*—the strategy addressed 'Water for Domestic Needs and Irrigation' as a key Sector Vulnerability Profile. To enhance climate resilience in this area, several priority actions were identified:

- Increasing the availability of irrigation water
- Ensuring dam safety and reducing reservoir siltation to protect existing irrigation water sources
- Expanding access to water for public use
- Preventing industrial pollution of water resources.

Implementing the Natural Disaster Management Plan (1999), which covers preparedness, mitigation, and preventive measures; recovery, relief, rehabilitation, and reconstruction; management of floods, landslides, and cyclones; and improvements in meteorological monitoring, forecasting, and early warning systems.

6.2.2 National Climate Change Policy (2012)

The NCCP was developed to provide guidance and direction for all stakeholders to address the adverse impacts of climate change in an efficient and effective manner. This policy lays out a set of guiding principles followed by broad policy statements under *Vulnerability, Adaptation, Mitigation, Sustainable Consumption and Production, Knowledge Management,* and *General Statements.* The NCCP was brief and did not identify water as a cross-cutting sector. However, promoting integrated watershed and water resources management, as well as efficient water use through technologies and behaviors adaptive to changing weather patterns and trends, was identified as a policy statement under 'Adaptation'.

6.2.3 National Environment Policy (2022) and National Environmental Action Plan (2022–2030)

Both the National Environment Policy (NEP) (MoE 2022a) and the National Environmental Action Plan (NEAP) 2022–2030 (MoE 2022b) were adopted in 2022. The NEP clearly states that human-induced adverse impacts on the environment, such as climate change, could derail the process of sustainable development. It further notes that *"overwhelming evidence suggests that expansion of human economic activities creates increased pressure on ecosystems, leading to degradation of air, water, and land resources in rural and urban areas, with many adverse spill-over effects."* Therefore, land and water resources are identified as key policy concerns.

The NEP states that the main technical issues are low water productivity in agriculture and growing water pollution resulting from the excessive use of chemicals (such as fertilizers and pesticides). Additionally, there is increasing competition for land and water resources between the agricultural and non-agricultural sectors, particularly among industry, infrastructure, housing, and energy projects, which must be balanced with agricultural needs.

The policy statements under ‘Land and Water Resources’ include:

- Integrated management of land and water resources, national spatial strategy, and land-use planning
- Environmentally sensitive land and water resources
- Participatory management of land and water resources.

Since water is combined with land resources in the NEP, it can be inferred that the policy primarily considers water in the context of agriculture.

The 2023 update of the NCCP (named the National Policy on Climate Change—NPCC) addresses critical gaps in climate mitigation, adaptation, loss and damage, and climate financing—areas that have historically hindered the effective planning, implementation, and monitoring of Sri Lanka’s NDCs and NAP. The updated policy is aligned with the NEP 2022, the Paris Agreement, and the principles of the UNFCCC.

The NPCC highlights the intricate interplay between climate risks and socioeconomic factors, aiming to mitigate vulnerability and enhance national resilience. It provides a comprehensive framework for implementing climate action while ensuring alignment with broader environmental policies and promoting effective resource mobilization.

Specifically, the NPCC offers a cross-sectoral approach to addressing climate impacts, including those affecting the water sector. It highlights the need for adaptive water management practices, the protection of water-related ecosystems, and the development of robust climate data systems to support informed decision-making.

Policy Statement 8 of the NPCC acknowledges water management and conservation as part of the broader goal of “ensuring the sustainable management of natural resources, biological diversity, and ecosystem services.” Key strategies outlined under this statement include:

- 8.6 Enhance watershed, river basin, and groundwater management to reduce pollution and conserve natural resources.
- 8.7 Minimize the adverse effects of climate change, such as irregular rainfall patterns, rising temperatures, and sea level rise, on natural and artificial water bodies.
- 8.8 Strengthen integrated water resources management, promote efficient water use, and encourage water-sharing through climate-smart technologies, behavioral change, community engagement, and stakeholder coordination.

The key themes emphasized in the NPCC include the conservation of water resources, integrated water management, and building resilience through a combination of hard and soft solutions. These include:

- **Diversifying water sources**: Utilizing alternative sources such as desalinated water, recycled water, and stormwater can reduce dependency on traditional groundwater and surface water supplies.
- **Enhancing water quality monitoring**: Leveraging technologies such as artificial intelligence (AI) to monitor water quality and usage in real time, helping to identify and address issues proactively.
- **Detecting infrastructure vulnerabilities**: Advanced monitoring systems can help identify potential infrastructure failures before they escalate into major problems.
- **Repairing and retrofitting facilities**: Measures such as installing saltwater intrusion barriers and implementing aquifer recharge techniques can prevent coastal aquifer contamination and maintain freshwater quality.
- **Adopting climate-smart water technologies**: Promoting efficient and sustainable water use through innovative technologies supports long-term water security and resilience.

However, the NPCC also highlights several persistent challenges in water governance:

- **Policy Coherence**: While the policy provides a comprehensive framework, it lacks detailed, actionable guidance for implementation, particularly at the local level. This gap can lead to inconsistencies in application, such as conflicting sectoral policies, uneven enforcement, and differing regional regulations.
- **Stakeholder Engagement**: Effective water governance requires inclusive participation from all stakeholders, especially local communities. However, barriers to meaningful engagement remain, particularly for women and marginalized groups, limiting their influence in decision-making processes.
- **Recognition of Water's Economic Value**: The economic value of water is largely overlooked in current policies, aside from taxation for domestic and industrial use. This may be due in part to the high cost of water treatment. Despite Sri Lanka receiving sufficient annual rainfall and possessing adequate water sources to meet current demand, recurring seasonal water shortages persist. These shortages are primarily the result of poor coordination, inadequate infrastructure, and inefficient management practices.

6.2.4 National Adaptation Plan (2016–2024)

The NAP 2016–2024 outlines specific adaptation measures across nine priority sectors, including water. These sectors are:

- Food security (agriculture, fisheries, and livestock)
- Water resources
- Coastal and marine ecosystems
- Health
- Human settlements and infrastructure

- Ecosystems and biodiversity
- Tourism and recreation
- Export agriculture
- Industry, energy, and transportation.

In addition, the NAP identifies ten cross-cutting thematic areas critical to enabling and supporting sectoral adaptation efforts:

- Policy, legal, economic, and governance frameworks
- Institutional development and coordination
- International cooperation and partnerships
- Resource mobilization
- Research and development
- Technology transfer and standardization
- Capacity building of communities
- Education, training, and awareness
- Climate-induced disaster risk management
- Climate information management.

Within the water sector, the NAP highlights four key focus areas: (a) agricultural water use, (b) human consumption, (c) industrial and energy-related water use, and (d) watershed degradation. The primary adaptation measures recommended for these areas include improving water-use efficiency, protecting water quality, and enhancing resilience to climate-induced water stress (see Table 6.1).

The NAP (2016–2024) also highlights the need to build resilience in the water sector in response to climate impacts, ensure effective water distribution, and minimize salinity intrusion into water bodies. Under the "Human Settlements and Infrastructure" sector, the following water-related adaptation needs and options have been identified (see Table 6.2).

The NAP (2016–2024) also emphasized minimizing the impacts of rising temperatures and periodic water scarcity on energy, industry, and transportation. In the combined sectors of Energy, Transport, and Industry, the following water-related adaptation needs and options are proposed (see Table 6.3).

The NAP (2016–2024) proposed an institutional mechanism for monitoring and reporting the progress of adaptation activities under a National Steering Committee (see Fig. 6.1).

(This figure illustrates the coordination structure: a National Steering Committee oversees NAP implementation, supported by Sectoral Climate Cells, Provincial/Regional Climate bodies, a National Working Group, a Civil Society Organization Forum, and advice from the National Experts Committee on Climate Change Adaptation.)

Sectoral Climate Cells (SCCs) coordinate activities within each vulnerable sector. Each SCC includes representatives from government agencies, the private sector, community organizations, and NGOs. For example, in the water sector, there are two chairpersons—one for the drinking water sub-sector (Ministry of Water Resources) and one for the irrigation sub-sector (Ministry of Irrigation).

Table 6.1 Adaptation needs and options

No	Adaptation needs	Adaptation options
01	Enhance the resilience of systems for water resource management and utilize them to overcome the scarcities caused by climate change impacts	• Improvement of watershed management • Capacity development of storage facilities • Initiating research studies to assess climate impacts • Strengthening the monitoring of climate change impacts (changing pattern of variability) • Promote efficient practices of water management and use
02	Ensure the safety of water management facilities and minimize disturbances to the supply due to extreme weather events	• Strengthening the monitoring of climate change impacts (extreme events) • Establishment of an efficient climate information and communication system • Improvement of disaster risk preparedness and management • Capacity development of storage facilities
03	Minimize the impacts of sea level rise on water supply and management in the coastal zone	• Strengthening the monitoring of climate change impacts (sea level rise) • Improve saltwater intrusion protection measures in coastal areas and wetlands

Source Ministry of Mahaweli Development and Environment (2016)

Table 6.2 Adaptation needs and options in human settlements and infrastructure

Adaptation needs	Adaptation options
Enhance the capacity of infrastructure in urban areas and settlements	• Extend the capacity of drainage and sewerage systems to avoid periodic flooding and overcrowding • Encourage the use of drainage infrastructure to recharge groundwater systems
Promote water safety and efficient utilization of surplus water	• Ensure the safety of domestic water for settlements (e.g., household water treatment and safe storage through point-of-use practices) • Strengthen drinking water surveillance systems • Promote domestic water recycling and reuse • Promote the use of rainwater harvesting systems to collect water in surplus periods for use during dry periods

Source Ministry of Mahaweli Development and Environment (2016)

Table 6.3 Adaptation needs and options in energy, transport, and industry

No	Adaptation needs	Adaptation options
1	System improvements and diversification of energy and power generation	• Reduce fluctuations in hydropower generation by enhancing system management • Develop and implement watershed management plans for major and mini-hydro reservoirs • Use short-term and seasonal weather forecasts to guide generation planning • Explore alternatives to optimize the use of hydropower facilities, such as pumped-storage hydroelectric systems

Source Ministry of Mahaweli Development and Environment (2016)

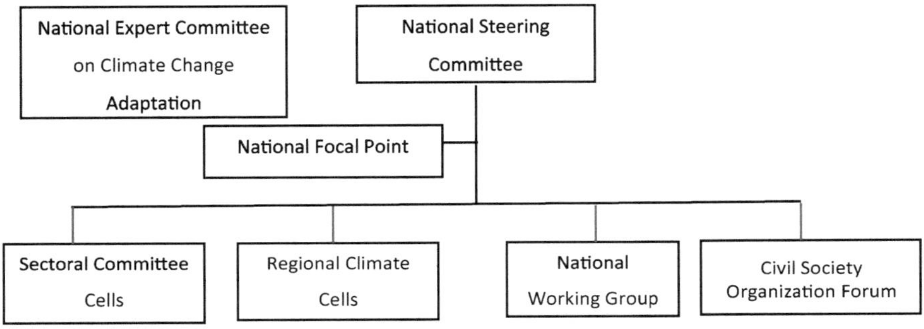

Fig. 6.1 Institutional mechanism for monitoring and reporting of NAP (2016–2024)

Regional Climate Cells (RCCs) implement actions specific to geographic areas. RCCs include representatives from Provincial Councils, local government bodies, and District and Divisional Secretariats. RCCs are established by the Ministry of Environment in accordance with regional requirements.

The National Working Group (NWG) functions as a consortium of national agencies and non-state actors, coordinated by staff from the Climate Change Secretariat (CCS). The NWG is responsible for implementing adaptation actions that address cross-cutting national issues.

The Civil Society Organization (CSO) Forum is a coordination platform comprising partner organizations that support grassroots-level initiatives. It facilitates the implementation and coordination of community-based adaptation activities at both sectoral and cross-sectoral levels, as proposed in the NAP.

The National Focal Point (NFP) is responsible for the overall coordination of the NAP. The Climate Change Secretariat carries out this role under the Ministry of Environment.

The National Steering Committee (NSC) serves as the primary supervisory body for implementing the NAP. It consists of selected members from SCCs, RCCs, the NWG, and CSOs, appointed by the Ministry of Environment to ensure broad representation across sectors and regions. Key responsibilities of the NSC include providing overall leadership for NAP implementation, monitoring implementation progress, guiding periodic reviews of the plan, and making decisions on course corrections as needed.

The National Expert Committee on Climate Change Adaptation (NECCCA) is a pre-existing advisory body composed of technical experts supporting the Ministry of Environment. It continues in an advisory role and expands its functions with the launch of the NAP. All institutional bodies involved in the NAP have access to NECCCA's technical advice through the NFP and NSC. Technical issues arising during implementation are referred to NECCCA and addressed in its periodic meetings or by designated individual experts, depending on the subject matter.

Key Challenges in NAP Implementation:

- **Implementation Gaps**: Despite the comprehensive nature of the National Adaptation Plan, significant gaps remain in implementation, particularly in rural areas where capacity and resources are limited.
- **Monitoring and Evaluation**: The plan lacks a robust monitoring and evaluation framework to track progress and assess the effectiveness of adaptation measures.

Considering the mandates of relevant institutions and the practicality of monitoring progress, the revised NAP (2025) aligns with the 13 sectors identified in the Nationally Determined Contributions (NDCs). This alignment is based on the NDC 2021 and the NDC Implementation Plan (2021–2030). The updated NAP will cover the period from 2025 to 2034, and stakeholder consultations are currently underway.

In the water sector, key strategies—beyond those outlined in the original NAP (2016)—include:

- Enhanced watershed and river basin management, along with the development of additional water storage infrastructure
- Improved management of groundwater resources
- Expansion of the domestic water supply
- Implementation of water demand management strategies in both the domestic and irrigation sectors
- Deployment of early warning systems in flood-prone areas.

In addition, nine provincial-level adaptation plans have been developed and are scheduled for publication in November 2025. These plans aim to address region-specific climate challenges by proposing targeted adaptation interventions suitable for each province. The following draft institutional coordination mechanism (Fig. 6.2) has been adopted to support progress reporting for both the provincial and national adaptation plans.

In this revised framework, Regional Climate Cells have been replaced by the Provincial Climate Board (PCB)/ Provincial Climate Committee (PCC) and the Provincial Climate Unit (PCU). The Chief Secretary of each province chairs the PCB together with relevant District Secretaries, Co-chairing to ensure central and provincial governance streams work in tandem. The PCB comprises of other key provincial line agencies and district level central agencies. The PCU, located within

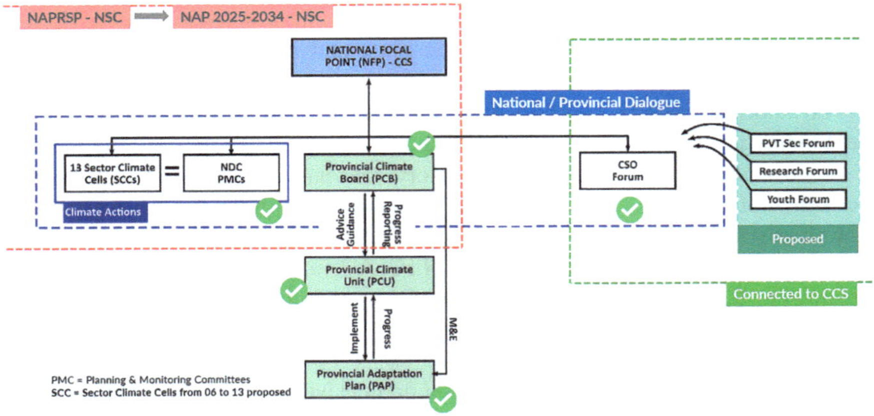

Fig. 6.2 An institutional coordination mechanism for progress reporting for both the provincial and national adaptation plans. *Source* Draft NAP 2025–2034

the Provincial Council and led by the Deputy Chief Secretary (Planning), is staffed by members of the Provincial Secretariat and is responsible for supporting the PCB for decision making, implementing, and monitoring of climate adaptation actions within the province. The Sector Climate Cells remain unchanged and continue to function as Sectoral Planning and Monitoring Committees under the NDC framework.

6.2.5 Updated Nationally Determined Contributions (NDC) 2021

Sri Lanka submitted its first Nationally Determined Contributions (NDCs) in 2016, followed by an updated version in 2021. Under the updated NDCs (2021–2030), Sri Lanka committed to reducing its greenhouse gas (GHG) emissions by 4% unconditionally and 10.5% conditionally, compared to the business-as-usual (BAU) scenario. These commitments reflect the country's obligations under the Paris Agreement to both mitigate emissions and enhance climate resilience.

The updated NDCs outline detailed commitments across 15 key sectors, covering climate change mitigation, adaptation, loss and damage, and means of implementation.

The six mitigation sectors include:

- Electricity (power generation and end-use)
- Transport
- Industry
- Waste
- Forestry

- Agriculture (including livestock).

The nine adaptation sectors—identified due to their vulnerability to extreme weather events and slow-onset climate impacts—are:

- Agriculture
- Livestock
- Fisheries
- Water (including irrigation and drinking water sub-sectors)
- Biodiversity
- Coastal and marine ecosystems
- Health
- Urban planning and human settlements
- Tourism and recreation.

Additionally, **Loss and Damage** has been identified as a separate area requiring focused interventions.

Water security in Sri Lanka is becoming increasingly vulnerable due to erratic weather patterns and rising demand. In response, the government has launched initiatives such as the Mahaweli Water Security Improvement Programme and the Climate Resilience Improvement Project, with a focus on Integrated River Basin Management (IRBM) to balance competing water needs. In the updated NDCs, water and irrigation have been merged into a single sector, while other water uses—such as industry, hydropower, and fisheries—are addressed in their respective sectors. Environmental water concerns are partly included under biodiversity NDCs, and agriculture NDCs emphasize water use efficiency. In the updated NDCs, adaptation in the water sector is structured into two sub-sectors: domestic and irrigation water use. The NDCs prioritize improving water efficiency, promoting Integrated Water Resources Management (IWRM), and enhancing climate resilience in infrastructure. However, challenges remain in aligning national goals with local governance and ensuring coordinated implementation. A total of ten NDCs support the water sector, including IRBM; five focus on domestic water (e.g., groundwater monitoring and climate-resilient supply) and four target irrigation through system upgrades, efficiency improvements, and flood early warning systems (see Table 6.4).

In 2023, Sri Lanka published the **Nationally Determined Contributions Implementation Plan (2021–2030)**, which serves as a national action plan and monitoring tool for achieving the targets in the NDCs submitted in 2021. It also functions as an investment roadmap, outlining national priorities for sustainable economic development. The plan acknowledges the gendered nature of water resource management, emphasizing the central role of women in household water use, agriculture, and disaster resilience. Given their responsibilities and vulnerabilities, women are key stakeholders in water management. The water sector's ten NDCs are supported by 75 sub-actions, with gender considerations integrated into their implementation.

Monitoring and Evaluation Structure of NDC Implementation: Sri Lanka has established a structured institutional framework (Cabinet Paper No. 21/1259/321/012) to implement and monitor its NDCs. Each NDC sector is supported by a Sectoral

Table 6.4 Ten NDCs supporting the water sector

NDC	Action
1	Implement integrated river basin management (IRBM) in 15 prioritized river basins
2	Monitor ground and surface water in drought-prone regions, including Northern, North Central, and Northwestern provinces
3	Promote climate-resilient water supply schemes
4	Promote the use of wastewater for gardening, sanitary, construction, etc., to reduce demand for treated water
5	Install salinity barriers in the three rivers (Kelani Ganga, Kalu Ganga, Malwathu Oya) affected by saline intrusion
6	Build the capacity of water sector personnel and raise public awareness on climate resilience
7	Restore, rehabilitate, and upgrade 25 major/medium reservoirs, 300 minor irrigation systems, and 200 km of irrigation canals
8	Promote alternative water sources for domestic and supplementary irrigation use
9	Improve water management in 40 irrigation schemes
10	Assess and mitigate river flood risks, and implement early warning systems in five priority basins (Kelani Ganga, Attanagalu Oya, Kalu Ganga, Kirindi Oya, and Malwathu Oya) on a pilot basis

Source Ministry of Environment (NDC Implementation Plan, 2023)

Planning and Monitoring Committee (PMC), chaired by the relevant Ministry Secretary, to integrate climate actions into sectoral development plans and facilitate access to funding. A National Steering Committee (NSC), chaired by the Secretary of the Ministry of Environment and comprising key government agencies, ensures cross-sector coordination, tracks progress, and resolves implementation challenges. The Ministry of Environment, as the UNFCCC focal point, operates through the Climate Change Secretariat (CCS) and National Expert Committees, leading climate reporting, greenhouse gas inventory development, and coordination of adaptation and mitigation efforts across the country.

Key Challenges in NDC Implementation: The NDC Implementation Plan calls for integrated action, but institutional fragmentation in water management hinders effective coordination and policy consistency. Funding gaps—especially at the local level—and limited focus on low-cost, water-saving technologies further challenge implementation. Public awareness of water conservation remains low, and key practices such as rainwater harvesting have seen limited progress. Additionally, policy updates need to reflect current climate forecasts, and technical personnel require capacity building to integrate climate considerations into infrastructure and sectoral planning.

6.3 Water Governance Challenges

- **Institutional Fragmentation**: Water governance in Sri Lanka faces significant challenges due to fragmentation of institutional responsibilities across multiple agencies. This fragmentation results in inefficiencies and poor coordination, particularly in climate adaptation efforts.
- **Outdated Policy and Legal Framework**: The policy and legal framework is outdated and ill-equipped to address modern climate challenges. Key laws, such as the Water Resources Board Act (1964) and the Irrigation Ordinance (1946, last amended in 1983), do not reflect current needs for integrated water management or climate resilience. These frameworks define institutional roles but lack coherence and adaptability to evolving environmental and governance demands.
- **Stakeholder Participation and Equity**: Effective water governance must address the diverse needs and capacities of all stakeholders, including local communities, farmers, and urban populations, to ensure equitable distribution of resources. However, a disconnect often exists between policymakers and local communities, resulting in unequal access to water resources and the marginalization of vulnerable groups.
- **Data and Information Systems**: Accurate and timely data are essential for informed water management and climate adaptation. In Sri Lanka, significant gaps persist in climate data collection, dissemination, and utilization. These gaps limit the ability of decision-makers to formulate evidence-based policies and respond effectively to changing climate conditions.
- **Financial and Technical Constraints**: Achieving the objectives of the NDC Implementation Plan, NCCP, and NAP requires considerable financial and technical resources. However, limited funding and a shortage of technical expertise—especially at the local level—continue to hinder effective implementation of climate adaptation measures and improvements in water governance.

6.4 Adaptation Strategies in Water Governance

- **Integrated Water Resources Management (IWRM)**: IWRM is a central strategy promoted in both the NDC Implementation Plan and the National Climate Change Policy. It emphasizes the coordinated development and management of water, land, and related resources to maximize economic and social welfare without compromising ecosystem sustainability. Effective implementation of NDCs in Sri Lanka requires addressing institutional fragmentation, strengthening stakeholder engagement, and improving access to reliable data.
- **Policy Reforms and Institutional Strengthening**: To improve water governance, Sri Lanka must undertake comprehensive policy reforms that integrate climate change considerations into water management laws and regulations. Strengthening institutional capacity at national, provincial, and local levels is essential

to ensure better coordination and effective implementation of climate adaptation strategies.

- **Community-Based Adaptation**: Community-based adaptation involves local communities in the planning and decision-making process, ensuring that adaptation strategies are locally relevant and culturally appropriate. This approach is especially valuable in rural areas of Sri Lanka, where traditional water management practices can be combined with modern adaptation measures to enhance resilience.
- **Use of Technology and Innovation**: Technological advancements—including remote sensing, geographic information systems (GIS), early warning systems, and digital investment platforms—present significant opportunities to enhance water governance. These tools improve monitoring, forecasting, and planning for climate impacts. Innovations such as the Climate Smart Governance (CSG) Dashboard and the AWARE platform, developed by the International Water Management Institute (IWMI), provide valuable support for data-driven decision-making, investment planning, and the long-term sustainability of climate adaptation projects.

6.5 Discussion

Current policies and legislation in Sri Lanka do not fully integrate climate change into water governance. Institutional fragmentation persists, with many agencies operating in silos and limited coordination of water-related activities. This has led to overlapping adaptation efforts and inefficiencies. A key gap is the absence of a comprehensive National Water Master Plan that clearly outlines activities and responsibilities for the conservation and management of water resources.

Although IWRM principles are referenced in various water-related policies, including the NAP and NDCs, water is primarily addressed in terms of its role in agriculture (food security) and drinking water (human health).

The NPCC (2023) and NAP (2016–2024) emphasize the importance of sustainable water management, climate resilience, and balancing competing water demands. However, policy coherence remains a major issue. Water management for agriculture, urban use, and industry often lacks alignment with environmental and climate policies. Additionally, the absence of clear guidelines for local-level implementation results in inconsistent enforcement and varied outcomes across regions.

Local capacity constraints are another significant barrier. Many local authorities lack the necessary technical expertise, funding, and trained personnel to implement national strategies, particularly in remote or climate-vulnerable areas. Balancing competing water demands among agriculture, industry, domestic use, hydropower, and environmental needs is increasingly complex amid growing water scarcity. Integrated approaches like IRBM are essential but often face resistance due to their complexity and coordination challenges.

Financial and resource limitations further hinder progress. While several initiatives support water management and resilience, funding for large-scale infrastructure and adaptation remains inadequate. Innovative financing models, public–private partnerships, and increased international support are urgently needed.

The urgency for climate-resilient infrastructure is growing as rainfall patterns become more unpredictable and the frequency of floods and droughts increases (Alahacoon & Edirisinghe, 2021). Despite policy emphasis on adaptation, implementation is delayed by gaps in data, weak monitoring systems, and the lack of localized adaptation plans.

Water pollution remains a persistent challenge, driven by agricultural runoff, industrial waste, and urban pollution. Although environmental concerns related to water are acknowledged in biodiversity NDCs, water quality regulations are often weak at the local level. Enforcement is further hindered by limited institutional capacity and resistance from certain sectors to adopt sustainable practices.

Stakeholder engagement and gender equity are also underdeveloped. Community involvement, local government participation, and private sector engagement need to be strengthened. Gender-responsive approaches are particularly important, as women often bear primary responsibility for household water management and are disproportionately affected by climate-related impacts. Unfortunately, these considerations are often overlooked in practice.

Technological tools, such as the Climate Smart Governance Dashboard and the AWARE platform (developed by IWMI), present opportunities for data-driven decision-making and investment planning. These platforms can support both national-level strategic planning and local-level adaptation.

6.6 Conclusion

Sri Lanka faces considerable challenges in managing its water resources effectively, particularly in light of climate change and increasing demand across sectors. While the NCCP and the NEP provide a solid framework, significant implementation gaps persist, especially in terms of policy coherence, local capacity, financial mobilization, and managing competing water demands. Water quality protection and the integration of gender considerations also require greater focus.

To ensure effective implementation, national and local policies must be better aligned, local capacities strengthened, and resources mobilized efficiently. Integrating climate change considerations into all water-related development planning, supported by digital tools and scientific evidence, is essential. A comprehensive National Water Master Plan, along with supportive legislation, is needed to guide the conservation, sustainable use, and governance of water resources.

Improving stakeholder engagement at all levels and adopting innovative financing and integrated management approaches will be key to building resilience and ensuring equitable, sustainable access to water for all sectors. Addressing these issues will help secure Sri Lanka's water future and support broader sustainable development goals.

References

Alahacoon, N., & Edirisinghe, M. (2021). Spatial variability of rainfall trends in Sri Lanka from 1989 to 2019 as an indication of climate change. *ISPRS International Journal of Geo-Information, 10*(2), 84. https://doi.org/10.3390/ijgi10020084

Government of Sri Lanka. (1964). *Water Resources Board Act No. 29 of 1964*. Government of Sri Lanka.

Government of Sri Lanka. (1983). *Irrigation Ordinance No. 32 of 1946: A law to consolidate the law relating to irrigation*. Government of Sri Lanka.

Ministry of Environment. (2010). *National climate change adaptation strategy for Sri Lanka 2011–2016*. Ministry of Environment.

Ministry of Environment. (2021). *Updated nationally determined contributions (NDCs) under the Paris agreement on climate change*. Ministry of Environment.

Ministry of Environment. (2022a). *National environment policy*. Ministry of Environment. https://env.gov.lk/web/images/pdf/policies/National_Environment_Policy_-_English.pdf

Ministry of Environment. (2022b). *National environmental action plan: Pathway to sustainable development (2022–2030)*. Ministry of Environment. https://env.gov.lk/web/images/downloads/policies/NEAP_2022.pdf

Ministry of Environment. (2023a). *National policy on climate change 2023*. Ministry of Environment.

Ministry of Environment. (2023b). *Nationally determined contributions implementation plan (2021–2030)*. Ministry of Environment.

Ministry of Mahaweli Development and Environment. (2016). *National adaptation plan for climate change impacts in Sri Lanka (2016–2024)*. Ministry of Mahaweli Development and Environment.

Ms. Hasula Wickremasinghe works at the Ministry of Environment. She has worked at the Biodiversity Secretariat for ten years and coordinated the preparation of the National Red List (fauna groups) in 2012 and was instrumental in developing the Action Plan for Palaeobiodiversity in Sri Lanka. She has worked on the Mitigation and Adaptation sectors at the Climate Change Secretariat for the past ten years. She assisted in submitting the Updated Nationally Determined Contributions (NDCs) in 2021 and in the development of the NDC Implementation Plan and the updated National Policy on Climate Change in 2023. She was also instrumental in designing a Climate Smart City for Kurunegala and assisted in implementing the National Adaptation Plan Readiness Support Project. Hasula has a degree in Biosciences from the Open University of Sri Lanka, an MSc in Environment Sciences from the University of Colombo, and a Postgraduate Diploma in Archaeology from the University of Kelaniya.

Chapter 7
Water Governance in Sri Lanka: Present Status

S. S. K. Chandrasekara and P. H. Sarath Gamini

Abstract Sri Lanka has been renowned for its hydraulic civilization since the era of the ancient kings. Historically, an advisory committee of experts provided guidance to kings on governing water resources, and communal participation, along with a compulsory labor system known as "Rajakariya," was used to manage these resources collaboratively. However, the colonial administrations abolished these traditional water governance methods (based on norms, customs, and traditions) and introduced formal legislation and institutions for water governance. Over 50 pieces of legislation and 40 institutions have since been established to manage the country's water resources. The current water governance framework has, to some extent, adopted a multiscale, polycentric structure. However, many of these laws, being outdated, do not adequately address climate change and variability. Therefore, it is necessary to amend existing legislation to incorporate measures that mitigate the effects of climate change and variability. Furthermore, raising awareness and building capacity regarding the economic value of irrigation and domestic water could serve as a water-saving strategy to help mitigate the impacts of climate change and variability. Ultimately, strong political will and transparent policies are required to improve Sri Lanka's existing water governance.

Keywords Climate change · Water governance · Water resources · Water policy · Awareness

S. S. K. Chandrasekara (✉)
Department of Agricultural Engineering, Faculty of Agriculture, University of Peradeniya, Peradeniya, Sri Lanka
e-mail: sewwandhich@agri.pdn.ac.lk

P. H. Sarath Gamini
Independent Consultant and Former Project Director, National Water Supply & Drainage Board, Kandy, Sri Lanka
e-mail: phsgamini@gmail.com

U. A. Amarasinghe et al. (eds.), *Water Governance for Climate Change Adaptation in Sri Lanka*, Global Issues in Water Policy 13,
https://doi.org/10.1007/978-3-032-04950-6_7

7.1 Introduction

Governance can be generally defined as the process of making decisions and the process by which those decisions are implemented (or not implemented). It involves analyzing the performance of stakeholders in decision-making and assessing the structures that facilitate or hinder the implementation of decisions. According to the Stockholm International Water Institute (2015), water governance encompasses the laws, acts, policies, regulations, and administrative processes through which decisions are made and implemented for managing water resources. These mechanisms involve relevant legal institutions and aim to ensure stakeholder participation in the development and management of water resources, as well as the provision of equitable and efficient water services to users.

Continuous amendment of existing governance mechanisms is crucial for water resource management for several reasons: (a) water, a basic human need, is becoming a scarce or limited resource, (b) The variability of water supply, exacerbated by climate change, poses a threat to reliable access, (c) Environmental degradation due to anthropogenic activities, (d) Over-exploitation of water resources.

The Integrated Water Resources Management (IWRM) concept, introduced in the Dublin Principles of 1992, aimed to provide financial and technological assistance to improve water governance practices in developing countries. Nonetheless, only a few governance reforms have been successful. Furthermore, research and interventions in water governance have received little attention (Bakker, 2003).

7.2 Historical Water Governance

Sri Lanka is renowned for its ancient hydraulic civilization. Historically, kings made rules and regulations in consultation with superior advisory committees. The State, village assemblies, monastic institutions, and individuals owned water storage facilities (reservoirs, tanks, anicuts, canals) for irrigation, domestic, and other uses. Ancient kings built large tanks and canals, which were owned by the State. The king was the source of law for water management and development, and these laws—bound by customs and traditions (often influenced by religion)—upheld justice. As the State grew, laws were inscribed to cover various situations; disobeying rules or damaging reservoirs was regarded as theft, punishable by fines paid to the king. The controversial compulsory labor system, known as the *Rajakariya* system, as well as hired labor, were used for the operation and maintenance (O&M) of ancient irrigation systems (including tank and tank cascade systems). Referred to as Vari Sal, this labor force was paid from the King's treasury. The king himself decided on the allocation of labor and the construction of reservoirs, with technical assistance from officials such as the irrigation engineer (*Ananika*), the officer in charge of channels (*Adikaya*), and reservoir inspectors (*Vavajaruma*). In the 10th century, inscriptions mention the existence of a department in charge of twelve great reservoirs, termed *Dolos Maha Vatan*.

On a village level, small tanks and canals were constructed and maintained by the villagers on a voluntary, communal basis (Kenyon et al., 2004). The village-level administrative officer (*Gamika/Gamladda*) convened assemblies to maintain law, order, and general administration at the village level. Decisions were made participatorily by villagers in these assemblies. Inscriptions from the 3rd century AD (such as *Samanthapasādikā* and *Thimbirivāva*) indicate that private or family-owned small reservoirs, ponds, and channels were managed by their owners (referred to as *Vapi Hamika*) (Siriweera, 2002).

This ancient administrative system began to change after the Portuguese, Dutch, and British established colonial administrations along Sri Lanka's coasts. The British administration abolished the *Rajakariya* system—perceived as a form of slavery—and in doing so dismissed the strong farmer entitlements and obligations that underpinned it. Moreover, *Rajakariya* was seen as an obstacle to the colonial goal of establishing a commercial economy. The colonials introduced written rules and regulations for governance in place of those based on customs, norms, and traditions. These written regulations were frequently updated to redefine maintenance responsibilities for water and agricultural systems. However, frequent restructuring created contradictions and, as a result, implementation at the field level often failed. It is believed that this may have led to the abandonment of many historical water resource systems in Sri Lanka. The British administration, recognizing the importance of the village-level officer for irrigation upkeep, re-established that post in 1879, prior to establishing the present Irrigation Department.

7.3 Present Water Governance in Sri Lanka

The British colonial administration enacted the first written legislation on water management in Sri Lanka—the Irrigation Ordinance No. 32 of 1856. Its primary objectives were to

- legalize traditional irrigation practices,
- Abolish the compulsory labor system, and
- Recommend water extraction specifically for paddy cultivation.

Since then, approximately 50 pieces of legislation and 40 institutions have been established to manage and develop the country's water resources. These legal frameworks address nearly every aspect of water management, including the following:

- Land acquisition and declaration
- Conservation of protected areas
- Soil conservation
- Flood control and protection

- Water pollution control
- Water allocation among different user groups
- Water conservation
- Recognition of customary water rights
- Water trading (buying and selling)
- Water ownership
- Planning, construction, maintenance, and development of water bodies
- Integration with other natural resource sectors
- Stakeholder participation
- Advisory roles to ministers
- Inter-sectoral coordination
- International cooperation
- Conducting research and experimental studies
- Setting water-related standards
- Formulating national policies.

7.4 Influence of Climate Change and Variability on Water Resources of Sri Lanka

Sri Lanka, like many other tropical islands, is vulnerable to climate change and climate variability. Climate change and variability affect water resource availability, leading to spatial and temporal water scarcity. This variability and emerging scarcity affect all sectors of water use, including agriculture, hydropower, industry, health of human and ecosystems, and settlements. Governance interventions are necessary to develop adaptive mechanisms that reduce the risk of climate change impacts. However, it is doubtful whether climate change and its impacts were considered in the development or amendment of water management legislation and institutional strategies. This chapter analyzes the challenges faced by current water governance mechanisms in adapting to climate change in Sri Lanka.

7.5 Multiscale Polycentric Governance in the Water Sector

Multiscale polycentric governance (MPG) is considered a promising approach for mitigating and adapting to climate change while conserving, managing, and developing existing water resources. Carlisle and Gruby (2019) defined a polycentric governance system as a framework for conflict resolution through competitive and cooperative relationships among multiple decision-making centers. Chandrasekara et al. (2021) illustrated that water governance in Sri Lanka is (a) multiscale (operating at national, provincial, district, divisional, and field levels) and (b) polycentric, with multiple semiautonomous decision-making centers.

The Water Management Panel (WMP) is the most effective mechanism for allocating irrigation water in Sri Lanka. Chandrasekara et al. (2021) described the WMP's multiscale polycentric components in decision-making related to bulk water allocation for irrigation and other uses. The WMP allocates bulk water for the Mahaweli irrigation systems for the two main cultivation seasons—Maha (major season, September to March) and Yala (minor season, May to end of August). This process involves participatory decision-making with stakeholders across different scales and sectors to allocate available water for irrigation after considering:

- Water requirements for other users (e.g., hydropower and domestic supply)
- The start/end dates of cultivation/water allocation and the amount of water required for each Mahaweli system
- Crop type (paddy vs. other field crops, or a combination of both)
- The percentage of area to be cultivated.

Established in 1985, the WMP is primarily mandated to allocate, manage, and monitor water resources for areas under the jurisdiction of the Mahaweli Authority of Sri Lanka (MASL). The Director General of MASL chairs the WMP, and its membership also includes the Director General of the Irrigation Department (ID), the Director General of the Department of Agriculture (DoA), the Chairman of the Ceylon Electricity Board (CEB), and the Government Agents/District Secretaries of the Mahaweli Basin and adjoining basins (Malwathu Oya, Kala Oya, Kantale, Maduru Oya, Yan Oya, and Mundeni Aru).

The Water Management Secretariat (WMS), established within MASL, provides technical guidance to the WMP. The Director of WMS serves as the secretary to the WMP. Once the WMP makes its decisions, the WMS oversees the implementation of water allocation among users. The WMS organizes Kanna (pre-seasonal) meetings twice a year, prior to each main cultivation season, to facilitate active participatory decision-making with all relevant stakeholders (including farmer representatives) on water sharing for the upcoming season. Before decisions are finalized in these meetings, the DoA and associated institutions inform participants about changes in rainfall and temperature patterns, forecast potential pest and disease outbreaks (along with recommended remedies), and advise on suitable crop varieties and water-conservation methods. These seasonal forecasts take into account observed climate variability and climate change. The decisions made by the WMP provide sustainable short-term (seasonal) solutions for water conservation and sharing, mitigating the impacts of climate variability. However, long-term water-saving strategies for irrigation and other sectors are also necessary.

The Central Coordination Committee for Irrigation Management and the National Water Supply and Sanitation Steering Committee were tasked with creating awareness and building the capacity of professionals in irrigated agriculture and domestic water supply, respectively. However, frequent changes in the government ministries related to the water sector have caused these committees to cease functioning effectively since 2002 (Chandrasekara et al., 2021). Therefore, it is essential to raise awareness and provide training for professionals on climate change adaptation in the irrigation sector. Furthermore, curbing the monopolistic tendencies of certain water-related institutions could facilitate more collaborative efforts to adapt to climate change impacts. Including newly established bodies such as the Climate Change Secretariat of Sri Lanka (mandated to address climate change mitigation and adaptation) in WMP deliberations would also inform decision-making. This would help strengthen the effective governance of bulk water management in Sri Lanka, particularly in the face of climate change.

7.6 Status of Existing Legislation in the Context of Climate Change

Almost all water-related legislation in Sri Lanka addresses water conservation in some form, either directly or indirectly. Agencies such as MASL, ID, DoA, the Department of Agrarian Development (DAD), the National Water Supply and Drainage Board (NWSDB), Municipal Councils (MC), Provincial Councils (PC), Urban Councils (UC), and Pradeshiya Sabhas (Divisional Secretariats, DS) have direct responsibilities under their respective legislative mandates. Many of the institutions and laws governing water resources were formulated decades ago, based on the needs and conditions that prevailed at the time of their establishment. Table 7.1 lists the primary water-related institutions and their governing legislation that were established before the 21st century.

Furthermore, some institutions have direct mandates to conserve soil or reserved lands, which ultimately contributes to water conservation (see Table 7.2).

During the colonial period, the widespread perception of abundant water resources, combined with a lack of awareness about climate change, meant that water conservation was neither prioritized in legislation nor included in development agendas. As a result, most existing laws now require amendments to incorporate strategies for mitigating and adapting to the impacts of climate change.

7.7 Necessity of Capacity Building and Awareness on Water Saving as a Climate Change Mitigation Approach

The swift implementation of climate change mitigation strategies in Sri Lanka is crucial. Conserving water and related resources is one of the most appropriate strategies to mitigate the effects of climate change and variability. However, before amending legislation, it is necessary to implement capacity-building and public awareness programs that emphasize the importance of water conservation and recognize water as a valuable economic resource.

Chandrasekara et al. (2021) described the evolution of water policy proposals in Sri Lanka and the reasons for their failure. The main hindrance was political resistance to the concept of pricing irrigation water, which led to public rejection of the idea of

Table 7.1 Water-related institutions in Sri Lanka and their governing legislation established before the 21st century

No	Institutions/agency	Legislation
1	Department of Agriculture (DoA)	Department of Agriculture Ordinance No. 37 of 1921
2	Land Commissioner General's Department (LCGD)	Land Development Ordinance No. 19 of 1935
3	Irrigation Department (ID)	Irrigation Ordinance No. 32 of 1946
4	Urban Council (UC)	Urban Councils Act No. 22 of 1955
5	Water Resources Board (WRB)	Water Resources Board Act 29 of 1964
6	National Water Supply and Drainage Board (NWSDB)	National Water Supply and Drainage Board Law No. 02 of 1974
7	Mahaweli Authority of Sri Lanka (MASL)	Mahaweli Authority of Sri Lanka Act No. 23 of 1979
8	Department of Agrarian Development (DAD)	Agrarian Services Act No. 58 of 1979, Agrarian Development Act No. 46 of 2000
9	Central Environmental Authority (CEA)	National Environmental Act, No. 47 of 1980
10	Coast Conservation Department (CCD)	Coast Conservation Act No. 57 of 1981
11	Pradeshiya Saba (PS)	Pradeshiya Sabha Act No. 15 of 1987
12	Municipal Council (MC)	Municipal Councils (Amendment) Act No. 19 of 1987
13	Provincial Council (PC)	Provincial Councils Act No. 42 of 1987
14	Geological Survey and Mines Bureau (GSMB)	Mines and Minerals Act No. 33 of 1992

"selling" state-owned water. Therefore, it is imperative to raise awareness and build the capacity of relevant officials to recognize the value of pricing bulk water to reduce overuse and wastage, thereby promoting water conservation. Furthermore, adjusting the tariff structure for domestic water use could encourage water conservation at the household level. However, attributing an economic value to water is a contentious step that necessitates prior consultation with all relevant stakeholders. This process is likely to be slow and will require financial support to implement the necessary awareness programs.

Scientific research and technological interventions can also play a vital role in water conservation. Chandrasekara et al. (2023) identified several strategies to reduce blue water usage, including:

- Optimize the use of blue water.
- Implement domestic defluoridation units.
- Promote rainwater harvesting.
- Utilize desalination.
- Adopt adaptive agricultural practices to reduce water consumption.

Table 7.2 Institutions and legislation with direct or indirect mandates for water conservation

Description of the conservation		Mandated institution
Water conservation	Agricultural lands	DoA, DoAD
	Water supplies	NWSDB, MC, UC, DS
	Major irrigation schemes	MASL, ID
	Minor irrigation schemes	DoAD
Conservation of reservations	State lands	PC
	Major irrigation schemes	MASL, ID
	Minor irrigation schemes	DoAD
	Water bodies	WRB, ID, MASL, NWSDB
	Streams	DoA, LCGD, MASL
	Lagoons, lakes and ponds	LCGD
	Groundwater	LCGD
	Tanks, reservoirs and bunds	GSMB, LCGD, MASL
	Ocean and seashore	CCD, LCGD
	Draining trenches in flood	ID
Soil conservation	State Lands	PC, CEA, LCGD
	Privately owned lands	DoA
	Land prone to flood and drought	DoA
	Water bodies and streams	LCGD, DoA,
	Major irrigation schemes	MASL, ID
	Minor irrigation schemes	DoAD
	Ocean (Foreshore and seabed)	CCD, LCGD

- Modify power consumption patterns.

Additionally, much of Sri Lanka's water-related infrastructure has exceeded its intended lifespan, resulting in significant levels of non-revenue water loss.

The 13th Amendment to the Constitution (1987) devolved certain water sector governance responsibilities to provincial councils. However, Samad et al. (2017) highlighted several challenges that undermine effective provincial water governance, including:

- Overlapping responsibilities between central government agencies and provincial administrations
- Unclear delineation of authority
- Limited institutional capacity
- Inadequate funding to fulfill statutory obligations.

Moreover, the absence of clear and transparent national or sectoral water policies in Sri Lanka hampers donor engagement and financial support for water-related initiatives (Samad et al., 2017).

7.8 A Way Forward for Water Governance

Currently, Sri Lanka's water governance framework has addressed many aspects. However, it requires tidying up to enhance coordination and collaboration between sectors and institutions, as well as reforms to address climate change and variability. Although the WMP is a good example of an existing multiscale, polycentric governance mechanism in the water sector, it is necessary to incorporate long-term mitigation and adaptation measures to support sustainable resource use. Furthermore, transparency and clearly defined boundaries of responsibility must be strengthened within the polycentric framework to achieve more effective multiscale water governance in Sri Lanka. Gaps and overlaps in institutional responsibilities within Sri Lanka's water governance framework could be resolved by strengthening multiscale polycentric governance mechanisms (Chandrasekara & Gunawardena, 2011).

Political will for the development of the country's water resources is essential. Water sector reforms have been ongoing since the 1980s, with a focus on strengthening grassroots-level water user organizations. Moreover, many water institutions remain static, concentrating on rehabilitating existing infrastructure. However, deficiencies in resource management and financial constraints (Samad et al., 2017) can be overcome by implementing transparent planning policies and allocating appropriate budgets to the water resources sector.

Increasing public awareness and building capacity are vital precursors to legislative reform, and these efforts should be pursued while maintaining institutional continuity and avoiding adverse political interference. In addition, academic and research communities must collaborate with foreign and local authorities to scientifically assess the effectiveness of various strategies for mitigating the impacts of climate change and variability. These strategies include introducing novel approaches, assessing current practices, and amending existing policies from a governance perspective. It is also essential to disseminate the knowledge gained from academic and research initiatives to the public—especially to youth—through social media platforms, which are effective in raising awareness among the younger generation.

References

Bakker, K. (2003). *Good governance in restructuring water supply: A handbook*. Federation of Canadian Municipalities.

Carlisle, K., & Gruby, R. L. (2019). Polycentric systems of governance: A theoretical model for the commons. *Policy Studies Journal, 47*(4), 927–952. https://doi.org/10.1111/psj.12212

Chandrasekara, S. S. K., & Gunawardena, E. R. N. (2011). Effectiveness of existing laws and regulations to prevent encroachments of stream reservations. *Tropical Agricultural Research, 22*(2), 134–143. https://doi.org/10.4038/tar.v22i2.2822

Chandrasekara, S. S. K., Chandrasekara, S. K., Sarath Gamini, P. H., Obeysekera, J., Manthrithilake, H., Kwon, H., & Vithanage, M. (2021). A review of water governance in Sri Lanka: The lessons learnt for future water policy formulation. *Water Policy, 23*, 255–273. https://doi.org/10.2166/wp.2021.152

Chandrasekara, S. S. K., Pathmarajah, S., Obeysekera, J., & Vithanage, M. (2023). Water scarcity—a concealed phenomenon in Sri Lanka: A mini review. *ACS ES&T Water, 3*(6), 1454–1462. https://doi.org/10.1021/acsestwater.2c00398

Kenyon, P., Pollett, C., & Wills-Johnson, N. (2004). *Sustainable water management practices: Lessons from ancient Sri Lanka* (Discussion Paper No. 5/04). Institute for Research into International Competitiveness.

Samad, M., Aheeyar, M., Royo-Olid, J., & Arulingam, I. (2017). *The political and institutional context of the water sector in Sri Lanka: An overview.* Publications Office of the European Union.

Siriweera, W. I. (2002). *History of Sri Lanka from earliest times up to the sixteenth century.* Dayawansa Jayakody.

Stockholm International Water Institute. (2015). *Improved water governance.* Retrieved December 10, 2023, from http://www.siwi.org/priority-area/water-governance/

Sewwandhi Chandrasekara is a lecturer at her alma mater—the Department of Agricultural Engineering, Faculty of Agriculture, University of Peradeniya. Recently, she received her Ph.D. in Hydraulic Engineering at Sejong University, South Korea. She has experience as a research scientist in local and international institutions, including APEC Climate Center, South Korea. Her expertise focuses on water resources management, including hydro-climatological assessments, agricultural water management, and analysis of irrigation and water policies. She has many publications on her expertise. Currently, she serves as the country coordinator for Cap-Net Lanka, a network dedicated to capacity development in sustainable water management in Sri Lanka and worldwide.

Part III
Water Governance and Climate Change Adaptation: Sectoral Perspectives

Chapter 8
Water Governance Challenges for Climate Change Adaptation in the Major and Medium Irrigation Sector

Herath Manthrithilake and Lal Mutuwatte

Abstract This chapter examines how improved management practices could be has been incorporated into various adaptation interventions in major and medium irrigation systems, emphasizing surface and groundwater use and the modifications required to withstand growing challenges due to external climate and economic risks. Some of the crucial governance challenges to enhance the sector's climate resilience include: improving use of rainfall, surface drainage, and seepage flows to address inequitable irrigation distribution across many smallholdings; initiating bulk water allocation to encourage farmer organizations to manage irrigation at the field level; promoting conjunctive management of surface and groundwater to ensure crop water requirements are met during critical growth periods with supplemental groundwater irrigation; encouraging groundwater use to enable farmers to diversify crops, especially during the Yala season; revising existing laws and regulations to promote crop diversification, particularly in areas currently allocated for paddy cultivation; enhancing collaboration among various institutions and across scales to support a more productive agricultural sector, and considering participatory water management models to address the increasing uncertainty of water availability due to climate change.

Keywords Major and medium irrigation · Hydropower · Groundwater use · Paddy cultivation · Crop diversification · Governance · Institutions

H. Manthrithilake (✉) · L. Mutuwatte
International Water Management Institute, Colombo, Sri Lanka
e-mail: h.manthri@cgiar.org

L. Mutuwatte
e-mail: l.mutuwatte@cgiar.org

U. A. Amarasinghe et al. (eds.), *Water Governance for Climate Change Adaptation in Sri Lanka*, Global Issues in Water Policy 13,
https://doi.org/10.1007/978-3-032-04950-6_8

8.1 Introduction

Major and medium irrigation systems are vital to Sri Lanka's food security. These systems irrigate approximately 70% of the country's paddy fields and contribute about 80% of national rice production, the staple food (DCS, 2024).

Major irrigation systems typically serve command areas greater than 405 hectares (1,000 acres), while medium systems cover areas between 81 and 405 hectares (200–1,000 acres) (Irrigation Department, 1978). However, all such systems are now officially classified as major irrigation systems (Rice Research and Development Institute [RRDI], 2024). Their primary water sources infrastructure include large or medium tanks (reservoirs), and anicuts, some of which are supplemented by trans-basin diversions. Over 379 major and medium irrigation schemes are managed by Sri Lanka's Department of Irrigation and the Mahaweli Authority.

Major irrigation reservoirs, predominantly located in Sri Lanka's dry zone, are increasingly recognized for their role in climate adaptation (World Bank, 2014). These systems help buffer against extreme weather events such as floods and droughts. The total storage capacity of major reservoirs is 3.9 billion cubic meters, representing 8% of the country's average annual surface runoff. This storage meets a substantial portion of the water requirements for paddy cultivation. Furthermore, although not yet fully quantified, groundwater recharge facilitated by reservoirs and canal irrigation plays a significant role in supporting agriculture in the surrounding command areas.

Major irrigation systems have received increasing attention in policy (and investment) due to their importance for national food security (Eriyagama et al., 2010). Notably, the Climate Resilient Improvement Project (CRIP)—a USD 152 million initiative funded by the World Bank and the Government of Sri Lanka—aims to enhance the climate resilience of key river basins and large irrigation schemes. CRIP activities included identifying climate-vulnerable hotspots, developing river basin investment plans focused on climate change impacts, and upgrading irrigation and associated infrastructure to cope with future climate variability.

Many climate adaptation projects incorporate both physical infrastructure and institutional reform components. However, the extent to which institutional reforms improve coordination across sectors and governance remains unclear. This chapter examines the historical development of major and medium-scale irrigation in Sri Lanka, assesses the integration of governance into various climate adaptation initiatives, and explores the challenges that must be addressed to enhance system resilience in the face of growing climatic and economic risks.

8.2 Water Governance for Irrigation

8.2.1 Historical Context

In 1842, the Colebrooke-Cameron Commission abolished the 'Rajakariya' - the existed feudal administration system (Shah et al., 2013), a move aimed at strengthening loyalty to the British colonial authorities (Casinader et al., 2018). However, colonial authorities soon realized the adverse impacts of this reform on food security, social stability, and revenue collection. In response, the first Irrigation Ordinance, No. 32 of 1856, was enacted. This Ordinance legalized customary irrigation practices and established priority conditions for water abstraction for paddy cultivation, marking the "new era of irrigation." Over time, the Ordinance was amended to adapt to changing socio-economic, political, and environmental conditions, meeting the evolving needs of the irrigation sector. Some key amendments to the Irrigation Ordinance include:

- **The 1856 Irrigation Ordinance** addressed issues of water rights and dispute resolution, formalizing the customary rules governing irrigation practices.
- **The Paddy Cultivation Ordinance No. 9 of 1867** made it compulsory for landowners to pay, in installments, for repairing or improving irrigation works; in cases of default, penalties were applied.
- **The Irrigation and Paddy Cultivation Ordinance No. 23 of 1889** authorized the formation of Central and Provincial Irrigation Boards (CIB and PIB). It also established an Irrigation Fund managed by the CIB, with government contributions. The Government Agent (GA) could release funds for irrigation works maintained by landowners based on estimates prepared by the CIB.
- **The Irrigation Ordinance No. 16 of 1906** abolished the CIB, PIBs, and the Irrigation Fund. In 1906, the Governor assigned all irrigation schemes to the Director of the newly established Department of Irrigation. The Director of Irrigation (DI) was empowered to call meetings of landowners and to appoint officials for operations and management (O&M) of irrigation schemes, to distribute water, and to perform related tasks. The GA sets rates for irrigation works according to region. Landowners could appoint and pay a village Irrigation Headman (in kind or money) to perform maintenance. If government support was unavailable, landowners shared maintenance costs via a lump sum per acre annually. Village committees (under the Village Communities Ordinance) were authorized to fund villagers to cultivate crown lands. The government provided a grant for paddy cultivation, and violations were adjudicated by a village tribunal as specified by law.
- **The Irrigation Ordinance (Amendment) No. 45 of 1917** empowered the GA to fix irrigation rates and authorized the DI to issue abstraction permits for irrigation from public water sources. The DI could appoint an arbitrator (in a court of arbitration) to resolve disputes. If landowners failed to maintain the irrigation system, the GA (in consultation with the DI) could correct defects and recover

costs from the landowners. The GA also had authority to establish rules to protect irrigation infrastructure and conserve water resources.

- **The Irrigation Ordinance No. 32 of 1946** established District Agricultural Committees (DACs) under the GA. DACs included senior representatives responsible for agriculture and marketing, with the DI as an ex officio member. DACs also included representatives of interest groups. An advisory committee was established for each scheme, chaired by the GA, with the DI and at least 11 elected landowner members. This amendment defined the role and responsibilities of the Village Irrigation Headman. Major and minor irrigation schemes were separated: all major schemes were vested with the DI and maintained using state funds. Rural courts were strengthened to handle irrigation offenses.
- **The Irrigation Ordinance No. 48 of 1968** abolished the positions of Village Irrigation Headman and Irrigation Agent, transferring their powers to newly established Cultivation Committees and their representatives. The Commissioner of Agrarian Services (CAS) was empowered to carry out duties under the Irrigation Ordinance, in accordance with the Paddy Land Act No. 1 of 1958. Additionally, responsibilities previously held by the GA were transferred to the CAS.
- **The Irrigation Act (Amendment) No. 23 of 1983** reintroduced the Irrigation Fund, now funded by fines and penalties from irrigation offenses as well as contributions from landowners and encroachers. The GA was authorized to evict encroachments and remove obstructions along watercourses, and to recover costs of eviction, removal, and repair from those responsible. Dispute resolution was shifted from rural courts to magistrate courts.
- **The Irrigation Act (Amendment) No. 34 of 1990** expanded the definition of "Government Agent" (GA) under Section 118 of the principal enactment to include Assistant GAs, the Provincial Irrigation Secretary, and the Divisional Secretary of the Division.
- **The Irrigation Act (Amendment) No. 13 of 1994** replaced the "Cultivation Committee" (CC) with Farmer Organizations (FOs), transferring the CC's responsibilities (under Part IV, Section 23) to FOs. This amendment exempted an FO's command area from irrigation tax in exchange for performing O&M work on distributary canals, and handed water distribution in those canals to the FOs. It also proposed establishing Project Management Committees for each scheme, as specified by the Secretary of the Ministry of Irrigation.

Subsequent amendments and acts have responded to changing political priorities, providing continuous legal support for the irrigation sector. The Mahaweli Authority Act No. 23 of 1979 and the Agrarian Services Act No. 58 of 1979 established the legal framework for irrigation activities. The Mahaweli Authority was granted legal authority to construct and maintain large irrigation schemes identified under the Mahaweli Master Plan.

8.2.2 Present Context

Major irrigation systems, which have long distributary canals and numerous control structures, provide water on a pre-planned schedule to serve wide command areas. These systems are primarily managed by Sri Lanka's Irrigation Department (ID) and/or by the Mahaweli Authority of Sri Lanka (MASL). A few large and medium schemes in the Northern and Northwestern Provinces are under Provincial management. The DI and the MASL manage 381 major irrigation schemes in the country (DI, 2020) (Table 8.1).

The ID and the Irrigation Management Division of the Ministry of Irrigation, under the Integrated Management of Agricultural Settlements (INMAS) program, jointly manage 52 of the major irrigation schemes. Major and medium irrigation schemes account for 54.3% of the total irrigated area (Salman et al., 2022). The MASL manages irrigation and undertakes multiple socioeconomic and environmental activities across nine systems (schemes) to provide water for both the Yala and Maha seasons (Saddhananda, 2022).

Table 8.2 lists the irrigation systems managed by the MASL of Sri Lanka, along with their command areas, cultivated areas, and cropping intensities.

Water management in major irrigation systems is a complex process involving the storage, diversion, conveyance, regulation, measurement, distribution, and delivery of water to each farm gate. Below the farm gate, farmers manage on-farm irrigation, ensuring that the right amount of water is applied at the right place and time, and removing excess water through drainage. The DI is responsible for collecting and removing this drainage water from irrigation schemes.

Within the Department of Irrigation, the Water Management Division oversees scheme operations and maintenance, as well as water distribution in DI-managed systems. Regional Directors and Divisional Irrigation Engineers handle field-level tasks. The Irrigation Management Division (IMD) schedules and conducts pre-seasonal and seasonal planning meetings, jointly reviews and adjusts cultivation

Table 8.1 Portfolio of Schemes under the Department of Irrigation

Portfolio factors	Values
Major reservoirs (with a command area of more than 405 ha)	73
Medium reservoirs (with a command area between 81 and 405 ha)	171
Total reservoir capacity	3,940 Mm^3
No. of Anicuts	113
No. lift irrigation schemes	5
Drainage schemes	19
Total no. of key structures	381
Total command area	304,953 ha

Source DI 2020 (http://www.irrigationmin.gov.lk/major-and-medium-irrigation-schemes/index.php)

Table 8.2 Irrigation systems managed by the MASL of Sri Lanka

Name of the irrigation system	Azweddumized extent (irrigable command area) in (ha)	Extent cultivated annually (ha)	Cropping intensity (%)
System B	20,006	39,370	196
System C	22,406	44,623	199
System D	1,227	2,690	219
System G–Elahera	6,721	11,706	174
System H	25,390	44,042	173
Huruluwewa	9,100	15,474	170
System L	980	1,956	199
UdaWalawe	20,000	39,631	198
Rambaken Oya	905	1,841	203
Mahaweli	106,735	201,334	188

Source Mahaweli Statistical Handbook (2022), Mahaweli Authority (2022)

plans with farm organization (FO) leaders, and coordinates tasks with field staff. This ensures continuous interaction with farmers at both the field canal and distributary canal levels. DI-IMD officials jointly manage water allocations, monitor reservoir operations, and evaluate scheme performance.

In MASL-managed systems, the system's water management officials are responsible for all aspects of water management. The total 'Mahaweli Water Complex' includes hydropower stations and inter-basin water transfers, with a detailed allocation procedures. This procedure meets the water needs of hydropower stations, irrigation schemes, drinking water supply, and environmental requirements, and it supplements DI-managed irrigation schemes. The Ceylon Electricity Board (CEB) oversees the hydropower stations.

The irrigation water allocation procedure used in MASL system is unique and employs a bottom-up approach. In pre-seasonal planning meetings, representatives of all relevant stakeholders for each irrigation scheme (including both Mahaweli systems and linked DI schemes) meet to reach consensus on seasonal water allocation plans. They determine the start and end dates of water releases and the extent of land to be cultivated in each scheme. The Mahaweli Water Management Secretariat then uses these decisions, along with other hydrological data, to estimate water allocations through a simulation model. The model incorporates 30 years of historical data (rainfall, reservoir levels, river flows) and treats 'irrigation scheme failure' as a final decision-making criterion. The "Irrigation failure" is defined as the model's inability to allocate sufficient water for any irrigation scheme to meet at least 75% of its total demand in any 5 of those 30 years. Three key parameters—the extent of cultivation, the start date of land preparation, and the paddy variety (3–4.5 months)—can be adjusted to fine-tune the allocations.

No modeling is used for schemes not supplied with Mahaweli water; in those cases, water allocation is based on the experience of farmer leaders and DI officials.

8.2.3 Water Delivery to Small Land Parcels

Water distribution to small paddy parcels presents a significant governance challenge. More than 70% of paddy-cultivated lands are smaller than 0.8 ha, and about 25% are less than 2 ha (Economic Census, 2013/14, 2018), DCS (2019). Although land preparation should align with the onset of rainfall over the local command area, farmers typically adhere to pre-agreed dates with farmers' organizations (FOs) for water releases. The highest water use and wastage occur during land preparation. The current practice of providing full canal supply during land preparation period requires a revision, with a greater focus on rainfall, timing, and managing surface drainage and seepage flows to farms.

One approach to improve water delivery to small land parcels is 'Bulk Water Allocation', which encourages farmers to manage their water allocation at the field level (Aheeyar et al., 2012). This system was successfully implemented in the Mahaweli System H. In 'Bulk Water Allocation', the FOs sign an agreement with MASL officials at the start of the season, specifying the quantity of water available to the FO at the distributary canal outlet. The FO is entitled to receive this entire volume on an agreed schedule and is responsible for distributing it among farmers based on their actual needs. These needs vary according to factors such as location, field topography, and farmers' readiness, and that volume often not exceed the 'normative' "water right", based on land area. Because FOs are familiar with actual local conditions, they can adjust distribution, typically reducing the volume wasted. Farmers were guaranteed to the access to 'any saved water' in the following season, although this proved problematic due to limited upstream storage capacity. Although, this system proved more efficient, it was not adopted beyond the Mahaweli System H for unknown reasons.

Generally, irrigation authorities deliver water up to the farm gate. In DI-managed schemes, on-farm water management falls to the hands of farmer, which is more akin to the role of the National Water Supply and Drainage Board (NWSD) for household drinking water. The difference lies in the quantity of water and the type of conveyance system involved. However, in MASL managed areas, the officials as well as FOs are directly involved.

In both the Mahaweli and DI-managed areas, a "Water Master" employed by the FO is responsible for distributing water among individual farms. The role of the Water Master differs between the Bulk Water Allocation process and DI-managed areas. Under Bulk Water Allocation, demand management is inherent, as only a definite quantity of water is "owned" by the FO. In DI-managed areas, however, there is often an undeclared assumption that effectively there is an unlimited water availability, and the Water Master has no control over water uses by farms. This illustrates the dilemma of "owned" versus "common" resource use: in the latter case, disciplining demand requires either raising awareness for behavior change or using other mechanisms such as financial instruments.

Another demand management approach often used is rotational supply. When available water is insufficient for continuous distribution, both MASL and DI impose a "rotational" supply schedule, where individual farms receive water only after a specified number of days. This decision is enforced administratively, by imposing the 'rotation period'.

However, inequitable distribution can occur in both MASL and DI-managed systems. Equitable distribution means providing irrigation water in sufficient quantities at the right time and place, which requires maintaining adequate flow rates throughout each distribution channel. The leading causes of inequitable water distribution include dilapidated canals due to inadequate maintenance, management system inefficiencies, insufficient water supply, and sequential delivery. In an inequitable system, upstream farmers have access to more water, while downstream users suffer from shortages, leading to conflicts between head-enders and tail enders.

8.2.4 *On-Farm Water Management*

Long and staggered land preparation is a major issue in efficient water management for paddy cultivation in Sri Lanka. According to the Department of Irrigation's Design Guide (1978), the 'water duty' for rice cultivation is estimated at 1.5–2.0 meters per hectare (3.0–5.0 acre-feet per acre, in British units). The 'Water duty' is defined as the volume of water required to meet crop water needs and other demands, such as land preparation, seepage, and runoff. The Rice Research and Development Institute (RRDI) estimates average paddy water requirements to be 1,200–1,800 mm per hectare (Table 8.3). Meeting these water requirements with precise land preparation and pre-agreed 'start' and 'end' dates for water releases, is a significant challenge.

In Sri Lanka, paddy is cultivated either by broadcasting germinated seeds or by transplanting nursery-grown seedlings. Direct-seeded paddy requires more water but less labor. In both methods, intermittent irrigation is practiced. About 375 mm of water per hectare (15 inches per acre) is used for land preparation, including raising groundwater levels. Approximately 87.5 mm per hectare (3.5 inches per acre) per week is supplied for crop growth, except during the last two weeks before harvest. The most crucial period is the two-week flowering stage, during which water requirements can rise to 175 mm per hectare (7 inches per acre). After a dragged land preparation,

Table 8.3 Water requirement of paddy (mm/ha) for different agro-ecological zones and crop durations:

Agro-ecological zone	Yala (4 months)	Maha (4 months)	Yala (3½ months)	Maha (3½ months)	Yala (3 months)	Maha (3 months)
Dry Zone	1550	1350	1475	1325	1425	1275
Intermediate Zone	1425	1325	1375	1300	1325	1250
Wet Zone	1350	1325	1300	1275	1275	1250

Source Rice Research and Development Institute (RRDI 2020)

meeting critical water needs of the 'paddy crop schedule' is a major issue in paddy cultivation.

The lack of flow measurements in canals and rainfall monitoring systems, exacerbates retail water distribution problems. Accurate information on canal flows and rainfall is crucial for estimating the supplementary water to be delivered through the canal network. In the absence of these measurements, farmers lack confidence in the 'estimated' water duty, which is based on the reservoir's water balance at the end of the season.

Groundwater use 'in and around' canal command areas, particularly during dry periods, is limited. A study by the International Water Management Institute (IWMI, 2017), based on field observations (of a drainage canal command area and monitoring wells), found that groundwater was nearly at the surface, following a prolonged dry period and a failed cropping season. This highlights the potential use of recharged groundwater, as a supplementary water source. In normal rainy periods, groundwater should supplement canal water for on-farm management. During dry periods, it could serve as the primary irrigation source, offering a clear adaptation strategy for major irrigation systems in the dry zone.

8.2.5 Improving Water Use Efficiency

Low 'water use efficiency' (WUE) is a significant issue in major irrigation systems. In these schemes, WUE is the product of delivery efficiency and application efficiency. Delivery efficiency is defined as the ratio of water released from the source to the amount delivered at the farm gates. Delivery inefficiencies arise from evaporation, seepage, leakage, and overflow in canals and distributary systems, as well as non-beneficial use by weeds and shrubs along the canals. Poor infrastructure maintenance, outdated irrigation technologies, neglect of canal flow rates, uncoordinated water allocation, and limited financial and human resources further reduces the delivery efficiency.

Application efficiency is defined as the ratio of a crop's consumptive water requirement to the actual water supply at the farm. In paddy cultivation, inefficient irrigation methods (such as continuous and intermittent flooding) are the leading causes of low application efficiency. Factors such as small landholding size, timing of land preparation, cultivation method (direct seeding vs. transplanting), weed control, and topography also affect application efficiency (DOA, 2015).

Improving WUE from current levels (around 30%) to 55% could reduce water withdrawals by about 30% (Eshani and Stakhiv, 2017). Supply management (reducing over-allocation and adjusting distribution plans) combined with demand management (using administrative, technical, and financial instruments) could increase WUE. Reducing non-beneficial water use—such as by maintaining a nearly continuous supply instead of large floods—can also improve WUE. Reducing standing water depth and maintaining the soil near saturation (rather than deep flooding), or using alternative wetting and drying methods (such as the System of

Rice Intensification - SRI) (FAO and WFP, 2022), could maintain water depths of 5–10 cm and increase WUE and water productivity by roughly 20% (Lakshani et al., 2023). Adopting dry-seeding technology for paddy cultivation can also significantly reduce water withdrawals.

8.2.6 Operation and Maintenance

Sri Lanka's canal irrigation systems currently operate on a "built-operate-rehabilitate" cycle, which needs urgent reform. The O&M is often given a lower priority. Hence, these systems frequently suffer from inefficient water distribution and deteriorating infrastructure. As a result, they require costly rehabilitation in every 10–15 years, usually funded by large loans from bilateral or multilateral donors.

Efficient operation requires professionally qualified technical staff. However, existing staff are spread across multiple responsibilities, including developing new water resources (feasibility studies, design, and construction of new reservoirs and schemes), reservoir management, flood forecasting, dam safety, and rehabilitation of existing schemes. Meanwhile, an increasing emphasis on financial targets has led to a low priority for investment in skill development and O&M, both of which are critical for maintaining long-term performance.

As seen in other countries, a well-designed asset management program involving beneficiaries can enhance O&M. Policymakers, the Ministry of Finance, the DI, and the MASL should prioritize these approaches to enhance O&M. One potential institutional solution is to separate resource development activities from irrigation management, allowing technical staff to focus on O&M, drainage management, and improving irrigation efficiency. Sri Lanka should explore and adapt new management models to improve its canal irrigation systems.

8.3 Conclusions

Major irrigation systems are vital to Sri Lanka's food security. They account for about 80% of the country's rice production. Water stored in reservoirs and tanks, as well as return flows from canal irrigation, support other sectors, particularly the domestic water supply in the dry zone. The indirect livelihood benefits of major irrigation agriculture—through backward and forward linkages—extend beyond the regions, where it is produced. Therefore, enhancing the sector's adaptive capacity to climate variability is imperative, and effective water governance will play a crucial role in this process.

The following governance solutions could address these challenges:

- **Use rainfall and return flows more effectively**. By focusing on irrigation timing (especially aligning with the onset of rainfall and available return flows), farmers—particularly those at the tail end—can better meet crop water requirements without relying on an unreliable supply

- **Initiate bulk water allocation**. Encourage farmer organizations to manage water at the field level. In this approach, FOs distribute water according to actual needs, improving on-farm water use efficiency.
- **Implement deliberate conjunctive management of surface and groundwater**. Conjunctive water use in major irrigation systems should be planned explicitly. This ensures that crop water requirements are met during critical growth periods via supplemental groundwater irrigation. Reusing surface return flows can also increase the water use efficiency of major irrigation (currently low).
- **Promote groundwater use and crop diversification**. Encouraging groundwater use can entice farmers to diversify crops, especially during the Yala season. A reliable water supply is essential for adopting advanced irrigation technologies (such as drip or sprinkler) for high-value crops, which would enhance economic water productivity.
- **Revise laws and regulations to support crop diversification**. Update existing regulations to promote diversification, particularly in areas currently dedicated to paddy. This policy shift would increase the resilience and productivity of agricultural lands.
- **Enhance collaboration through polycentric governance**. Strengthen collaboration among institutions and scales to ensure a productive agricultural sector. Major irrigation schemes should adopt the polycentric governance approach practiced by MASL-managed systems. Pre-season planning meetings with a bottom-up approach can improve communication and coordination between farmers and institutions.
- **Adopt hybrid water management models**. For example, consider approaches like *Bethma* (marketable water rights and rotational supply) to address uncertainty in water availability due to climate change. Such water rights should be used to improve productivity and safeguard the food security of vulnerable farming communities.

References

Aheeyar, M. M. M., Padmajani, M. T., & Bandara, M. A. C. S. (2012). *Farmer participation in irrigation system management: Achievements and drawbacks*. Hector Kobbekaduwa Agrarian Research and Training Institute.

Casinader, N., Wijeyaratne, R. D. S., & Godden, L. (2018). From sovereignty to modernity: Revisiting the Colebrooke-Cameron Reforms–transforming the Buddhist and colonial imaginary in nineteenth-century Ceylon. *Comparative Legal History, 6*(1), 34–64.

Ceylon Electricity Board. (2022). Annual Report.

DCS (Department of Census and Statistics). (2019). Household Income and Expenditure Survey—2019, Final Result (January to December 2019). Department of Census and Statistics (DCS) & Ministry of Economic Policies and Plan Implementation. Battaramulla. Cited

2019. http://www.statistics.gov.lk/Resource/en/IncomeAndExpenditure/HouseholdIncomeandExpenditureSurvey2019FinalResults.pdf
DCS. (2024). *Paddy statistics*. Department of Census and Statistics (DCS). Battaramulla. December 2021. http://www.statistics.gov.lk/Publication/Economic-Statistics-2021
DOA (Department of Agriculture). (2015). *Pest management recommendations*. Department of Agriculture of Sri Lanka. Peradeniya. Cited 2015. https://doa.gov.lk/wp-content/uploads/2020/06/Introduction_to_book.pdf
Economic Census 2013/14. (2018). General Report—Economic Census 2013/14, Agricultural Activities, Sri Lanka. Department of Census and Statistics of Sri Lanka & Ministry of National Policies and Economic Affairs. Battaramulla. Cited June 2018. http://www.statistics.gov.lk/Economic/Final_Report_Agri.pdf
Eriyagama, N., Smakhtin, V., Chandrapala, L., & Fernando, K. (2010). *Impacts of climate change on water resources and agriculture in Sri Lanka: A review and preliminary vulnerability mapping*. International Water Management Institute (51 pp.). (IWMI Research Report 135). https://doi.org/10.3910/2010.211
Eshani, N., & Stakhiv, E. (2017). Reservoir operations under climate change: Storage capacity options to mitigate risk. *Journal of Hydrology*. https://doi.org/10.1016/j.jhydrol.2017.09.008
FAO and WFP. (2022). Special Report – FAO/WFP Crop and Food Security Assessment Mission (CFSAM) to the Democratic Socialist Republic of Sri Lanka. September 2022. Rome. https://doi.org/10.4060/cc1886en
GOSL (Government of Sri Lanka). (2018). Annual report, 2018. Department of Agrarian Development, Ministry of Agriculture, Government of Sri Lanka.
Harding, T. J., & Franks, T. R. (1988). The matrix management of irrigation projects: Integrated management of major irrigation schemes in Sri Lanka. *Agricultural Administration and Extension, 28*(2), 133–146. https://doi.org/10.1016/0269-7475(88)90062-1
Irrigation Department. (1978). *Design guide for irrigation canals*. Irrigation Department.
Irrigation Department. (2022). *Scheme Register*, (2nd ed.). Irrigation Department.
IWMI. (2017). *Groundwater use in irrigation*. Report submitted to the Asian Development Bank.
Lakshani, M. T., Deepagoda, T. C., Li, Y., Hansen, H. F. E., Elberling, B., Nissanka, S. P., Senanayake, D. M., Hamamoto, S., Babu, G. S., Chanakya, H. N., & Arunkumar, P. G. (2023). Impact of water management on methane emission dynamics in Sri Lankan paddy ecosystems. *Water, 15*(21), 3715.
Mahaweli Statistical Handbook. (2022). *Planning and monitoring unit*. Mahaweli Authority of Sri Lanka.
Rice Research and Development Institute (RRDI). (2020). *Published by the water management and GIS division, Peradeniya*. Department of Agriculture, Government of Sri Lanka.
RRDI. (2024). *Rice in Sri Lanka-water management: Irrigation systems in Sri Lanka*. Rice Research and Development Institute, Bathalagod, Sri Lanka.
Saddhananda, K. W. S. (2022). *Technical background of the paddy crop cutting survey in Sri Lanka*. Department of Census and Statistics, Democratic Socialist Republic of Sri Lanka.
Salman, M., Suzuki, H., Ahmad, W., Giusti, S., Ali, A., Rathnayake, W. M. U. K., Sirisena, D. N., Senanayake, D. M. J. B., Herath, W. M. T. M., Meegasthenna, J., Ponnampalam, Y., Bandulasena, W. M., De Silva, A., Bandara, D., Nandharathne, A. B. D. T., Sooriyaarachchi, A. T., & Pathmarajah, S. (2022). Efficient agricultural water use and management in paddy fields in Sri Lanka – National outlook. *Rome, FAO*. https://doi.org/10.4060/cc2778en
Shah, T., Samad, M., Ariyatane, R., & Jinapala, K. (2013). Ancient small-tank irrigation in Sri Lanka: Continuity and change. *Economic and Political Weekly, 48*, 58–65.
World Bank. (2014). *Sri Lanka—Improving climate resilience project*. World Bank Group. http://documents.worldbank.org/curated/en/256411468334291922/Sri-Lanka-Improving-Climate-Resilience-Project

Herath Manthrithilake was a Senior Researcher (currently, an Emiratus Scientist and Advisor) at the International Water Management Institute (IWMI). He represented IWMI in Sri Lanka, India (Hyderabad), and Central Asia (Tashkent), and has led multiple research programs with a variety of stakeholders. Before IWMI, he worked for the Mahaweli Authority of Sri Lanka (MASL) – a premier water resources development agency of the country, for over 27 years, in different capacities. Since retirement from IWMI, he has been engaged as a visiting lecturer by the Post Graduate Institutes of Rajarata (Ag. Faculty) and the Engineering faculties of Peradeniya and Ruhuna Universities. He has published over 70 scientific papers and reports on a variety of topics related to water management and has made presentations on different aspects of this subject at national, regional, and international conferences. He obtained his Ph.D. from the Moscow Civil Engineering Institute (1979-82) of the Soviet Union. He is a graduate of Moscow Friendship University, USSR (1971-76).

Dr. Lal Mutuwatte is a hydrologist specializing in numerical modeling for water resources management. He has over 25 years of experience in GIS, remote sensing, water resources management, and numerical modeling. He has worked in multiple countries across Asia and Africa, conducting hydrological assessments, analyzing climate change impacts, and managing water resources, with a focus on agricultural applications. He has contributed to over 50 research articles in various high-quality scientific journals. In 2010, 2014, 2015, 2018, and 2019, he was awarded the prestigious Sri Lanka Presidential Award for his research work. He has been a visiting lecturer at various universities in Sri Lanka. Lal holds a Ph.D. in water resources management and hydrological modeling from the University of Twente, the Netherlands. His research interests include working with remote sensing and numerical modeling to study the impacts of climate change on water resources in river basins and regions.

Chapter 9
Governance Issues for Sustainable Management of Village Irrigation in the Dry Zone of Sri Lanka

Upali Imbulana and Mohamed Aheeyar

Abstract Over 70 percent of small tanks in the Dry Zone are hydrologically linked in cascades and are highly vulnerable to climate change impacts. Analyzing the cascade as a system comprising social, technical, and ecological domains reveals that the current governance modality cannot fully realize its development potential and requires transformative adaptation for enhanced climate resilience. Analysis shows that the polycentric governance elements that prevailed in ancient village irrigation management helped sustain these systems, and similar principles apply to the current context. A mix of traditional knowledge and modern technology, a community-centered institutional mechanism, strong policy and legal support, and linking tank cascades with river basin management plans are recommended for realizing the full development potential of the cascades and enhancing their climate resilience.

Keywords Sri Lanka · Climate change · River basin · Village irrigation · Tanks · Cascade system · Irrigation · Management committees

9.1 Introduction

9.1.1 The Role of Village Irrigation in Water Resources Management

Village irrigation systems (VIS) were a key component of the ancient hydraulic civilization of Sri Lanka. Despite the collapse of this civilization in the thirteenth century, VISs remained the center of social organization and management of local

U. Imbulana (✉)
Water Resources Engineer, Colombo, Sri Lanka
e-mail: upali.imbulana@gmail.com

M. Aheeyar
International Water Management Institute, Colombo, Sri Lanka
e-mail: m.aheeyar@cgiar.org

U. A. Amarasinghe et al. (eds.), *Water Governance for Climate Change Adaptation in Sri Lanka*, Global Issues in Water Policy 13,
https://doi.org/10.1007/978-3-032-04950-6_9

resources by the village community at the subsistence level (Kenyon et al., 2004) until the British colonial times. Traditionally, the villagers managed these systems, which consisted of small reservoirs (tanks), and diversions (anicuts). VISs function not only as production systems but also provide social, economic, and cultural functions. VISs include isolated tanks and those organized in a "cascade", a chain of small reservoirs connected hydrologically using a natural drainage pattern (Panabokke, 2009). Over 70 percent of small tanks in major river basins in the dry zone are clustered in these cascades. This connectivity in a cascade enables the reuse of water several times, which is a good water management strategy for addressing recurring droughts (Madduma Bandara, 2009). The cascade system also helps to maintain groundwater in relatively larger surface areas. Studies also demonstrate their ability to mitigate floods (Perera et al., 2020). Therefore, the cascades have the potential to adapt to current climate change impacts characterized by high-intensity rains and longer dry periods.

Since cascade systems consist of hydrologically interconnected tanks, ad hoc interventions targeting isolated tanks have often created problems in the past, affecting the sustainability of the entire cascade system. This paper proposes measures to improve the sustainability of village irrigation systems (VIS) with a focus on cascades. The proposal is based on an analysis of past lessons and the current status of VIS and cascades.

9.2 A Synopsis of Village Irrigation Systems

There are 17,194 VIS-14,421 tanks and 12,773 anicuts in working order (DAD, 2023). VISs account for 37% of total lands under irrigation (DCS, 2023). The total number of farmers benefiting from VIS (excluding small anicut systems) is over 300,000 (DAS, 2000). A database prepared by the UNDP and the Department of Agrarian Development (DAD) in 2017 indicates a density of approximately one tank per 1.02 km^2 in the Northwestern Province (NWP) and one tank per 2.16 km^2 in the North Central Province (NCP), including both working and non-working tanks.

Approximately 25% of the paddy lands cultivated in 2021/22 are under minor irrigation (DCS, 2023), contributing to rural food security. When non-agricultural benefits, including domestic water supply, fisheries, livestock needs, and environmental services, are also considered, village tanks perform a significant social and economic function.

Small tanks are prevalent in approximately 70 river basins (Panabokke et al., 2002). Approximately 90% of the cascade systems are in 27 river basins (Dharmasena, 2019). Madduma Bandara (1985) defined the cascade as "a connected series of village irrigation tanks organized within a micro (or meso-) catchment of the dry zone landscape, storing, conveying and utilizing water from an ephemeral rivulet." Panabokke (cited in Sakthivadivel et al., 1997) recommended that "micro-catchment" be replaced by "meso-catchment" and "ephemeral rivulet" be replaced by either "second-order inland valley" or "first-order ephemeral stream." A modification introduced by Dharmasena (2019) describes them as "an ecosystem where water and land resources are organized within the micro-catchments of the dry zone landscape, providing basic needs to human, floral and faunal communities through water, soil, air, and vegetation with human intervention on a sustainable basis." Perera et al. (2020), focusing on the hydrological principles behind the formation of tank bunds in a cascade, define it as "an autogenic runoff harvesting technique consisting of strategically located series of bunds in shallow valleys of a small catchment for storing water while increasing the ability to regulate peak flows experienced over time." In summary, the various definitions of cascades cited above suggest that they provide distinct social, technical, and ecological benefits.

9.3 Current Issues and Challenges

9.3.1 Environmental Degradation

Ancient land use planning recognized the connectivity among land, water, and human activities, with due regard for nature-based solutions, and this connectivity supported the ecosystem services provided by the cascades. Recent observations indicate that the erosion of rainfed uplands and home gardens contributes to silting in tanks, decreasing the productivity and water infiltration rates of eroded lands (JICA & Nippon Koei, 2000). The encroachment of tank catchments, spill tail canals, and drainage canals in cascades by farmers disrupts the erosion, evaporation, flood, and water-logging control functions of the cascade ecosystem.

Research indicates that tank water quality decreases during dry periods (Cooray et al., 2019). Water-related health issues, including chronic kidney disease of unknown etiology (CKDu), impact climate resilience, reduce community engagement in economically productive activities, and exacerbate poverty.

9.3.2 Rehabilitation and Maintenance of Village Irrigation Systems

The rehabilitation of irrigation systems has gained prominence since the mid-1980s, and numerous rehabilitation projects focusing on specific geographic areas and village irrigation systems have been implemented in the country (World Bank, 1981). However, post-rehabilitation maintenance remains a concern, affecting the sustainability of rehabilitation investments. A deteriorated sense of community ownership of the rehabilitated irrigation infrastructure provided little incentive for its maintenance (Abeyratne, 1990). World Bank (1999) raised concerns about the sustainability of the rehabilitation benefits without increased financial support for operation and maintenance (O&M).

Decreased income from the irrigated command and the emergence of alternative sources of income have contributed to the inadequate priority for VIS maintenance. Fragmentation of land, complex land tenure patterns, low profitability of paddy cultivation, and water scarcity in the dry season (sometimes resulting from unauthorized increases in irrigated command during the wet season compared to the designed command area) are noted to reduce income from the irrigated area. Alternative income sources include groundwater-supported diversified cropping outside the irrigated command, livestock, and inland fisheries. A major strategy for coping with the loss of income from village irrigation, particularly among male farmers, is engaging in off-farm activities (Aheeyar, 2015; Gunatilake, 2015). Anecdotal evidence suggests that such migration leads to poor maintenance of irrigation systems.

9.3.3 Inadequacies of the Current Water Governance System

Current governance arrangements are inadequate to address the major issues listed above, despite numerous government institutions providing various services required by the community. Moreover, it is often found that the responsibility for various services, such as rehabilitation, water management, and agricultural extension, is divided among different government organizations that are legally mandated to provide these services. While inadequate maintenance and environmental degradation of the catchments affect the sustainability of village irrigation, there is no mechanism in place to engage the entire set of water users to address these issues. Although many rehabilitation projects promise institutional changes, infrastructural development often takes precedence during implementation. One of the lessons learned from rehabilitation is that it should be preceded by institutional and agricultural policy reforms (World Bank, 1981).

Considering the relevance of governance to many of the issues challenging the sustainability of village irrigation, we studied the evolution of their governance to learn lessons useful for improving the current situation.

9.3.4 Evolution of Water Governance in Village Irrigation

Historical Buddhist literature suggests that private individuals, the community, and state-owned irrigation works coexisted before the thirteenth century (Abeywardana et al., 2018; Gunawardana, 1971). The state was responsible for building major irrigation works. However, it did not maintain ownership of all major works. Buddhist monasteries-controlled sections of irrigation work obtained through donations and purchases, and generally maintained their irrigation works (Abeywardana et al., 2018).

A powerful bureaucracy, guided by numerous rules, customary regulations, and sanctions, managed VISs through *Gamsabawa* (Village Councils). The community-led indigenous institutions were responsible for the construction, operation, and maintenance of tanks (Panabokke, 2009). Such institutions formulated methods to raise the incomes of the community through fisheries and aquatic plants, as well as mechanisms to share water based on property rights within the tank's jurisdiction. They also managed the ecosystem associated with irrigation infrastructure. The concept of an equitable right to water was central in all decisions on irrigation and cropping systems (Leach, 1980).

The ancient cascades contained tanks serving various purposes, including agriculture, groundwater recharge, wildlife needs, silt trapping, and an ecosystem with several environmentally important components. The sustainability of the cascade system depended on ensuring these distinct functions of the different tanks and their ecological components. Therefore, it is reasonable to assume that the ancient village irrigation management system extended beyond the tank to the cascade. This management system could have been informal, with a "cellular" rather than centrally administered structure (Leach, 1959). Although the cascade management system is not well-documented, a few ancient inscriptions indicate that water management based on the "Cascade Tank-Village System" was introduced in the third century BC and underwent uninterrupted expansion and development until the 13th Century AD. The fully developed cascades were governed by a hierarchical yet durable governance system, under which the King, community, and leaders had distinct responsibilities, accompanied by fees and sanctions for water use (Melles & Perera, 2020).

The hydraulic civilization in Sri Lanka's Dry Zone collapsed after the thirteenth century and was never restored. The disintegration of the irrigation governance structure, due to foreign invasions, the spread of malaria, and climatic reasons (Gunawardana, 1971; Wijesundere & Ramasamy, 2017), has contributed to this collapse. However, the ancient irrigation governance depended on several decision-making centers that were reasonably independent. These polycentric elements in the ancient governance system enabled the continuity of village irrigation infrastructure during the post-hydraulic civilization period (Abeywardana et al., 2018), which may have been administered by temples and local nobles (Kenyon et al., 2004).

Sweeping changes to the traditional governance system were implemented during British rule, following the Colebrook Reforms (1829–1832), which weakened the traditional governance system and reduced its judicial powers (Abeywardana et al., 2018). Although several reasons are cited for these actions, making traditional institutions dysfunctional without providing an alternative arrangement has contributed greatly to the deterioration of irrigation systems (Roberts, 1980; Silva & Vidanapathirana, 1984), and has disrupted the holistic management of the tank, paddy, and forest areas including pasture lands, and associated agriculture-based economic activities (Panabokke et al., 2002).

The resulting severe impacts on the rural economy (Perera, 1955) and national economy (Silva & Vidanapathirana, 1984) prompted the authorities to improve domestic peasant agriculture in the later part of the nineteenth century (Farmer, 1957). However, Abeyratne (1990) noted that government interventions in VISs since the mid-nineteenth century adversely affected their being "farmer-managed." Although some governance arrangements were introduced occasionally, they did not help improve sustainable cascade management.

The above discussion demonstrates that the polycentric elements in ancient governance facilitated the continuity of village irrigation when a single center of decision-making malfunctioned. However, recently introduced governance arrangements in village irrigation did not retain all such elements. Therefore, an analysis of the current state of governance was conducted to establish a foundation for developing an improved governance system.

9.4 Opportunities for Improvement of the Governance System

9.4.1 Current State of Cascade Systems Using Social-Technical-Ecological Systems (STES) Approach

The STES primarily consists of three domains: social, technical, and ecological. Each of these domains interacts to influence the state of a cascade system.

9.4.2 Social Domain

The laws and regulations include the following:

- **Agrarian Development Act (2000)**: Covers issues such as tenant farmers' rights, Agrarian Tribunals (for resolving disputes), the use of agricultural lands, and the formation and registration of Farmer Organizations (FOs), including their legal authority, meetings, minor irrigation development, and management within FO areas. This Act defined irrigation schemes commanding less than 200 acres (80 ha) as "Minor Irrigation Schemes" (FAO, 2023).
- **Irrigation Ordinance** LAWNET, 2016): Addresses matters such as the District Agricultural Committee, meetings of proprietors, construction and maintenance of irrigation works, special provisions applicable to major and minor irrigation works, protection of irrigation works, and dealing with offenses, among others.
- **Other laws**: Applicable to village irrigation, covering land administration, environmental management, flood management, provincial statutes, etc.
- **Customs and traditions**: The customs and traditions of the community, which should be considered when analyzing the system, also fall in the social domain.

The institutions relevant to the village irrigation system include:

- **Department of Agrarian Development (DAD)**: Deals with the implementation of matters relating to the Agrarian Development Act.
- **Provincial Councils (PC)**: Have authority to deal with technical matters such as agriculture and irrigation within the provinces. Established by the 13th Amendment to the Constitution in 1987, some PCs have gazetted their irrigation functions and are legally empowered to conduct technical functions, including disaster risk reduction, irrigation works, agriculture, and water supply.
- **Irrigation Department**: Responsible for irrigation works, particularly when a tank serving a command area larger than 80 hectares is located in a cascade in an inter-provincial river basin.
- **District Secretaries and Divisional Secretaries**: Empowered by several laws related to irrigation to carry out functions related to land and environmental conservation.
- **Other institutions**: Provide services to the public in areas such as drinking water, groundwater, fisheries, livestock, agriculture, credit, insurance, micro-credit, environmental conservation, and disaster management.
- **Community organizations**: Including farmer organizations, women's organizations, fisheries societies, and domestic water user organizations. These organizations are affiliated with government institutions and primarily operate at the village level. Their role in ecosystem management is vague or given low priority in their operational arrangements.
- **Private sector**: Plays a major role in providing agricultural inputs and markets for produce.
- **Paddy Marketing Board**: Purchases paddy at a guaranteed price and helps stabilize prices.

9.4.3 Ecological Domain

As per Dharmasena (2010), the traditional tank ecological domain included (among others):

- Upstream tree belt and sediment filter.
- Soil ridges, water holes, and small tanks to trap sediment and provide water for wild animals and cattle.
- Drainage canal.
- Downstream reservation (interceptor, hamlet buffer, assigned portion of land) for protection from birds and pests.

The efficient functioning of these ecosystem components, within the constraints imposed by the current socio-economic environment, is crucial in addressing the impacts of climate change. For example, the tree belt in the upstream reservation of the traditional VIS acts as a wind barrier, reducing evaporation from the tank surface. It also facilitates vegetation growth among the trees, controls runoff, and improves groundwater recharge. This vegetation and artificial soil ridges trap sediment inflow to the tank. Controlling sedimentation helps maintain the tank's water retention capacity and mitigate the effects of floods and droughts. Aquatic plants in the upper periphery of the tank trap nutrients flowing with agricultural drainage (Mahatantila et al., 2008). The dead storage of a tank retains some water after the irrigation season, helps recharge groundwater during the dry periods, and maintains the water level of nearby wells. The upstream tanks in forest areas provide water for wildlife, thereby minimizing human–wildlife conflicts. The upstream reservations are also potential grazing grounds for livestock.

Furthermore, the downstream reservation acts as an interceptor to retard the flow of undesirable salts to the command area. Properly functioning drainage canals prevent waterlogging and soil salinization and facilitate water reuse. The tank and the ecosystem provide fish, fuelwood, fruits, vegetables, and materials for cottage industries (Dharmasena, 2010).

9.4.4 Technical Domain

The technical domain includes the standards and tools to carry out technical functions, such as:

- **Technical guidelines and design standards:** Provided by the Irrigation Department and other mandated institutions.
- **Circulars:** Issued by government institutions for specific tasks, such as de-silting, climate-smart agriculture guidelines, and agricultural extension methodologies.
- **Mathematical models, weather forecasts, and other technical tools:** Facilitate improved decision-making in water management.

Main infrastructure components include the tank bund, sluices, spillways, and riprap for large bunds. Farmers generally operate and maintain these infrastructures with specific inputs from relevant government institutions, especially when the cost of infrastructural improvements exceeds the community's capacity.

The state of cascade management in the context of three domains.

Figure 9.1, modified from Imbulana et al. (2023), illustrates how the three domains interact and influence each other in a cascading manner within a pluralistic system. This interconnectedness implies that isolated interventions in a domain can sometimes be detrimental to the system's efficient functioning, as verified by previous studies. Abeyratne (1990) noted that inadequate coordination between implementing institutions and beneficiaries poses a threat to the sustainability of the benefits of a rehabilitation project. Aheeyar and Smith (1999) noted that adopting a sectoral approach to solving complex irrigation problems has adverse effects on the environment and recommended a multidisciplinary and integrated approach to irrigation development. The Climate Resilient Integrated Water Management Project's (CRIWMP) preliminary investigation surveys in Anuradhapura in 2017 revealed that tapping spill water from an upstream tank by downstream farmers without proper control structures, water-sharing agreements, and disaster response mechanisms has led to conflicts among farmers and increased flood risk.

Although technical tools, including computer models, are available to make decisions on sharing water equitably among users, there is no institutional mechanism to exploit such opportunities for the entire population in the cascade. For example, the field observations of the CRIWMP revealed that four cascades in Vavuniya have been connected by tapping drainage and utilizing spill water, thereby enabling increased water reuse. However, the lack of an institution to utilize technical tools and make water-sharing decisions at the cascade level hinders the optimal realization of this potential. Therefore, sectoral decision-making within a complex legal and bureaucratic system, characterized by a confusing array of overlaps and gaps, continues in an unproductive manner. It constrains women and youth from contributing to the management of water and associated resources and prevents different community-based water user organizations from taking collective action for optimal water sharing.

This implies that the current governance modality is insufficient for the cascade to achieve its development potential or to be sustainable, and this is one of the root causes of vulnerability.

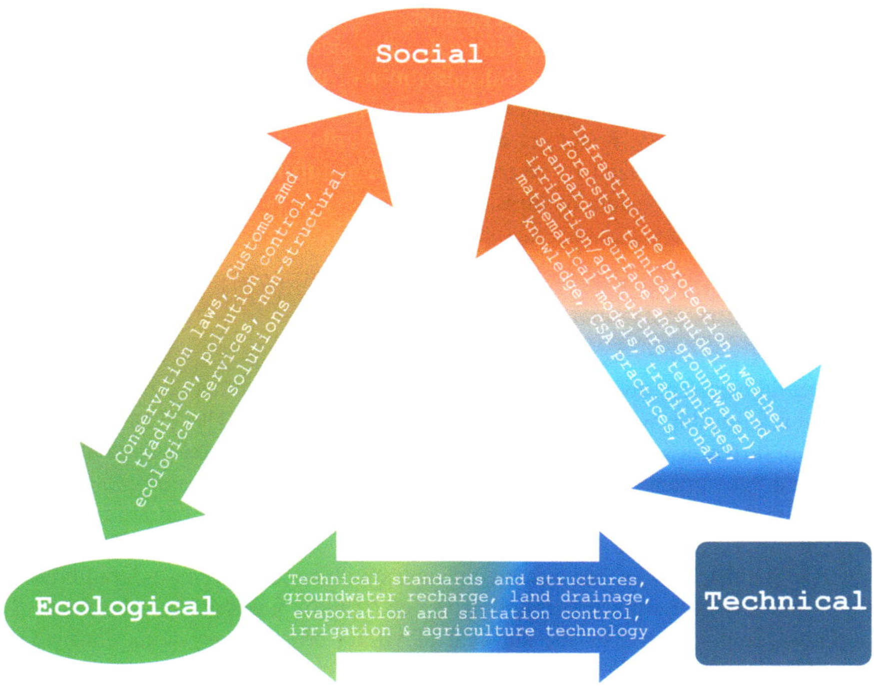

Fig. 9.1 Interactions among different domains of a cascade system. *Source* Authors'

9.5 Need for Transformative Adaptation of Cascade Systems

Fedele et al. (2019) describe the strategies that enable the community to maintain the climate change–affected system in an unchanged state as coping strategies. Incremental adaptation strategies make small-scale adjustments to the system without altering its fundamental characteristics, so that people are resilient to the current impacts. However, the potential for coping strategies or incremental adaptation to mitigate future vulnerability is limited because the fundamental characteristics of the system remain unchanged. Several irrigation rehabilitation projects implemented since the 1980s contain a heavy proportion of incremental adaptation elements. Although many such projects have substantially achieved their physical targets, it is questionable whether they have effectively addressed the root causes of vulnerability in village irrigation.

Furthermore, unpredictability in seasonal weather patterns resulting from climate change increases the reliance on accurate weather forecasts. Many technical standards need modification due to changes in flood frequency and other hydrological parameters. Considering the impact of climate change on crops, climate-smart agriculture practices (CSA) are becoming increasingly relevant. Above all, the cascades need to be managed in an integrated manner.

Therefore, conventional coping strategies and incremental adaptation to climate change are unlikely to reduce the vulnerabilities of the communities in village irrigation systems. Instead, a systemic transformation is required that addresses the root causes of vulnerability. Fedele et al. (2019) describe "transformative adaptation" as a strategy characterized by restructuring, path-shifting, innovation, multiscale, system-wide, and persistent features.

9.6 Need for Polycentric Governance

A polycentric system involves multiple governing authorities that do not have a hierarchical relationship with each other (Morrison et al., 2019). These authorities may operate at different scales while making mutual adjustments. Oberlack et al. (2018) noted that "governance systems are polycentric if they involve multiple arenas of decision making, which operate with some degree of autonomy but are interlinked through processes of cooperation, coordination, or conflict." Gelcich (2014) notes that a polycentric governance system, through smaller-scale successes, could reduce the probability of failure throughout a region. This is particularly relevant to Sri Lanka, as demonstrated by the survival of village irrigation systems during failures of the central government, due to the polycentric elements inherent in ancient governance systems.

In ancient times, cascades operated as mostly isolated and self-sufficient systems that depended substantially on local resources and were mainly community-managed.

They served multiple uses, and village life depended on the complementary services of rain-fed and irrigated agriculture. However, the analysis of the current social domain reveals that numerous government institutions, at national, provincial, and district levels, are involved in managing the cascades. Essential needs, such as drinking water, health, agricultural inputs, marketing, protection from wild animals, and land use, are provided by institutions, including the private sector. The market forces govern the community's income. The farmer community is also affected by human–elephant conflict, and it needs the support of government institutions to maintain their lives and livelihoods.

A polycentric governance system can bring the actors operating at different scales and within social, ecological, and technical domains to a common forum. Abeyratne (1990) and Kekulandala et al. (2020) emphasized the need for an improved governance arrangement, considering the multifunctional nature of cascade systems, to avoid potential conflicts in resource use and infrastructure management. Madduma Bandara (2009) emphasized the need for a management structure that encompasses the entire cascade to realize the full potential benefits of cascade water management. While there may have been elements in the ancient governance system that made the cascades function efficiently, mechanisms such as compulsory labor cannot be reactivated. Therefore, the current situation calls for a systemic transformation of cascade governance.

Accordingly, we examine a few recent interventions for their polycentric governance and transformative adaptation elements, as well as any barriers to their sustainability.

9.7 The Conceptual Approach of Recent Development Projects

9.7.1 Adoption of a "Cascade Approach"

Panabokke et al. (2002) noted that altering the hydrology of a few tanks in a cascade affects the entire cascade. As such, the hydrological assessment of the cascade should precede interventions on selected tanks. Sakthivadivel et al. (1997) noted that the shift of emphasis from tank to cascade happened in the mid-1980s. Meaningful attempts to incorporate this concept into development planning took some more time. One such example is the cascade-based small tank rehabilitation project in the Anuradhapura District, implemented by Plan Sri Lanka from 2004 to 2010. Subsequently, The Hong Kong and Shanghai Banking Corporation Limited (HSBC), in partnership with the International Union for Conservation of Nature (IUCN), implemented a cascade-focused project to enhance rural livelihoods and environmental services in the Kapirikgama cascade in the Anuradhapura District (MMDE, 2016). This project gave due consideration to the cascade ecosystem, which could be considered a further advancement from the cascade-hydrology model (Dharmasena, 2019).

The Climate Resilient Integrated Water Management Project (CRIWMP), implemented from 2017 to 2024 by the Ministry of Irrigation with technical assistance from the UNDP and funding support from the Green Climate Fund (GCF), is a pioneering effort to identify cascade-based village irrigation systems as a planning, development, and water management unit within a river basin. The studies leading to the formulation of the project identified poverty, the incidence of CKDu, and floods and droughts as some of the main issues contributing to the vulnerability of Dry Zone farmers to climate change. The interventions included upgrading irrigation systems, restoring cascade ecosystem components and their maintenance arrangements, converting weather advisories into improved agricultural advisories for cultivation planning, modeling complex water balances in cascade systems, and promoting community-supported agriculture (CSA), drinking water management, and groundwater management.

The Climate-Smart Irrigated Agriculture Project (CSIAP), funded by the World Bank, implemented activities in 11 districts from 2019 to 2024. The project aims to improve the productivity and climate resilience of the smallholder agriculture sector in selected vulnerable areas (World Bank, 2019). Novel approaches adopted by the project include river basin–level hydrological assessments, considering future climate vulnerability, and identifying "hot spot" areas, including cascades. A hotspot area is identified based on exposure and vulnerability to increasing climate variability, as determined by drought and flood impacts, as well as current and future climate vulnerability. The interventions focused on improving agricultural production, marketing, and irrigation systems through rehabilitation, operation, and maintenance.

9.7.2 *Institutional Arrangements Proposed for Cascade Management*

A common intervention with polycentric governance elements in both CSIAP and CRIWMP is the formulation of a Cascade Management Committee (CMC). The CRIWMP's project proposal recommends establishing cascade-level water management committees to optimize the utilization of water resources, promote multi-stakeholder planning, conflict resolution, and improve water-sharing between individual tanks and anicuts for multiple uses, thereby enhancing climate resilience (MMDE, 2016). The CSIAP creates a CMC to ensure that individual tanks are managed in the broader context of the system. The committees, comprising key stakeholder agencies and users, are expected to be the main entry point for water management within a watershed (World Bank, 2019). In either approach, the CMC has the potential to bring production entities, such as command areas, home gardens, livestock, and fisheries, together in a common forum, thereby increasing the resources available for maintaining water infrastructure. The CMC can facilitate the participation of women and youth in the resource management process. When fully implemented, it will promote polycentric governance by engaging national, provincial, district, and local-level actors in a single forum. It also has the potential to resolve issues arising from different institutions managing water within a cascade. However, this institutional framework is only partially implemented now due to several reasons, including inadequate legal and policy support.

9.7.3 Integrated Management Model Proposed by the CRIWMP

A unique feature of the CRIWMP is the development of an integrated management plan to manage water and associated natural resources at the cascade level. The work of several scholars in the field of tank-cascade development influenced this plan. The plan includes seven thematic areas (Imbulana et al., 2023) (Fig. 9.2):

- Upgrading the village irrigation system in a climate-resilient manner.
- An agriculture development plan with CSA principles.
- Operation and maintenance of VIS and disaster response.
- Environmental management and catchment conservation.
- Drinking water management.
- Groundwater management.
- An institutional framework for cascade management.

9.7.4 Barriers to Implementing Functional CMCs

An inadequate enabling environment (policy and legal backing) was one of the major barriers to establishing functional CMCs. Although it is possible to engage key actors in the agriculture sector through existing mechanisms, such as seasonal cultivation meetings and the District Agriculture Committee, current policies do not support the participation of different water users outside the agriculture sector in such forums. Some of the potential barriers to facilitating transformative adaptation in village irrigation among different tank-based communities are the large number of government institutions providing services to the community, administrative boundaries that do not coincide with hydrological boundaries, competition within the community for scarce resources, inefficient village-level community organizations, and inadequate social cohesion.

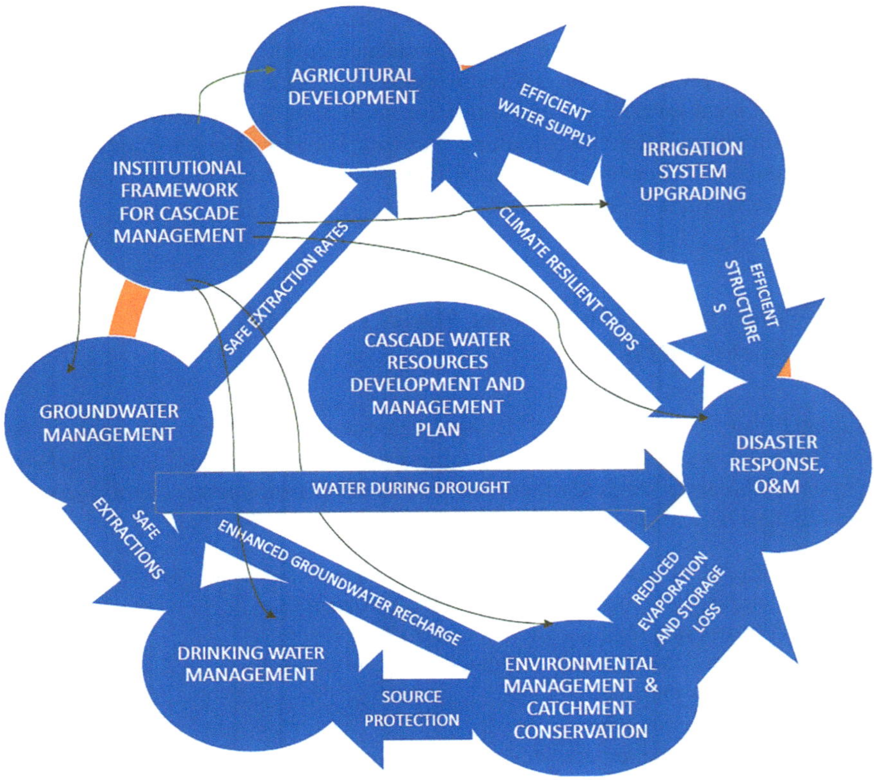

Fig. 9.2 The concept of the integrated management plan of CRIWMP

9.8 Conclusions and Policy Recommendations

The discussion demonstrates that a tank cascade has inherent advantages that support its management as an integrated system. Technologically, hydrological linkages between tanks make the cascade an ideal unit for water management. As an STES providing multiple services, it has the potential to promote integrated water resources management with a "bottom-up" approach.

The scientific basis of some traditional elements of a cascade suggests that the sustainable development of cascades should be carried out using an appropriate mix of traditional knowledge and modern technology. Analyzing the cascades in an STES framework enables understanding of the current status of the cascade, its functional gaps, and overlaps. It facilitates making more specific and targeted inputs from social scientists, economists, environmentalists, engineers, and other stakeholders.

The current interventions in VIS have transformative adaptation options. These include.

Development and management plans that incorporate surface and groundwater resources, multiple water uses, new technologies including CSA, and restoration of selected traditional features into integrated solutions.

A new institutional setup that encourages interested members of the community, especially the youth, to engage in commercial farming and introduce improved business practices and partnerships with due regard to environmental management and the user rights of other farmers.

Measures to address issues such as inadequate representation of women in decision-making, improved vertical and horizontal linkages across the community and government institutions, and under-representation of tenant farmers and non-agricultural water users in the current institutional setup.

Grouping the production of a larger number of smallholders, which can influence markets and make it easier to achieve the cascade's potential to mitigate droughts and floods.

Linking off-farm incomes and attracting external financial support to cascade system management.

Formal recognition of cascade-based community institutions by government policies and legislation, which will directly address some barriers to sustainable operations and facilitate detailed investigations to achieve the full potential of a cascade system. The goals, objectives, scope, and accountability aspects of such an organization should be designed in consultation with government institutions responsible for agriculture, irrigation, forestry, wildlife, and drinking water. This will lead to a polycentric governance model at the cascade level, in which the community's role will increase with its economic development. Different cooperative business models, such as producer groups, cooperatives, and farmer companies, as well as partnerships with the private sector, can enable the economic empowerment of the cascade community with a sound business plan.

A considerable number of isolated tanks that exist outside cascades should not be excluded from the development process. Besides, the impact of interventions such as improved water management, cropping patterns, and groundwater recharge extends beyond the cascade boundaries. As such, any water resource development should be carried out with a proper hydrological assessment that extends beyond the cascade. A sub-basin (consisting of several cascades, isolated tanks, and anicuts) can be used as a development planning unit, while the cascade can be used as a water and agriculture management unit. Such an arrangement can optimize in-basin water resource use and reduce adverse impacts resulting from the excessive utilization of cascade water resources by downstream users and the environment.

The river basin management plans should recognize the hydrology, livelihoods, and natural resources of cascades. Understanding cascade hydrology will help river basin planners assess the potential for in-basin water resource development. If the concept of cascade management is recognized in river basin management, it will constitute a bottom-up approach to river basin management and integrated natural resources management.

This, however, will require strong policy and legal support. Furthermore, the political will to adopt long-term, sustainable solutions is also crucial for success. It will be challenging because the current global economic downturn can make immediate and short-term solutions more attractive. Meaningful discussions among researchers, decision-makers, donors, and subject-matter specialists are also essential to formulate strategies that overcome the identified barriers. However, the preceding discussion highlights that cascade systems will not be sustained without addressing the root causes of vulnerability through a transformative adaptation approach. Therefore, the commitment of all interested parties is required, and accountability for implementing the governance changes envisioned during the planning stage of development projects should be asserted.

The following recommendations emerge from the above analysis:

- Policy support, followed by legal recognition, should be provided to community-based institutions such as cascade management committees at the cascade level. This will enable increased bottom-up approaches in water and natural resources management and facilitate investigations into ill-defined subject areas in cascade management, as well as complex socio-political and economic issues and challenges.
- Water resources development should be planned with due consideration for a river basin-level water resources assessment, which can reveal the cumulative impacts of any modification to water use throughout the basin. Such an analysis can identify whether there is a need for inter-basin water transfers.
- If the entire river basin cannot be adopted as the spatial unit for water resources planning and development, a sub-basin (sub-watershed) can be selected for that purpose. The cascade shall be considered as the primary spatial unit for water management in river basins where cascades are predominantly present.
- Water resources management policies should recognize the importance and relevance of polycentric governance and transformative adaptation to make cascade systems sustainably climate-resilient and managed by empowered communities.

References

Abeyratne, S. (1990). *Rehabilitation of small-scale irrigation systems in Sri Lanka, state policy and practice in two systems*. International Irrigation Management Institute (IIMI).

Abeywardana, N., Bebermeier, W., & Schütt, B. (2018). Ancient water management and governance in the dry zone of Sri Lanka until abandonment, and the influence of colonial politics during reclamation. *Water, 10*, 1746. https://doi.org/10.3390/w10121746

Aheeyar, M. M. M., & Smith, L. E. D. (1999). The impact of farmer participation on water distribution performance in two irrigation schemes in Sri Lanka. *Sri Lanka Journal of Social Sciences, 22*(1&2), 27–43. https://doi.org/10.4038/sljss.v0i0.7428

Aheeyar, M. M. M. (2015). *Resilience of irrigated farmers to climate change effects*. Survey Report submitted to the Water and Climate Change Resilience Programme (WACREP), Sri Lanka Water Partnership. Colombo.

Cooray, T., Wei, Y., Zhong, H., Zheng, L., Weragoda, S. K., & Weerasooriya, R. (2019). Assessment of Groundwater Quality in CKDu affected areas of Sri Lanka: Implications for drinking water treatment. *International Journal of Environmental Research and Public Health, 16:*1698. https://doi.org/10.3390/ijerph16101698. www.mdpi.com/journal/ijerph

Department of Agrarian Services (DAS). (2000). *Data book for village irrigation schemes of Sri Lanka—island summary, water management division.* Department of Agrarian Services, Sri Lanka.

Department of Agrarian Development (DAD). (2023). DAD website. Retrieved July 6, 2023, from https://www.agrariandept.gov.lk/web/index.php?option=com_content&view=article&id=56&Itemid=148&lang=si

Department of Census and Statistics (DCS). (2023). DCS website. Retrieved July 6, 2023, from http://www.statistics.gov.lk/Agriculture/StaticalInformation/new

Dharmasena, P. B. (2010). *Essential components of traditional village tank systems.* https://www.researchgate.net/publication/303944355_Essential_Components_of_Traditional_Village_Tank_Systems

Dharmasena, P. B. (2019). Cascaded tank-village system: Present status and prospects. In B. Marambe, J. Weerahewa, & W. S. Dandiya (Eds.), *Agricultural research for sustainable food systems in Sri Lanka: A historical perspective* (Vol. 1, pp. 63–75). Springer Nature Singapore Pte Ltd. https://www.springer.com/gp/book/9789811521515

Farmer, B. H. (1957). *Pioneer peasant colonization in ceylon: A study in Asian agrarian problems.* Oxford University Press.

Fedele, G., Donate, C. I., Harvey, C. A., Hannah, L., & Hole, D. G. (2019). Transformative adaptation to climate change for sustainable social-ecological systems. *Environmental Science and Policy, 101*, 116–125.

Food and Agriculture Organization (FAO). (2023). *Agrarian development act, no. 46 of 2000.* Retrieved July 6, 2023, from https://faolex.fao.org/docs/pdf/srl43285.pdf

Gelcich, S. (2014). Towards polycentric governance of small-scale fisheries: Insights from the new “management plans” policy in Chile. *Aquatic Conservation: Marine and Freshwater Ecosystems, 24*, 575–581.

Gunatilake, M. M. (2015). Drought and household food security in rural Sri Lanka: A case study. *International Journal of Scientific Research and Innovative Technology, 2*(7).

Gunawardana, R. A. L. H. (1971). Irrigation and hydraulic society in early medieval ceylon.*Past & Present* (53), 3–27. Oxford University Press on behalf of the past and present society. Retrieved February 27, 2023, from https://www.jstor.org/stable/650279

Imbulana, U., Aheeyar, M., Amarasinghe, U. A., & Amarnath, G. (2023). *Governance issues for sustainable management of village irrigation in the Dry Zone of Sri Lanka* (p. 21). International Water Management Institute (IWMI). CGIAR Initiative on Climate Resilience.

Japan International Cooperation Agency (JICA), Nippon Koei Co., Ltd. (2000). *The study for the potential realization of irrigated agriculture in the dry and intermediate zones of Sri Lanka.* Appendix—I, Environmental Study. JICA Library Portal Site. Retrieved July 1, 2021, from https://libopac.jica.go.jp/images/report/11600434_12.PDF

Kekulandala, B., Jacob, B., & Cunningham, R. (2020). Management of small irrigation tank cascade systems (STCS) in Sri Lanka: Past, present and future. *Climate and Development.* https://doi.org/10.1080/17565529.2020.1772709

Kenyon, P., Pollett, C., & Wills-Johnson, N. (2004). *Sustainable water management practices: Lessons from ancient Sri Lanka.* Discussion Paper No. 5/04. https://ssrn.com/abstract=1002254

Lawnet. (2016). *Irrigation.* Ministry of Justice. https://www.lawnet.gov.lk/irrigation/

Leach, E. R. (1959). Hydraulic society in Ceylon. *Past and Present, 15*(1), 2–26. https://doi.org/10.1093/past/15.1.2. (Published: 01 April 1959)

Leach, E. R. (1980). Village Irrigation in the dry zone of Sri Lanka. In E. W. Coward (Ed.), *Irrigation and agricultural development in Asia: Perspectives from the social sciences.* Cornell University, Ithaca.

Madduma Bandara, C. M. (1985). Catchment ecosystems and village tank cascades in the dry zone of Sri Lanka: A time-tested system of land and water resource management. In J. Lundqvist,

U. Lohm, & M. Falkner (Eds.), *Strategies for river basin management* (pp. 99–113). Reidel Publishing Company.
Madduma Bandara, C. M. (2009). *Village tank cascade systems of Sri Lanka: A traditional technology of water and drought management*. Retrieved May 1, 2023, from https://www.indiawaterportal.org/articles/village-tank-cascade-systems-sri-lanka-traditional-technology-water-and-drought-management
Mahatantila, K., Chandrajit, R., Jayasena, H. A. H., & Ranawaka, K. B. (2008). Spatial and temporal changes of hydrogeochemistry in ancient tank cascade systems in Sri Lanka: Evidence for a constructed wetland. *Water and Environmental Journal, 22*(2008), 17–24. https://doi.org/10.1111/j.1747-6593.2007.00077.x
Melles, G., & Perera, D. (2020). Resilience thinking and strategies to reclaim sustainable rural livelihoods: Cascade tank-village system (CTVS) in Sri Lanka. Retrieved March 14, 2023, from https://www.researchgate.net/publication/344618687_Resilience_Thinking_and_Strategies_to_Reclaim_Sustainable_Rural_Livelihoods_Cascade_Tank-Village_System_CTVS_in_Sri_Lanka
Ministry of Mahaweli Development and Environment (MMDE). (2016). *Technical feasibility report*. Strengthening the resilience of small holder farmers in the dry zone to climate variability and extreme events through an integrated approach to water management. Annex II—feasibility study, Green Climate Fund Funding Proposal. MMDE, Colombo.
Morrison, T. H., Adger, W. N., Brown, K., Lemos, M. C., Huitema, D., Phelps, J., Evans, L., Cohen, P., Song, A. M., Turner, R., Quinn, T., & Hughes, T. P. (2019). The black box of power in polycentric environmental governance. *Global Environmental Change, 57*, Article 101934. https://doi.org/10.1016/j.gloenvcha.2019.101934
Oberlack, C., Boillat, S., Brönnimann, S., Gerber, J. D., Heinimann, A., Speranza, C. I., Messerli, P., Rist, S., & Wiesmann, U. (2018). Polycentric governance in telecoupled resource systems. *Ecology and Society, 23*(1), 16. https://doi.org/10.5751/ES-09902-230116
Panabokke, C. R., Sakthivadivel, R., & Weerasinghe, A. D. (2002). *Evolution, present status and issues concerning small tank systems in Sri Lanka*. International Water Management Institute (IWMI).
Panabokke, C. R. (2009). *Small village tank systems of Sri Lanka: Their evolution, setting, distribution and essential functions*. Special Publication July 2009. Hector Kobbekaduwa Agrarian Research and Training Institute 114, Wijerama Mawatha. Colombo 07, Sri Lanka.
Perera, L. H. H. (1955). *Ceylon under western rule*. Macmillan and Company Limited.
Perera, K. T. N., Wijayaratna, T. M. N., Jayatillake, H. M., Manatunge, J. M. A., & Priyadarshan, T. (2020). Hydrological principle behind the development of series of bunds in ancient tank cascades in small catchments, Sri Lanka. *Water Practice & Technology*, **15**(4). IWA Publishing. https://doi.org/10.2166/wpt.2020.088
Roberts, M. (1980). Traditional customs and irrigation development in Sri Lanka. In E. W. Coward (ed.), *Irrigation and agricultural development in Asia: Perspectives from social science*. Cornell University Press.
Sakthivadivel, R., Fernando, N., & Brewer. J. D. (1997). *Rehabilitation planning for small tanks in cascades: A methodology based on rapid assessment*. Research Report 13. International Irrigation Management Institute (IWMI), Colombo, Sri Lanka.
Silva, M., & Vidanapathirana, A. (1984). Major irrigation and settlement systems in Sri Lanka: Some policy issues in operation and maintenance. *Sri Lanka Journal of Agrarian Studies, 5*(2), 1–11.
Wijesundere, D. A., & Ramasamy, R. (2017). Analysis of historical trends and recent elimination of malaria from Sri Lanka and its applicability for malaria control in other countries. *Frontiers in Public Health, 5*, 212. https://doi.org/10.3389/fpubh.2017.00212
World Bank. (1981). *Staff appraisal report, Sri Lanka, Village irrigation rehabilitation project*. Report No. 3363-CE, The World Bank.
World Bank. (2019). *International development association*. Project appraisal document on a proposed credit in the amount of U$125.00 Millions to the democratic socialist Republic of

Sri Lanka for a climate smart irrigated agriculture project, February 13, 2019. Document of the World Bank. Report No: PCBASIC0113664. International Development Association.

Mr. Upali Imbulana is an engineer with about 42 years of experience in irrigation management and water resources development. He is a graduate of the University of Moratuwa and holds an MSc in Irrigation Engineering from Utah State University. He is also a member of the Institution of Engineers, Sri Lanka, and the American Society of Civil Engineers. His experience in the public sector includes being attached to the Irrigation Department (15 years), handling the operation and maintenance of irrigation systems, design, training, and research. He served another 15 years in the Ministry of Irrigation in various positions and retired as Additional Secretary (Technical). His experience outside the public sector includes serving IIMI (now IWMI) as a research officer, working in the private sector as a consultant for the North Central Province Canal Project, GWP (South Asia), and being a UNDP consultant for the Climate Resilient Integrated Water Management Project implemented by the Ministry of Irrigation.

Mr. Mohamed Aheeyar is an agricultural economist. He has 30 years of experience in designing and executing multidisciplinary research and training projects in the agrarian and natural resources management, with special focus on socio-economic and institutional issues. Aheeyar has excellent analytical, report-writing, and communication skills, as well as extensive experience organizing and conducting seminars, workshops, and symposia at the local and regional levels. He has worked with government-sector organizations and has experience in working with donor agencies, NGOs, and the private sector. Currently, he is a regional researcher at the International Water Management Institute.

Chapter 10
Climate Change and Water Governance Challenges in the Aquaculture, Fisheries, and Livestock Sector

Upali S. Amarasinghe and Saman B. Athauda

Abstract Culture-based fisheries and milk production can enhance the food production of small-scale inland fishers and livestock farmers in Sri Lanka, who currently account for an 80% share of food production, and thereby improve their food, nutritional, and livelihood security. Water, a key resource for culture-based fisheries and milk production in small-scale farming systems, is under stress and is increasingly affected by climate change. However, water governance poses a significant challenge to the expansion of culture-based fisheries and livestock production among smallholder farmers. Improved coordination and integration of inter- and intra-institutional policies, as well as commitment from regional administrators, are crucial for integrating culture-based fisheries and milk production into smallholder farming.

Keywords Sri Lanka · Climate change · Aquaculture · Fisheries · Livestock · Water governance

10.1 Introduction

As vital food production systems both globally and nationally, the aquaculture, fisheries, and livestock sectors should be emphasized in the context of the water–food–nutrition nexus (Salem et al., 2022). Globally, the consumptive use (depletion from evapotranspiration) of renewable freshwater resources for aquaculture is relatively low, accounting for approximately 0.83% of total water use (Boyd & McNevin,

U. S. Amarasinghe (✉)
Department of Zoology and Environmental Management, University of Kelaniya, Kelaniya, Sri Lanka
e-mail: zoousa@kln.ac.lk

S. B. Athauda
Department of Animal Science, Faculty of Agriculture, University of Peradeniya, Peradeniya, Sri Lanka

U. A. Amarasinghe et al. (eds.), *Water Governance for Climate Change Adaptation in Sri Lanka*, Global Issues in Water Policy 13,
https://doi.org/10.1007/978-3-032-04950-6_10

2015). When green water (i.e., rainwater) falling on agricultural fields and aquaculture ponds is also accounted for, the global consumptive use of freshwater for aquaculture is approximately 3.2% (Verdegem & Bosma, 2009).

On a global scale, agriculture accounts for 70% of blue water use (i.e., surface and groundwater). Approximately 30% of this is used by livestock production. Most consumptive water use in agriculture is for feed production, and only about 2% is for drinking and servicing (Heinke et al., 2020). The total water footprint can vary significantly depending on the animal species, the food product, and the type of farming system involved. For example, the water footprint increases from 1,020 L per kg (L/kg) of milk to 5,500 L/kg for goat meat, 6,000 L/kg for pig meat, 10,400 L/kg for sheep meat, and 15,400 L/kg for beef (Mekonnen & Hoekstra, 2010). The total water footprint also varies greatly depending on the animal farming system. However, intensive animal production has a large water and carbon footprint. Thus, when selecting a farming system, careful consideration should be given to economic and production aspects, the water resources required, and the environmental impacts.

Livestock contributes to nearly 40% of total agricultural output in developed countries and about 20% in developing countries, supporting the livelihoods of at least 1.3 billion people worldwide and providing about 34% of global food protein. The contribution of livestock to agricultural GDP varies significantly across countries, averaging approximately 25% in low- and middle-income countries, compared with nearly 50% in high-income countries (Baltenweck et al., 2020).

10.2 Contribution of Inland Fisheries, Aquaculture, and Livestock to Economic and Social Development

10.2.1 Contribution to the Global Economy

Historically, in inland Asia, the harvest of wild fish from extensive networks of rivers and floodplains provided a significant fish supply (Brummett et al., 2013). With conventional food and fish supplies plateauing—especially in the marine fisheries sector—narrowing the gap between supply and demand for fish has become more critical. However, the intensification of inland aquaculture is confronted with resource limitations, including land, water, and inputs such as feed. Inland fisheries have nonetheless been gaining recognition in development initiatives, encouraging research to enhance knowledge of their importance (Youn et al., 2014).

At the farm level in Africa and Asia, livestock production generates income through the sale of milk, meat, and products such as wool, hides, and skins. The economic benefits of the livestock sector extend beyond the farm, employing individuals across various allied industries throughout the value chain, including agricultural inputs and services, animal health and breeding services, equipment and machinery, and even banks and insurance companies.

10.2.2 Contribution to the Sri Lankan Economy

Sri Lanka's reservoirs, covering 206,000 hectares (Jayasinghe & Amarasinghe, 2018), can support productive inland fisheries that provide affordable animal protein to rural communities. With both marine and inland capture fisheries reaching sub-optimal levels, alternative methods for increased fish production are needed. One alternative is culture-based fisheries (CBF), which stock small- and medium-sized water bodies with fish that grow on naturally produced food resources. This stock-and-recapture strategy is most effective when communally managed (Amarasinghe & Nguyen, 2010).

CBF has expanded due to government support and scientific inputs (Amarasinghe & Nguyen, 2010). CBF development, initiated in 2023, has shown positive impacts happening fish production and improved living standards for fisher communities using medium (200–800 ha) and large (>800 ha) reservoirs (Pushpalatha et al., 2017). CBF in perennial reservoirs enhances production by stocking exotic species, while in small and medium tanks, the entire biomass is harvested during the dry season (Wijenayake et al., 2005). Inland fish production increased from 44,490 MT (12.9% of total fish production) in 2008 to 104,235 MT (31.4%) by 2021 (Fig. 10.1).

In Sri Lanka, the livestock sector contributes approximately 1.2% of the national GDP, with the dairy industry being the most critical investment area (Daniel et al., 2020). Approximately 750,000 cattle and buffalo are milking cows out of a population of 1.1 million cattle and 0.3 million buffalo. Goats (0.3 million) play a significant role in the dry zone. Swine (~0.08 million), predominant in the northwest to southern coastal areas, provide a lower-cost source of pork, while the sheep population is comparatively smaller than that of other livestock.

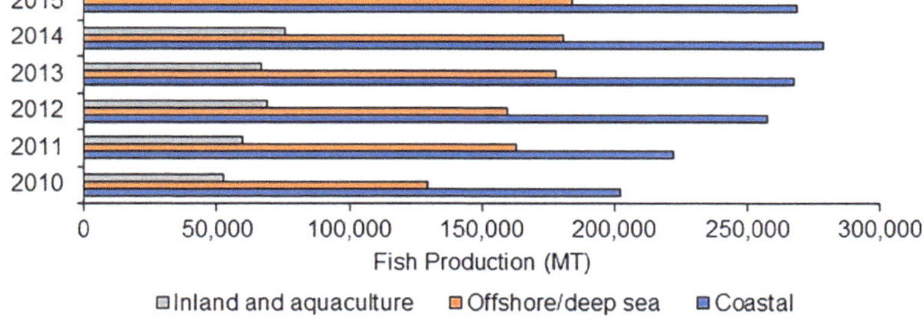

Fig. 10.1 Contribution of three sub-sectors of fish production in Sri Lanka during 2010–2021 (MoF 2022)

10.2.3 Contributions to Food and Nutritional Security

In Sri Lanka, per-capita fish consumption is approximately 14 kg per year, providing 50% of animal protein and 11% of total protein intake (MoF, 2022), which is higher than the global averages of 17% and 7%, respectively, from fish. The fisheries sector, which employs nearly 586,000 people (9.4% of whom are women) (MoF, 2022), supports about 2.3 million livelihoods (10.7% of the population). Promoting agricultural diversification through agroforestry and integrating crops and livestock that enhance natural nutrient recycling, along with farming systems that use lower external inputs, can help achieve food security and improve the nutritional well-being of people vulnerable to climate and economic shocks (Gunaratne et al., 2021).

10.3 Inland Fisheries, Aquaculture, and Livestock Production Systems in Sri Lanka

10.3.1 Aquaculture and Inland Fisheries—Production Systems

Sri Lanka's inland aquaculture production encompasses capture and culture-based fisheries, freshwater aquaculture, and coastal aquaculture (MoF, 2022). Shrimp farming, which dominates coastal aquaculture, began in 1985, and shrimp production (Penaeus monodon) increased to approximately 9,000 kg/ha/year by the early 1990s (Harkes et al., 2015). However, outbreaks of White Spot Syndrome Virus (WSSV) in 1996, 1998, and 2004 caused severe economic losses to shrimp farmers (Munasinghe et al., 2010). In 2019, the country introduced the Pacific white shrimp to revive the shrimp farming sector (Jayasinghe & Amarasinghe, 2021). Inland capture and culture-based fisheries can be further classified into two groups: (i) *culture-enhanced capture fisheries* (with irregular stocking in major (>800 ha), medium (200–800 ha), and minor (<200 ha) reservoirs, and in floodplains) (Fig. 10.2a); and (ii) *culture-based fisheries* (with regular stocking in non-perennial and minor perennial reservoirs, floodplains, and irrigation canals) (Fig. 10.2b).

Most inland capture fisheries production has come from major perennial reservoirs; however, in recent years, minor perennial reservoirs have become increasingly prominent (Fig. 10.2a). Culture-based fisheries (CBF) production, which is dominated by non-perennial reservoirs (Amarasinghe & Nguyen, 2010), has declined significantly in recent years (Fig. 10.2b). Increased culture-enhanced fish production in minor perennial reservoirs has apparently resulted from intensified stocking since 2016. Conversely, the decline of CBF production in non-perennial reservoirs resulted from inadequate stocking (see Fig. 10.3).

CBF in irrigation tanks and reservoirs is a promising and sustainable method for inland fish production. Small seasonal tanks are the most productive for CBF

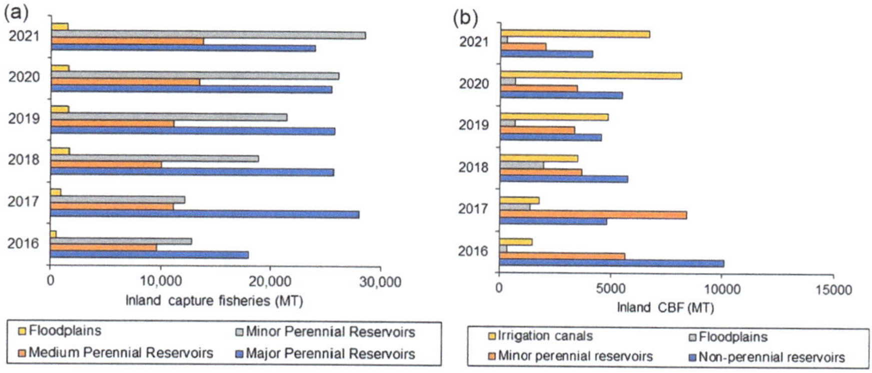

Fig. 10.2 Inland fish production (2016–2022) in Sri Lanka. **a** Culture-enhanced capture fisheries production; **b** Culture-based fisheries (CBF) production

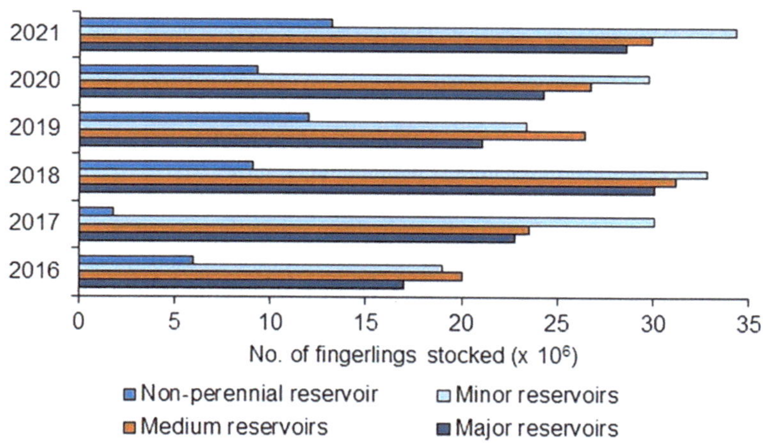

Fig. 10.3 Annual fish stocking in different types of reservoirs during 2016–2021. *Note* Due to the scarcity of fish fingerlings, stocking of non-perennial reservoirs is heavily constrained

(Amarasinghe & Nguyen, 2010). Stocking fish fingerlings is essential in small village tanks, which are filled during the inter-monsoonal rainy periods in November and January. As these tanks gradually dry up, the stocked fish should be harvested before the water reaches dead storage levels in August and October. Hence, the cropping calendar for CBF in non-perennial tanks should include appropriate timing for induced breeding of fish, post-larval and fry rearing, and fry-to-fingerling rearing (Wijenayake et al., 2021). The stocked fish rely on naturally occurring food sources in the tank ecosystem, with minimal supplementary feeding, which is an environmentally friendly strategy (De Silva, 2003). Common CBF species include hatchery-produced fingerlings of catla, rohu, mrigal, bighead carp, common carp, silver carp, and grass carp. Additionally, community-based organizations (CBOs) stock

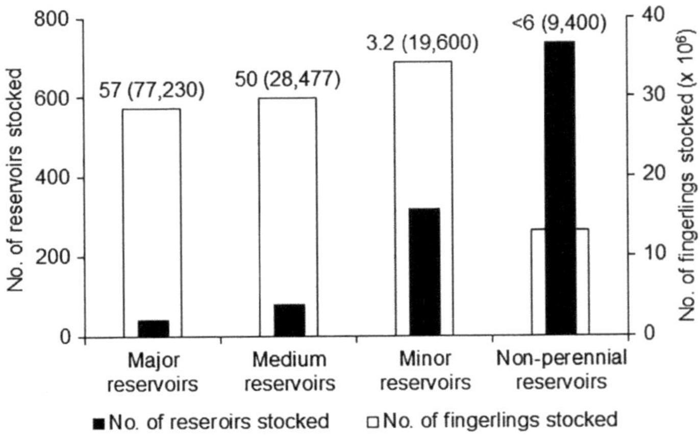

Fig. 10.4 Cumulative number of fingerlings stocked in each type during 2021 in different reservoirs *Note* Percentage of reservoirs stocked, and cumulative stock area of reservoir categories are given at the top of each column. The proportion of non-perennial reservoirs stocked was less than 6%

GIFT (genetically improved farmed tilapia) strains of Nile tilapia and freshwater prawn. CBF is a community effort, with leadership, empowered decision-making, and teamwork essential for success (Pushpalatha et al., 2020).

A major technical constraint to developing CBF in Sri Lankan reservoirs is the insufficient supply of fingerlings at the right time. When fingerlings are stocked in one type of reservoir, stocking in other types of reservoirs is limited (Fig. 10.4). Although small tanks are known to be highly productive, insufficient fingerling availability after they are filled by the monsoonal rains in December and January limits their ability to reach full production potential.

In Sri Lanka, although there are no apparent conflicts between inland/culture-based fisheries and other water uses (such as agriculture), agricultural uses often take priority. One exception is the incorporation of aquaculture and fisheries in small village reservoirs, as outlined in the Agrarian Development Act No. 46 of 2000 (Anon, 2000). This Act identified the rearing of fry to the fingerling stage in floating net cages in reservoirs as a major aquaculture production system. However, some adverse impacts have been reported, with frequent fish kills occurring due to poor water quality caused by the proliferation of cage culture (Abery et al., 2005). Scientific inputs are therefore needed to determine the carrying capacity for environmentally less disturbed cage culture in reservoirs. Studies have found that the impact of exotic species used for inland fisheries enhancement in these quasi-natural water bodies on the biodiversity of indigenous flora and fauna is not severe (De Silva & Funge-Smith, 2005).

10.3.2 Livestock—Production Systems

In Sri Lanka, the livestock sector accounts for approximately 1.2% of the national GDP. Livestock farming is prevalent in all regions of the country, with specific farming systems concentrated in certain areas due to cultural, market, and agroclimatic factors. Most dairy farms in extensive systems rely on pasture as their primary feed source, whereas the cut-and-carry method is used in both semi-intensive and intensive systems. In most commercial farms, there has been a trend toward increased use of concentrates by adopting partial or total mixed rations to achieve higher milk production per cow and higher stocking rates.

Mixed smallholder farming systems are environmentally friendly and resilient in rural communities. Small-scale crop-livestock integration has minimal climate impacts and offers environmental benefits (Daniel et al., 2020). Keeping small ruminants (e.g., goats) and poultry (e.g., chickens, quails, ducks, or turkeys) on a small scale in rural households can provide climate-resilient livelihoods. Pastoral farming systems are found in drier and fragile lowland areas, where rainfall is relatively low, unevenly distributed, and unreliable, making cultivation and sustainable livestock development challenging. Pastoralists rely heavily on forests or villus as resources and grazing areas for their livestock. Addressing complex issues of land tenure, rights, and access to water is essential to avert conflicts. Changes in land use from rural to urban also create significant problems for livestock farmers. Pastoralists' cultural practices encompass valuable indigenous knowledge for sustainable and inclusive land use management. In Sri Lanka, only a limited number of dairy products are processed locally, primarily by a handful of companies and some small-scale household producers. The leading product among them is yogurt; other main dairy products include ice cream, curd, ghee, liquid milk (pasteurized and flavored), cheese, and some sweets. However, opportunities exist to incorporate advanced technology into primary livestock products, allowing the sector to enter the downstream end of global value chains and thus accelerate economic development (FAO, 2018).

10.4 Climate Change Challenges

Climate change may impact aquaculture and fisheries through increased water temperatures, extreme climatic events, water stress, and changes in monsoonal patterns. In inland waters, elevated temperature can (i) change habitat quality—for example, causing eutrophication and algal blooms—which affects the carrying capacity for cage culture practices (De Silva and Soto, 2005), and (ii) reduce oxygen supply, leading to earlier spawning and smaller maximum body size (known as "stunting") (Amarasinghe & Pauly, 2021). Changes in monsoonal rainfall patterns can impact water availability in tanks, underscoring the importance of stocking fingerlings at the optimal time.

Unlike ruminant animals (cattle, buffalo, sheep, goats, etc.), which produce significant amounts of methane in the rumen through microbial fermentation (enteric fermentation) and from manure (FAO, 2018), farmed aquatic organisms do not emit methane and therefore are not direct contributors to global warming. Additionally, the manure produced in livestock farming must be managed, as it can result in significant methane emissions. This is a serious issue, given that the global warming potential (GWP) of methane is estimated to be about 23 times that of carbon dioxide. Although aquaculture has been criticized as polluting (e.g., Allsopp et al., 2008), its lack of methane emissions is often overlooked.

Climate change could also impact the captive breeding of major carp species, such as those found in China and India, in hatcheries. Reduced seed production in hatcheries due to low rainfall and changes in flood regimes has been reported elsewhere (Siddique et al., 2022). Elevated water temperature is also one of the reasons for the low performance of induced breeding of major carp in state-owned hatcheries.

Livestock is generally more climate-resilient than crop-based systems (Escarcha et al., 2018). Climate change can directly affect livestock through heat stress, increased morbidity and mortality, reduced feed intake, and decreased milk production. It can also indirectly impact livestock through changes in the quality and availability of feed and forage, as well as an increased incidence of animal diseases (FAO, 2018). Among the direct impacts are reduced feed intake, lower milk and meat production, decreased reproductive performance, adverse effects on immune function, and increased mortality. Rising temperatures will increase animal water consumption, irrigation water use for pasturelands, and feed and fodder needs (Cheng et al., 2022). Smallholder livestock keepers and pastoralists are among the most vulnerable to the impacts of climate change.

10.5 Water Governance Challenges

Inland fisheries development is a secondary use of reservoirs, which were primarily constructed in the past for irrigation and more recently for hydroelectric power generation. From a policy perspective, inland fisheries are often considered low priority in water management strategies for reservoirs. This is particularly critical for CBF development in small village reservoirs, whose command areas are less than 80 ha. These small reservoirs fall under the jurisdiction of the Agrarian Development Act No. 46 of 2000 (Anon, 2000) and its Amendment Act No. 46 of 2011 (Anon, 2011). Under the legal provisions of these Acts, farmers' organizations (FOs) are established to empower communities for water management, and fisheries and aquaculture aspects are incorporated into reservoir management. However, the lack of firm commitment at regional administrative levels presents a significant obstacle to the development of CBF.

The Irrigation Department manages medium and major reservoirs with irrigable areas exceeding 80 hectares, while fishers practice community-based fisheries (CBF) and are organized into Rural Fisheries Organizations (RFOs). Community-based aquaculture (for fingerling rearing in mini-nurseries) falls under the mandate of RFOs whenever a farming community adopts it. As CBF is considered an environmentally friendly form of aquaculture (De Silva, 2003), there is potential for CBF to gain global recognition through the implementation of the FAO Code of Conduct for Responsible Fisheries, which emphasizes an ecosystem approach to fisheries and aquaculture (Staples & Funge-Smith, 2009). Therefore, the Irrigation Department's recognition of water rights for fisheries organizations in medium and major reservoirs is a major prerequisite for the successful operation of CBF.

Several water governance challenges constrain Sri Lanka's livestock sector. These include:

- **Lack of coordination**—A lack of effective coordination and integration between institutions at both inter- and intra-institutional levels in policymaking.
- **Underutilized land**—Unused pockets of land along river basins that have potential for mixed farming remain idle.
- **Socio-cultural factors**—Socio-cultural considerations in water resource development and livestock system improvements are not sufficiently considered due to a lack of understanding.
- **Water use conflicts**—Agricultural water use and pastureland management are not harmonized, resulting in underutilization of extensive grazing areas.

10.5.1 Realizing the Opportunities

In Sri Lanka, inland fisheries management regulations focus on mesh restrictions in gillnet fisheries targeting exotic cichlid species, which prevents the exploitation of small indigenous species (SIS). Previous work has demonstrated the feasibility of introducing a small-meshed gillnet fishery (15 to 52 mm stretched mesh) for exploiting SIS without negatively impacting the existing fishery for larger species, such as cichlids (Amarasinghe et al., 2016; De Silva & Sirisena, 1987). Introducing a subsidiary fishery for these small fish resources would help provide supplementary income for fishing communities and an additional source of food fish from Sri Lankan inland waters.

There is a national need to prepare aquaculture zones based on strategic environmental assessments, which would help overcome land eligibility constraints and facilitate responsible aquaculture planning. Such an approach is crucial for minimizing the impacts of aquaculture on biodiversity and protecting ecosystems in compliance with relevant national legislation. It would help maintain healthy populations of wild species, prevent the escape of cultured species into the environment, and avoid harmful interactions with wild fish stocks (Jayasinghe & Amarasinghe, 2021).

Without effective and inclusive governance, increased competition for freshwater can exacerbate inequalities in access and lead to inefficient use. The share of consumptive water use (CWU) by different livestock production types often does not match their contribution to production. For example, Professor G.L.L.P. Silva (pers. comm.) noted that under an intensive management system, European dairy cattle kept outside their thermo-neutral comfort zone (below 68 °F) may incur an income loss of Rs. 150,000 per lactation. This loss is attributed to the temperature-humidity index (THI) exceeding the tolerance level of European dairy cattle (Narmilan et al., 2021).

Opportunities for increasing livestock production and strategies to consider include:

- **Enhance livestock management**—Through crop-livestock integration and the use of crop by-products for livestock feed.
- **Improve production efficiency**—Increase the efficiency of crop and livestock production and reduce negative impacts on blue water. For example, using early-maturing feed/fodder varieties may help synchronize maximal growth with periods of green water availability.
- **Adjust livestock diets**—Increase the proportion of fresh grass or silage in ruminant diets to reduce drinking-water needs.
- **Utilize alternative water sources**—Safely use non-conventional water sources (e.g., recycled wastewater and saline water) in feed crop and pasture cultivation, especially in water-scarce areas.
- **Breed drought-resistant crops**—Develop drought-resistant corn and soybean varieties (via genetic engineering or traditional breeding), or replace corn with sorghum, which produces more biomass without irrigation (giving sorghum a greater green water footprint).
- **Promote saline-tolerant agriculture**—Support sustainable food and feed production in increasingly saline environments, while restoring and protecting productive natural resources affected by saline intrusion and scarcity.
- **Reduce pollution**—Improve human, plant, and animal health by reducing water pollution from nutrients, pesticides, and other leached materials, and by preventing antibiotics and chemicals used in livestock, crop, and aquatic food production from entering natural water systems.

10.6 Conclusion

Compared to livestock production, the water footprints of aquaculture and fisheries are appreciably low. Although aquaculture in Sri Lanka is primarily restricted to shrimp farming in coastal regions, culture-based fisheries (CBF) in inland reservoirs have gained momentum in providing affordable animal protein to rural communities. Because CBF strategies generally do not depend on external inputs such as supplementary feeding, they are considered environmentally friendly fisheries enhancement methods. In inland reservoirs, there is a greater potential for CBF development in non-perennial and small village reservoirs. Nevertheless, the full potential of these water bodies is not realized due to inadequate stocking of fish fingerlings.

The major climate change challenge for CBF development in non-perennial reservoirs is the unpredictability of inter-monsoonal rains, which creates uncertainty in the timing of stocking. Shifts in inter-monsoonal rains in recent years make it challenging to plan seed production for stocking, which is crucial for the sustainability of CBF. Improved weather forecasting can help ameliorate this situation.

Although many cultured species may be less impacted by global warming under climate change scenarios, oxygen limitation due to increased temperatures might result in earlier maturity and smaller body size. This smaller size can be advantageous for aquaculture production, especially in cichlid species.

Even though there are no apparent conflicts between water resource users in agriculture, aquaculture, and fisheries, in many cases, aquaculture and inland fisheries are treated as low priority in water management. Although provisions exist to incorporate aquaculture and fisheries into village reservoirs, the lack of a firm commitment at regional administrative levels presents a serious setback for CBF development.

There is a national need to prepare aquaculture zones based on strategic environmental assessments, which would help overcome land eligibility constraints and facilitate responsible aquaculture planning. This is particularly important to minimize aquaculture's impacts on biodiversity and protect ecosystems in compliance with relevant national legislation. It would help maintain healthy populations of wild species, prevent the escape of cultured species into the environment, and avoid harmful interactions with wild fish stocks.

Developing an intersectoral policy framework with targeted adaptation strategies for climate change and water governance is essential to achieving efficient water resource use and environmental sustainability. The purpose of this intersectoral policy framework is to enhance agricultural production across crops, livestock, fisheries, and aquaculture. This is best achieved through effective coordination and integration of the various activities that rely on shared water and watershed areas. Within this general framework, the following recommendations are outlined to optimize the use of water resources in support of livestock development needs:

- **Evaluate water supply**—Critically assess the problems, needs, and potential of water supply in meeting the current and future requirements of livestock production systems.
- **Assess water use**—Conduct a comprehensive assessment of water use in livestock production systems along the production chain (including water used for feed production) to inform policy and practice.
- **Address root causes**—Implement policies and plans that address the underlying causes of environmental degradation rather than its symptoms. Urgent action in livestock development is required to meet food needs and preserve the resource base, combining education, capacity building, local institutional empowerment, zoning, and property rights.
- **Utilize temporary watercourses**—Utilize temporary watercourses by creating and improving water-retaining structures, such as artificial pools, to provide water for herds grazing on nearby dry pastures.

- **Develop groundwater sources**—Where surface water sources are insufficient for drinking, establish wells and boreholes in shallow aquifer areas identified by geological surveys.
- **Irrigate pocket lands**—Develop pocket lands suitable for irrigation along rivers, and improve the productivity of dry pasture or grazing lands (villages) using water-harvesting techniques.
- **Restore tank cascades**—Restore and improve the village tank cascade system in the Dry Zone of Sri Lanka to provide dairy farmers with a free water source (tank water) and a feeding area (upper inundation area and native shrublands).
- **Integrate crop and dairy farming**—Integrate crop farming with dairy farming to create synergies between the two systems for a sustainable and climate-resilient agricultural system.
- **Seek external support**—In addition to government support, seek external assistance to meet needs for integrated water supply infrastructure, improved rangeland management, effective veterinary services, research and extension, and credit and marketing services.

References

Abery, N. W., Sukadi, F., Budhiman, A. A., Kartamihardja, E. S., Koeshendrajana, S., & Budhiman and De Silva, S. S. (2005). Fisheries and cage culture of three reservoirs in West Java, Indonesia: A case study of ambitious developments and resulting interactions. *Fisheries Management and Ecology, 12,* 315–330. https://doi.org/10.1111/j.1365-2400.2005.00455.x

Allsopp, M., Johnston, P., & Santillo, D. (2008). *Challenging the aquaculture industry on sustainability* (p. 22). Greenpeace.

Amarasinghe, U. S., & Nguyen, T. T. T. (2010). Enhancing rural farmer income through fish production: Secondary use of water resources in Sri Lanka and elsewhere. In S. S. De Silva, & F. B. Davy, F.B. (Eds.), *Success stories in Asian aquaculture* (pp. 103–130). Springer Science, Netherlands. ISBN 978-90-481-3085-6 (Print) ISBN 978-90-481-3087-0 (Online). http://library.enaca.org/emerging_issues/success_stories/success-stories-asian-aquaculture.pdf. https://doi.org/10.1007/978-90-481-3087-0_6

Amarasinghe, U. S.,& Pauly, D. (2021). The relationship between size at maturity and maximum size in cichlid populations corroborates the Gill oxygen limitation theory (GOLT). *Asian Fisheries Science, 34,* 14–22. https://doi.org/10.33997/j.afs.2021.34.1.002

Amarasinghe, U. S., Ajith Kumara, P. A. D., & De Silva, S. S. (2016). A rationale for introducing a subsidiary fishery in tropical reservoirs and lakes to augment inland fish production: Case study from Sri Lanka supplemented with evidence from elsewhere. *Food Security, 8*, 769–781. https://doi.org/10.1007/s12571-016-0596-4

Anon. (2000). Agrarian Development Act No. 46 of 2000. Gazette of the Democratic Socialist Republic of Sri Lanka of August 18, 2000, Colombo.

Anon. (2011). Agrarian Development (Amendment) Act No. 46 of 2011. Gazette of the Democratic Socialist Republic of Sri Lanka of November 25, 2011, Colombo.

Baltenweck, I., Enahoro, D., Frija, A., & Tarawali, S. (2020). Why is the production of animal-source foods important for economic development in Africa and Asia? *Animal Frontiers, 10*(4), 22–29. https://doi.org/10.1093/af/vfaa036

Boyd, C. E., & McNevin, A. A. (2015). *Aquaculture, resource use, and the environment.* Wiley-Blackwell. https://doi.org/10.1002/9781118857915

Brummett, R. E., Beveridge, M. C. M., & Cowx, I. G. (2013). Functional aquatic ecosystems, inland fisheries, and the millennium development goals. *Fish and Fisheries, 14*(3), 312–324. https://doi.org/10.1111/j.1467-2979.2012.00470.x

Cheng, M., McCarl, B., & Fei, C. (2022). Climate change and livestock production: A literature review. *Atmosphere, 13*, 140. https://doi.org/10.3390/atmos13010140

Daniel, S., Mombauer, D., & Wijenayake, V. (2020). Livestock policy brief 1: Sri Lanka's livestock sector and climate change. Retrieved August 13, 2023, from https://slycantrust.org/knowledge-portal/home

De Silva, S. S. (2003). Culture-based fisheries: An underutilised opportunity in aquaculture development. *Aquaculture, 221*, 221–243. https://doi.org/10.1016/S0044-8486(02)00657-9

De Silva, S. S., & Funge-Smith, S. J. (2005). A review of stock enhancement practices in inland water fisheries of Asia. RAP Publication 2005/12 (p. 93). Asia-Pacific Fishery Commission, Bangkok, Thailand.

De Silva, S. S., & Sirisena, H. K. G. (1987). New fish resources of reservoirs in Sri Lanka: Feasibility of introducing a subsidiary gillnet fishery for minor cyprinids. *Fisheries Research, 6*, 17–34. https://doi.org/10.1016/0165-7836(87)90004-X

De Silva, S. S., & Soto, D. (2009). Climate change and aquaculture: Potential impacts, adaptation and mitigation. In K. Cochrane, C. De Young, D. Soto, & T. Bahri (Eds.), *Climate change implications for fisheries and aquaculture: Overview of current scientific knowledge* (pp. 151–212). FAO Fisheries and Aquaculture Technical Paper No. 530. Rome, FAO.

Escarcha, J. F., Lassa, J. A., & Zander, K. K. (2018). Livestock under climate change: A systematic review of impacts and adaptation. *Climate, 6*, 54. https://doi.org/10.3390/cli6030054

FAO (Food and Agriculture Organization) (2018). *World livestock: Transforming the livestock sector through the sustainable development goals* (p. 222). Rome. https://doi.org/10.4060/ca1201en

Gunaratne, M. S., Radin Firdaus, R. B., & Rathnasooriya, S. I. (2021). Climate change and food security in Sri Lanka: Towards food sovereignty. *Humanities and Social Sciences Communications, 8*, 229. https://doi.org/10.1057/s41599-021-00917-4

Harkes, I. H. T., Drengstig, A., Kumara, M. P., Jayasinghe, J. M. P. K., & Huxham, M. (2015). Shrimp aquaculture as a vehicle for climate-compatible development in Sri Lanka: The case of Puttalam Lagoon. *Marine Policy, 61*, 273–283. https://doi.org/10.1016/j.marpol.2015.08.003

Heinke, J., Lannerstad, M., Gerten, D., Havlík, P., Herrero, M., Notenbaert, A. M. O., Hoff, H., & Müller, C. (2020). Water use in global livestock production—opportunities and constraints for increasing water productivity. *Water Resources Research, 56*(12), e2019WR026995. https://doi.org/10.1029/2019WR026995

Jayasinghe, J. M. P. K., & Amarasinghe, U. S. (2018). Inland aquatic resources. In M. J. S. Wijeyaratne, A. H. M. Jayasuriya, & N. P. Wijayananda (Eds.), *Natural resources of Sri Lanka: Conditions, trends and prospects* (pp. 327–343). National Science Foundation of Sri Lanka.

Jayasinghe, J. M. P. K., & Amarasinghe, U. S. (2021). Coastal aquaculture in Sri Lanka: Opportunities and environmental challenges. *Journal of the Indian Society of Coastal Agricultural Research, 39*(2), 1–14. https://doi.org/10.54894/JISCAR.39.2.2021.110687

Mekonnen, M. M., & Hoekstra, A. Y. (2010). The green, blue and grey water footprint of farm animals and animal products. Value of Water Research Report Series No. 48, UNESCO-IHE, Delft, the Netherlands.

MoF (Ministry of Fisheries). (2022). Fisheries statistics 2022. Ministry of Fisheries, Colombo, p. 60. Retreived July 30, 2024 from https://www.fisheries.gov.lk/web/index.php?option=com_content&view=article&id=44&Itemid=153&lang=en

Munasinghe, M. N., Stephen, C., Abeynayake, P., & Abeygunawardena, I. S. (2010). Shrimp farming practices in the Puttalam district of Sri Lanka: Implications for disease control, industry sustainability, and rural development. *Veterinary Medicine International, 2010*, Article 679130. https://doi.org/10.4061/2010/679130

Narmilan, A., Puvanitha, N., Ahamed, A. S., & Senthirakumar, S. (2021). Relationship between temperature-humidity index and milk production of dairy cows in a tropical climate. *Asian Journal of Dairy and Food Research, 40*(3), 246–252. https://doi.org/10.18805/ajdfr.DR-213

Pushpalatha, K. B. C., Chandrasoma, J., Fernando, W. A. J. R., & Sanjeewa, K. D. (2017). Impacts and importance of the introduction of culture-based fisheries in three medium-sized perennial reservoirs in Sri Lanka. *Asian Fisheries Science, 30,* 139–151. https://doi.org/10.33997/j.afs.2017.30.3.001

Pushpalatha, K. B. C., Kularatne, M. G., & Amarasinghe, U. S. (2020). Institutional robustness of culture-based fisheries in perennial reservoirs of Sri Lanka. *Asian Fisheries Science, 33,* 91–105. https://doi.org/10.33997/j.afs.2020.33.2.002

Salem, H. S., Pudza, M. Y., & Yihdego, Y. (2022). Water strategies and water–food nexus: Challenges and opportunities towards sustainable development in various regions of the world. *Sustainable Water Resources Management, 8,* 114. https://doi.org/10.1007/s40899-022-00676-3

Siddique, M. A. B., Ahammad, A. K. S., Bashar, A., Hasan, N. A., Mahalder, B., Alam, Md. M., Biswas, J. C., & Haque, M. M. (2022). Impacts of climate change on fish hatchery productivity in Bangladesh: A critical review. *Heliyon, 8*(12), Article e11951. https://doi.org/10.1016/j.heliyon.2022.e11951

Staples, D., & Funge-Smith, S. (2009). Ecosystem approach to fisheries and aquaculture: Implementing the FAO code of conduct for responsible fisheries. FAO Regional Office for Asia and the Pacific, Bangkok. RAP Publication 2009/11.

Verdegem, M. C. J., & Bosma, R. H. (2009). Water withdrawal for brackish and inland aquaculture, and options to produce more fish in ponds with the present water use. *Water Policy, 11,* 52–68. https://doi.org/10.2166/wp.2009.003

Wijenayake, W. M. H. K., Jayasinghe, U. A. D., Amarasinghe, U. S., Athula, J. A., Pushpalatha, K. B. C., & De Silva, S. S. (2005). Culture-based fisheries in non-perennial reservoirs in Sri Lanka: Production and relative performance of stocked species. *Fisheries Management and Ecology, 12,* 249–258. https://doi.org/10.1111/j.1365-2400.2005.00447.x

Wijenayake, W. M. H. K., Najim, M. M. M., & Amarasinghe, U. S. (2021). Constraints in culture-based fisheries of Sri Lanka and strategies for future development. *Sri Lanka Journal of Aquatic Sciences, 26*(2), 81–95. https://doi.org/10.4038/sljas.v26i2.7589

Youn, S.-J., Tayler, W. W., Lynch, A. J., Cowx, I. G., Beard, T. D., Barley, D., & Wu, F. (2014). Inland capture fishery contributions to global food security and threats to their future. *Global Food Security, 3,* 142–148. https://doi.org/10.1016/j.gfs.2014.09.005

Prof. Upali S. Amarasinghe is a Professor Emeritus in Zoology at the University of Kelaniya, Sri Lanka. He has research experience for over 45 years on fisheries biology, inland fisheries, and management of small-scale fisheries. In recognition of his outstanding contribution to fisheries science, he was awarded two D.Sc. degrees. In July 2010, Bangor University, UK, awarded him a D.Sc. for his research on tropical artisanal fisheries management and a D.Sc. (Honoris Causa) by the University of Ruhuna in November 2019 for his outstanding contribution to fisheries biology and management. Professor Amarasinghe is a Fellow of the National Academy of Science of Sri Lanka and a Chartered Biologist of the Institute of Biology of Sri Lanka. Professor Amarasinghe is the Associate Editor of the Asian Fisheries Science journal, a member of the Academic Editorial Board of the Journal of Applied Ichthyology, and the Egyptian Journal of Aquatic Research. He won several life-time awards for his research on fisheries biology, which include (i) Third World Academy of Sciences (TWAS) Prize for the Young Scientist in the field of Biology in 1993, (ii) Committee of Vice Chancellors and Directors (CVCD) excellence award 2008 for the most outstanding scientist in biological sciences, and (iii) General Research Committee (GRC) Award of Sri Lanka Association for Advancement of Science in 2009 for the most outstanding scientist in biological sciences. He also won several national awards, including the President's Award for Scientific Publications and the Vice Chancellor's Award for Most Outstanding Researcher in the Faculty of Science at the University of Kelaniya.

Prof. Saman B. Athauda commenced his academic career at the University of Peradeniya in 1996 as a lecturer. After qualifying with an M.Sc. from the Asian Institute of Technology, Thailand, and a Ph.D. from James Cook University, Australia, he continued his academic pursuits. He was promoted to the positions of Senior Lecturer-Grade II and Senior Lecturer-Grade I in 2005 and 2011, respectively. Prof. Athauda has been involved in various activities related to teaching and research in the field of Animal Science, as well as disseminating knowledge through his publications. He has published several full research articles in SCI journals, with over 60 abstracts in refereed journals. He was a recipient of the National Research Council of Sri Lanka Merit Award in 2012 and 2014, and also served as a Reviewer for several international and national journals that are currently publishing. Apart from his scholastics works, he held various positions in National Organizations as the President of Sri Lanka Association of Animal Production (SLAAP) in 2013 and the President of ALUMNI Association of Faculty of Agriculture, University of Peradeniya (AAFAUP) in 2018 and presently as the Ex-co member of the Sri Lanka Institute of Agriculture (SLIAg) and Sri Lanka Association Fisheries and Aquatic Resources (SLAFAR).

Chapter 11
Water Governance Challenges in the Domestic, Industrial, and Power Sectors

Chandra Embuldeniya

Abstract Sri Lanka's non-agricultural sectors are expected to experience a significant increase in water demand. This chapter examines the challenges of climate change and water governance on non-agricultural water use, with a specific focus on the domestic, industrial, and power sectors. It emphasizes the need for cross-sectoral collaboration and integrated water management approaches to ensure sustainable water use across the domestic, industrial, and power sectors. It recommends adopting innovative technologies such as water recycling, treatment, and real-time monitoring systems; enhancing water-use efficiency through conservation practices, awareness campaigns, and capacity-building efforts; and improving water governance through transparent, participatory, and accountable systems.

Keywords Climate change · Water governance · Non-agricultural sector · Industry · Domestic sector · Hydropower · Wastewater treatment

11.1 Introduction

Climate change and water governance are interlinked challenges that require immediate attention to ensure the sustainable use of water resources. While agriculture is often the primary focus of water management efforts, non-agricultural sectors, including the domestic, industrial, and power sectors, also play a significant role in water usage and face unique challenges in the context of climate change. In Sri Lanka, increasing temperatures and changing rainfall patterns have put immense pressure on water resources, making it critical to address the impact of climate change on non-agricultural sectors.

C. Embuldeniya (✉)
Founder and Former Vice Chancellor, Uva Wellassa University of Sri Lanka, Badulla, Sri Lanka
e-mail: chandrajembuldeniya@gmail.com

Past Chairman, Technology Development and Innovations Arm, National Science Foundation, Colombo, Sri Lanka

U. A. Amarasinghe et al. (eds.), *Water Governance for Climate Change Adaptation in Sri Lanka*, Global Issues in Water Policy 13,
https://doi.org/10.1007/978-3-032-04950-6_11

The effective management of water resources in non-agricultural sectors requires an understanding of the challenges of water governance and the identification of appropriate adaptation strategies. It is essential to note that although the industrial and power sectors currently consume a smaller share of water resources, their demand is increasing rapidly. As industries expand and power generation capacities rise, the water requirements for these sectors are projected to escalate significantly in the coming years. This increasing demand, coupled with competition for scarce water resources, poses specific challenges for the industrial and power sectors in terms of water availability, quality, and sustainable use.

Moreover, as climate change impacts intensify in Sri Lanka, with altered rainfall patterns and greater uncertainty, the water dynamics further complicate the challenges faced by non-agricultural sectors. Climate change-induced water scarcity, changes in water availability, and extreme weather events pose additional risks and vulnerabilities for the industrial and power sectors.

In this context, it is crucial to address the impact of climate change on water use in non-agricultural sectors, where water demand is likely to increase and face specific challenges related to water availability, quality, and use. The domestic sector, for example, requires an adequate and safe water supply for drinking, sanitation, and hygiene, while the industrial sector needs water for various processes, including cooling and manufacturing. The power sector is also a significant water user, requiring water for power generation and cooling processes.

To ensure the effective management of water resources in non-agricultural sectors, it is essential to understand the water governance challenges and identify appropriate adaptation strategies. This requires an integrated and holistic approach that considers the needs and interests of all stakeholders, including the environment and marginalized communities. Cross-sectoral collaboration is increasingly recognized as a crucial component of effective water management, as it facilitates coordinated efforts and shared responsibility for achieving sustainable water management objectives.

Climate change is having a profound impact on water resources in Sri Lanka, with increasing temperatures and changing rainfall patterns exacerbating existing water challenges. The country is already facing water scarcity and frequent droughts, particularly in the dry zone, which is home to over a third of the population and many important economic activities. Climate change is expected to exacerbate these challenges, resulting in heightened water stress and increased competition among various sectors.

In the following sections, we will examine the specific water governance challenges and adaptation strategies for the domestic, industrial, and power sectors.

11.2 Climate Change Impacts on Water Resources

11.2.1 Changes in Rainfall Patterns

The impacts of changing rainfall patterns on water resources pose significant challenges for non-agricultural sectors, particularly the power and industrial sectors. At present, the country heavily relies on hydroelectric power generation, which is dependent on a consistent and adequate water supply. However, with changing rainfall patterns, there is a greater risk of reduced water availability for hydropower generation. Extended dry spells during the monsoon seasons can lead to lower water levels in reservoirs, impacting power production and potentially causing energy shortages.

Industries require a reliable and consistent water supply for their operations, including manufacturing processes and cooling systems. Extended dry periods can lead to water scarcity, which in turn affects industrial production and poses significant economic and operational risks. Conversely, intense rainfall events can result in flooding, causing damage to infrastructure, disrupting operations, and compromising water quality.

According to the Sri Lanka Climate Risk Country Profile, the country has already witnessed a decline in rainfall during the dry season, a trend projected to continue in the future. This decreased water availability has implications for both agriculture and domestic water use. However, it is important to note that non-agricultural sectors, including the power and industrial sectors, also rely on adequate water resources for their operations and sustainability.

In addition to reduced water availability, extreme rainfall events have become more frequent in Sri Lanka, leading to flash floods and landslides. These events not only pose immediate risks to human lives and infrastructure but also have long-term consequences for water quality and availability. Flooding can contaminate water sources, affecting both domestic and industrial water supplies and requiring additional measures for water treatment and management.

The impact of changing rainfall patterns on the power and industrial sectors in Sri Lanka highlights the need for proactive adaptation strategies. It is crucial for these sectors to enhance their water resilience by exploring alternative water sources, implementing efficient water management practices, and developing contingency plans to mitigate the risks associated with both water scarcity and extreme rainfall events.

Furthermore, extreme rainfall events can cause flash floods and landslides, which have become more frequent in recent years. In May 2017, Sri Lanka experienced its worst flooding in 14 years, affecting over half a million people and causing an estimated $2 billion in damages. The floods damaged water supply infrastructure and contaminated water sources, resulting in water scarcity in the affected areas.

The changing rainfall patterns in Sri Lanka due to climate change have significant implications for non-agricultural sectors, particularly the power and industrial sectors. The reliance on hydroelectric power generation and the need for a consistent water supply make these sectors vulnerable to water scarcity and the impacts of extreme rainfall events. Adapting to these challenges through effective water management strategies and resilience-building measures is essential for ensuring the sustainability and resilience of the power and industrial sectors in the face of climate change.

11.2.2 Sea-level Rise

Sri Lanka has a long coastline and numerous estuaries. Many of its major cities are located near the coast and are vulnerable to saltwater intrusion, which is likely to be exacerbated with sea level rise. Saltwater intrusion contaminates freshwater sources, rendering them unsuitable for human consumption and agriculture, and negatively impacts coastal ecosystems, including mangroves and coral reefs, which are crucial for maintaining water quality and biodiversity. A study by the International Water Management Institute (IWMI) found that saltwater intrusion could affect up to 15% of Sri Lanka's coastal aquifers. Saltwater intrusion is a major concern for water resources in Sri Lanka.

11.2.3 Extreme Weather Events

The country is already vulnerable to natural disasters, including floods, landslides, and droughts, exacerbated by extreme weather events. Climate change is expected to exacerbate these events, increasing their frequency and intensity. This could have severe implications for water availability, quality, and reliability in non-agricultural sectors such as tourism and industry.

The prolonged drought that occurred in Sri Lanka in 2016 was one of the most severe in the country's history and was attributed by the government to the influence of El Niño. It caused widespread crop failures, water shortages, and power cuts. The drought was followed by a poor monsoon season in 2017, which further impacted water availability and agriculture.

The 2017 floods and landslides in Sri Lanka resulted in over one hundred deaths, the displacement of hundreds of thousands of people, and considerable economic losses. These floods underscored the country's vulnerability to extreme rainfall events, which are anticipated to become more frequent due to climate change.

11.2.4 Adapting to Climate Change Impacts on Water Resources

To address the impacts of climate change on water resources, Sri Lanka is implementing adaptation measures in various sectors. These measures include enhancing water storage capacity through reservoir rehabilitation and construction, promoting rainwater harvesting, and implementing drought management and early warning systems. Sri Lanka requires $6.2 billion in climate-smart investments to improve solid and water-waste management in its cities, highlighting its potential for growth in sustainable sectors within the region (South Asia IFC Analysis, 2017). The Sri Lanka Climate Resilience Improvement Program (Kahandawa, 2017) supports such initiatives by providing financing for climate-resilient water supply projects.

Additionally, there is a need to increase public awareness about the impacts of climate change on water resources and the importance of water conservation. Education campaigns, capacity building, and community-based water management can help communities adapt to changing water availability and usage patterns.

11.3 Water Governance Challenges in the Domestic Sector

Sri Lanka faces several water governance challenges in the domestic sector, which includes households and municipalities. These challenges are becoming increasingly urgent due to population growth, urbanization, and climate change. This section examines some of the key challenges and potential solutions.

11.3.1 Water Scarcity

Water scarcity is a major challenge in urban areas where demand is growing rapidly. According to the Water Supply and Sanitation Improvement Project (WASSIP), about 44% of the population in Sri Lanka lacks access to a piped water supply (World Bank, 2015). In 2015, this was equivalent to approximately 8 million people needing a piped water supply. Furthermore, the supply of piped water is often intermittent, and the water quality is frequently poor. However, safe water is available to almost 85% of the population.

To address this issue, the government has invested in new water supply infrastructure, such as reservoirs and treatment plants. However, these projects often face delays and cost overruns, and there are concerns about their long-term sustainability. A more sustainable approach is to promote water conservation and efficient use, which can be achieved through public awareness campaigns, education, and the use of water-saving technologies.

Increasing efficiency in water use can certainly help meet the additional demand in the industry and power sectors. By implementing water conservation practices and adopting water-saving technologies, industries and power plants can optimize their water usage and reduce wastage. This includes measures such as recycling and reusing water, improving irrigation techniques, and employing advanced technologies that minimize water consumption.

Efficiency measures can help industries and power plants reduce their reliance on freshwater sources, thereby alleviating the pressure on water resources. By using water more efficiently, these sectors can maximize their output while minimizing the amount of water needed. This is particularly important in regions where water scarcity is a pressing concern.

However, it is important to note that increasing efficiency alone may not be sufficient to fully address the additional demand in the face of growing industrial and power sector requirements. It is also crucial to consider the broader context of water availability and the potential exacerbation of water scarcity due to climate change.

Climate change can exacerbate water scarcity by altering rainfall patterns, leading to more frequent and intense droughts and affecting the overall availability of water resources. As mentioned earlier, Sri Lanka has already experienced changes in rainfall patterns, with longer dry spells and increased uncertainty. These changes can limit the availability of water for both domestic and non-agricultural sectors.

Furthermore, climate change can lead to increased evaporation rates and higher temperatures, which can further exacerbate water scarcity. Higher temperatures can intensify water demand for cooling purposes in industries and power plants, while increased evaporation rates can reduce water availability from surface water sources.

11.3.2 Water Quality Deterioration

Another challenge in the domestic sector is the deterioration of water quality, which is caused by a range of factors, including industrial pollution, agricultural runoff, and improper sanitation practices. A substantial portion of Sri Lanka's surface water is contaminated with high levels of bacteria and other pollutants. To address this issue, the government has implemented various policies and regulations aimed at controlling pollution and promoting proper sanitation practices.

For example, the National Water Supply and Drainage Board (NWSDB) has established a Water Quality Surveillance Unit to monitor and control the quality of the water supply (NWSDB, 2021). Additional expenditures will be required to improve service quality. Of the 332 schemes managed by the NWSDB, about 60 require improvements to be made to their treatment facilities to respond to deterioration in their raw water quality. In addition, approximately 400 CBO-managed systems need new treatment facilities (Fan, 2015). Additionally, the government has launched a National Sanitation Campaign to promote proper sanitation practices and improve public awareness about the importance of water quality.

11.3.3 Increasing Water Demand in Urban Areas

As Sri Lanka's population grows and urbanizes, there is an increasing demand for water in urban areas. According to the WASSIP, urban water demand in Sri Lanka is projected to increase by 15–20% over the next decade (NWSDB, 2019). This increased demand places pressure on existing water supply infrastructure, exacerbating water scarcity.

To address this challenge, the government has implemented policies to promote water conservation and efficient use, such as the National Water Conservation and Efficiency Improvement Program. Additionally, the government has introduced reforms to the water pricing system, including the introduction of a block tariff, which charges higher rates for higher levels of consumption. This strategy is common among utility companies, particularly in the water and electricity sectors. A block tariff system charges customers a higher rate as they consume more of a resource, incentivizing consumers to use water more efficiently and generating revenue for new water supply infrastructure.

Role of Policy Frameworks, Water Pricing, Infrastructure Development, and Public Awareness

The challenges facing the domestic sector in Sri Lanka require a comprehensive approach that includes policy frameworks, water pricing, infrastructure development, and public awareness. Effective water governance is essential to ensuring sustainable and equitable access to water resources.

The NWSDB is responsible for managing water supply and sanitation services, as well as developing policies and regulations to promote sustainable water management. To improve water governance in Sri Lanka, there is a need for increased public awareness about the importance of water resources and the necessity of sustainable water management practices. This can be achieved through education and outreach programs, as well as the promotion of public participation in decision-making processes.

11.3.4 Successful domestic water management practices and initiatives

Several successful domestic water management practices and initiatives have been implemented, demonstrating the potential for sustainable and effective water governance.

- **Rainwater harvesting**: In many rural areas of Sri Lanka, rainwater harvesting has been implemented as a means of collecting and storing water during the rainy season. This water can then be used for domestic purposes during the dry season when water scarcity is more acute (Ranathunga et al., 2016).
- **Community-based management**: In some areas, community-based organizations have been established to manage water resources. These organizations

involve residents in the planning, operation, and maintenance of water supply systems, ensuring that they are tailored to the specific needs of the community (Banda & Pathmarajah, 2018).

- **Improved sanitation**: The Sri Lankan government has implemented various initiatives to enhance sanitation, including the provision of toilets and sewage treatment plants. These measures have not only improved public health but also reduced pollution of water resources (UNICEF, 2018).
- **Water pricing reform**: As mentioned earlier, the government has introduced reforms to the water pricing system to encourage water conservation and efficient use. These reforms have been successful in reducing water consumption and generating revenue for investment in new water supply infrastructure (MoCPWS, 2019).

11.4 Water Governance Challenges in the Industrial Sector

The industrial sector is a significant contributor to the Sri Lankan economy, accounting for approximately 27% of the country's GDP. However, the water-intensive nature of many industries poses significant challenges for water governance in Sri Lanka.

11.4.1 Water Intensive Industries

Industries such as textile manufacturing, food processing, and beverage production are particularly water intensive. These industries consume large quantities of water for various processes, including cleaning, cooling, and product manufacturing. As a result, they place a significant strain on local water resources, particularly in areas where water scarcity is already a concern. While freshwater use in agriculture has decreased, water use in both domestic and industrial activities has increased. Currently, 87.3% of freshwater is used for agriculture, 6.2% for domestic use, and 6.4% for industrial use.

11.4.2 Impact on Water Resources

The industrial sector's impact on water resources goes beyond water consumption. Industrial activities can also contribute to water pollution, affecting both water quality and aquatic ecosystems. Inadequate wastewater treatment facilities and poor pollution prevention measures further exacerbate this issue.

According to the 2018 State of the Environment Report, the industrial sector is responsible for approximately 30% of total water pollution in Sri Lanka, with the textile, food and beverage, and chemical industries being the largest contributors (CEA 2019). The discharge of untreated or inadequately treated wastewater from these industries can have severe environmental and health consequences, including contamination of drinking water sources and degradation of aquatic ecosystems. For example:

- In 2013, a factory in Rathupaswala, Sri Lanka, was accused of contaminating the local water supply with chemicals used in textile manufacturing. The pollution caused protests and clashes with police, with residents reporting health problems as a result of the contaminated water (Wipulasena, 2013).
- In 2017, a chemical factory in Ekala, Sri Lanka, was found to be illegally dumping industrial waste into a nearby canal. The waste contained heavy metals and other toxic substances, resulting in significant water pollution and environmental damage. The incident prompted action by the authorities and calls for stronger enforcement of environmental regulations.
- In 2021, a sugar factory in Sevanagala, Sri Lanka, was accused of polluting a nearby river with untreated wastewater. The pollution caused fish deaths and other negative impacts on the local ecosystem and led to protests by local residents and environmental activists (Colombo Gazette, 2021).

11.5 Role of Policy Regulations

To address these water governance challenges, the Sri Lankan government has implemented several policy regulations. These include:

- **The Industrial Effluent (Treatment) Regulations**: Provide standards for the discharge of effluent from industrial processes.
- **The National Water Supply and Drainage Board Act**: Regulates the abstraction and use of groundwater.
- **National Policy, Strategies and Institutional Framework for Water Resources Development, Conservation and Management (2019)**: Provides a framework for the sustainable management of water resources, including the protection of water quality and ecosystems.

These policies emphasize the importance of integrated water resources management. Ensuring the coordination and integration of various stakeholders and sectors involved in water management is a key challenge. The policies promote a holistic approach by considering social, economic, and environmental aspects.

- The National Environment Act provides the legal framework for the protection and management of the environment in Sri Lanka, including provisions for the prevention and control of water pollution and the conservation of water resources and ecosystems. It sets standards and regulations for the discharge of pollutants into water bodies, aiming to protect water quality and ecosystems. Effective implementation of these regulations is crucial to mitigate the impact of industrial, agricultural, and domestic activities on water resources.

- The Mahaweli Authority Act established the Mahaweli Authority, which is responsible for implementing integrated water resources management practices, including regulating water use and conserving water resources in the Mahaweli River Basin (MA n.d.). River basin management is a key challenge in water governance, as it involves balancing competing demands for water from various sectors. The Act establishes the Mahaweli Authority as the governing body responsible for managing water resources in the basin.
- The Irrigation and Water Resources Management Act regulates irrigation and water resources in Sri Lanka. It includes provisions for the development of irrigation schemes, the protection of water quality, and the conservation of water resources (Mahaweli Authority n.d.).
- Industrial effluent management is a significant water governance challenge. The Industrial Effluent (Treatment) Regulations address this challenge by providing standards and guidelines for the discharge of effluent, aiming to protect water quality and minimize pollution from industrial activities.
- Groundwater is a vital water source for various sectors, including industry and domestic use. The National Water Supply and Drainage Board Act regulates the abstraction and use of groundwater, helping to ensure its sustainable management and prevent overexploitation.

These regulations and policies play a crucial role in addressing water governance challenges in Sri Lanka by providing legal frameworks, guidelines, and mechanisms for the sustainable management and protection of water resources.

11.5.1 Water-Use Efficiency

Improving water-use efficiency is another key aspect of water governance in the industrial sector. This can be achieved through technologies such as water recycling and reuse, as well as by implementing measures to reduce water consumption.

According to a study by the United Nations Industrial Development Organization (UNIDO, 2018), water consumption in the Sri Lankan industrial sector can be reduced by up to 50% through the implementation of water-use efficiency measures. The study also found that the cost of implementing these measures could be recovered within two years through savings in water and energy costs.

11.5.2 Wastewater Treatment and Pollution Prevention

Wastewater treatment and pollution prevention measures are critical for sustainable industrial water management. The government has implemented initiatives such as the Cleaner Production Program, which promotes resource efficiency and pollution prevention in industries. The program encourages industries to adopt cleaner production processes and technologies to reduce water consumption and wastewater generation.

The success of these initiatives is evident in the textile industry, which has implemented measures such as the adoption of waterless dyeing technologies. Waterless dyeing presents a significant opportunity for Sri Lanka's textile industry to enhance sustainability, address water scarcity, and meet global eco-friendly demands. Technologies like supercritical CO_2 dyeing drastically reduce water and energy consumption, minimize effluent discharge, and align with the industry's shift towards greener practices, offering a competitive edge in the global market. The shift towards waterless dyeing and sustainable processing methods, such as spraying and foam finishing, demonstrates the industry's commitment to reducing its environmental impact (SGS, 2025).

11.5.3 Successful Industrial Water Management Practices and Initiatives

Several successful industrial water management practices and initiatives have been implemented in Sri Lanka. For example, the Coca-Cola Company has implemented a water stewardship program in its bottling plants, resulting in a significant reduction in water consumption and wastewater generation. According to the company's 2020 sustainability report, its operations in Sri Lanka achieved a 45% reduction in water consumption and a 63% reduction in wastewater discharge since 2013. The textile industry in Sri Lanka has also taken measures to reduce water consumption, including the adoption of waterless dyeing technologies (Ratnayake & Fernando, 2018).

Another successful initiative is the "Greening Sri Lanka's Hotels" project, which was launched in 2011 with the aim of promoting sustainable water and energy management in the hotel industry. The project has resulted in the implementation of various water-use efficiency measures.

11.6 Water Governance Challenges in the Power Sector of Sri Lanka

The power sector is one of the most significant non-agricultural water users in Sri Lanka, with a high demand for water in power generation and cooling processes. The water-intensive nature of the sector poses several challenges to water governance, including potential conflicts between water availability and energy needs. This section examines the specific water governance challenges faced by the power sector in Sri Lanka and explores potential solutions to promote sustainable water use.

11.6.1 Water Requirements for Power Generation and Cooling Processes

The power sector in Sri Lanka relies on a mix of energy sources, including hydropower, thermal power, and renewable energy. Hydropower generation is the largest contributor to the national power grid, accounting for about 50% of the total installed capacity. However, climate change has impacted the availability of water resources for hydropower generation, with changing rainfall patterns and extreme weather events affecting water inflows to reservoirs. The thermal power sector, which uses fossil fuels to generate electricity, is also water-intensive, requiring large quantities of water for cooling processes. This high demand for water in power generation and cooling can create conflicts with other water users and exacerbate water scarcity issues.

11.6.2 Impacts of Climate Change on Hydropower Resources

Hydropower currently accounts for approximately 30% of the country's electricity generation. However, the construction of hydropower dams and reservoirs can have significant impacts on water resources, such as altering river flow patterns and affecting downstream water availability. For example, the Upper Kothmale hydropower project was found to have reduced the downstream flow of the Kotmale River by 38%, affecting downstream agriculture and other water uses.

The impacts of climate change on hydropower resources in Sri Lanka are significant. Changes in rainfall patterns and extreme weather events have impacted water inflows to reservoirs, affecting the generation capacity of hydropower plants. Recent studies indicate that climate change is already impacting hydropower generation in Sri Lanka. For example, the country's three largest hydropower stations produced less than half of their maximum potential output in 2021, a decline attributed to rising temperatures, increased evaporation, and delayed monsoon rainfall—phenomena linked to climate change (Climate Advocates Sri Lanka, 2023). Similarly, during the 2016–2017 drought, hydropower production declined by approximately 40%, forcing greater reliance on fossil fuel imports (Dharshana, 2023). The power sector's reliance on hydropower highlights the need for adaptation strategies to ensure a reliable and sustainable power supply.

Climate change is expected to have significant impacts on hydropower resources in Sri Lanka, with changes in rainfall patterns and temperature affecting water availability and the reliability of hydropower generation. Climate change could reduce hydropower generation potential in Sri Lanka by up to 10% by 2050 (ADB, 2016).

Role of Policy Regulations, Water-Use Efficiency, Wastewater Treatment, and Pollution Prevention Effective policy regulations can play a critical role in promoting sustainable water use in the power sector. The NWSDB and the Central Environmental Authority (CEA) are the key regulatory bodies responsible for managing water resources and regulating water use in Sri Lanka. The NWSDB is responsible for water allocation and supply, while the CEA enforces environmental regulations and monitors water quality.

Water-use efficiency and wastewater treatment are also essential components of sustainable water use in the power sector. Improving water-use efficiency can reduce the demand for freshwater in power generation and cooling processes. Wastewater treatment can help minimize water pollution from power generation, reducing the sector's impact on water resources and the environment. For example, the Kelanitissa Power Station has implemented a wastewater treatment plant to treat its effluent before discharging it into the Kelani River, thereby helping to minimize the power station's impact on water quality and the environment (ADB, 2019).

11.6.3 Integrated Water and Energy Planning

Given the close linkages between water and energy, it is important to consider water availability and use in energy planning. One example is the Mahaweli Water Security Investment Program, a joint initiative between the Sri Lankan government and the World Bank that aims to enhance water security and promote sustainable hydropower development in the Mahaweli River basin. The program includes measures such as improving irrigation efficiency, promoting rainwater harvesting, and implementing climate-resilient hydropower projects.

11.6.4 Policies and Practices that Promote Sustainable Water Use

Several policies and practices promote sustainable water use in the power sector. The Sri Lanka Sustainable Energy Authority (SLSEA) is responsible for promoting sustainable energy development in the country, including renewable energy sources and energy efficiency measures. The SLSEA has implemented initiatives to promote sustainable water use in the power sector. For example,

- Dry-cooling technologies require much less water than conventional cooling systems. The Norochcholai coal power plant, which utilizes a hybrid cooling system, achieves a 90% reduction in water use.
- Renewable energy sources, such as solar and wind power, also have significantly lower water requirements compared to hydropower and thermal power.

The Ceylon Electricity Board (CEB), the primary power utility in Sri Lanka, has implemented a policy to utilize treated wastewater from nearby urban areas for cooling processes in its power plants. This initiative has reduced the demand for freshwater in power generation and minimized the sector's impact on water resources.

11.6.5 Cross-Sectoral Collaboration and Adaptation Strategies

There is a growing need for cross-sectoral collaboration and integrated water management approaches to address the challenges of climate change. Effective water governance in non-agricultural sectors with water-intensive operations requires coordinated efforts among various stakeholders, including policymakers, industry leaders, and civil society. For instance,

- The NWSDB has established partnerships with various stakeholders, including local governments, community-based organizations, and the private sector, to improve water service delivery and ensure sustainable water management (NWSDB, 2019). Such collaborations can help enhance access to safe and reliable water supply, improve wastewater management, and promote water conservation and reuse.
- The Water Partnership Program, a joint initiative of the Ministry of Irrigation and Water Resources Management and the United Nations Development Programme, has been promoting integrated water resources management through stakeholder engagement, policy reform, and capacity building. Similarly, the Climate Resilience Improvement Project, funded by the World Bank, supports the implementation of climate-resilient water supply systems in vulnerable regions of the country (Kahandawa, 2017).

Adaptation strategies, such as water-use efficiency, water recycling, and technological innovations, are also essential for mitigating water-related challenges in non-agricultural sectors. For instance,

- Promoting water-efficient technologies such as low-flow faucets and showerheads can reduce water consumption in the domestic sector, while implementing water recycling systems can help reduce industrial water demand.
- Similarly, the adoption of renewable energy technologies, such as solar or wind power, can reduce the water intensity of power generation.

According to a study conducted by the Ministry of Environment and Renewable Energy in Sri Lanka, the domestic sector is the largest consumer of water, accounting for 54% of total water demand. The industrial sector follows closely behind, consuming 23% of total water demand, while the power sector consumes 5% of total water demand (MERE, n.d.).

- The Central Environmental Authority (CEA) reports that most industries in the country do not have adequate wastewater treatment facilities, resulting in significant levels of industrial effluent being released into water bodies (CEA, n.d.).
- The Asian Development Bank (ADB) projected that the water crisis would worsen due to climate change, with an average annual rainfall decrease of 10–30% by 2050 (ADB, n.d.).
- The Ministry of Power and Renewable Energy in Sri Lanka estimates that the country will require an additional 3,000 MW of power generation capacity by 2030 to meet growing energy demand and acknowledges that climate change impacts on hydropower resources may pose significant challenges to meeting this increased demand.

These statistics highlight the urgent need for cross-sectoral collaboration and adaptation strategies to ensure sustainable water management in non-agricultural sectors. Collaborative initiatives and integrated water management approaches can help mitigate the impacts of climate change and address water governance challenges in a more coordinated and effective manner. For example, the Water Supply and Sanitation Improvement Project, implemented by the National Water Supply and Drainage Board in Sri Lanka, aims to enhance domestic water supply and sanitation services through a multi-stakeholder approach that involves the government, private sector, and local communities (NWSDB 2019).

11.7 Conclusions

This chapter has highlighted the significant water governance challenges facing non-agricultural sectors in Sri Lanka, which are exacerbated by the impact of climate change. The focus was on specific challenges faced by the domestic, industrial, and power sectors in terms of water availability, quality, and usage. The chapter emphasized the need for:

- Enhancing cross-sectoral collaboration and integrated water management approaches, which are essential for achieving sustainable water management goals in the country.
- Engaging stakeholders for effective policy interventions, which is crucial for addressing governance challenges effectively.

- Promoting public-private partnerships and community-driven initiatives to improve water governance in non-agricultural sectors. The government can play a critical role by creating an enabling policy and regulatory environment that promotes sustainable water management practices. Meanwhile, the private sector, non-governmental organizations, and community-based organizations can mobilize local communities, promote awareness, and facilitate capacity building to enhance the effectiveness of water governance initiatives.
- Conducting research studies on the socioeconomic impacts of climate change on water resources and related sectors for informed decision-making and policy development. Such studies provide critical insights into the challenges faced by non-agricultural sectors and inform the development of evidence-based policies.
- Promoting innovative water management technologies to enhance water-use efficiency at all levels.

Addressing water pollution issues in non-agricultural sectors. Measures such as enforcing water quality standards, promoting eco-friendly technologies, and raising awareness among stakeholders are necessary for preventing and controlling water pollution.

Strengthening monitoring and evaluation systems for water governance initiatives to assess their effectiveness and identify areas for improvement.

Overall, it is evident that effective water governance in non-agricultural sectors is crucial for achieving sustainable development in Sri Lanka. It requires collective efforts, stakeholder engagement, and effective policy interventions to address the challenges and ensure the efficient and equitable use of water resources.

References

ADB (Asian Development Bank). (2016). *Climate change impact and adaptation study for hydropower sector in Sri Lanka.* https://www.adb.org/sites/default/files/linked-documents/45876-001-climate-change-hydropower.pdf

ADB. (2019). *Strengthening water security in Sri Lanka's power sector.* https://www.adb.org/news/features/strengthening-water-security-sri-lankas-power-sector

ADB. (n.d.). *Climate risk country profile: Sri Lanka.* https://www.adb.org/sites/default/files/institutional-document/178320/sri-lanka-climate-risk-profile.pdf

Banda, N., & Pathmarajah, S. (2018). Community-based management of water resources in Sri Lanka: Lessons learned and the way forward. *Journal of Environmental Management, 180*, 323–331.

CEA (Central Environmental Authority) (n.d.). *Industrial pollution control in Sri Lanka.* http://www.cea.lk/web/en/industrial-pollution-control-in-sri-lanka

Dharshana, M. (2023). Sri Lanka's Persistent Power Predicament: A Comprehensive Overview, *LinkedIn*, https://www.linkedin.com/pulse/sri-lankas-persistent-power-predicament-comprehensive-dr-dharshana-xmejc

Fan, M. (2015). *Sri Lanka's water supply and sanitation sector: Achievements and a way forward.* Asian Development Bank. Retrieved June 28, 2023, from https://www.adb.org/sites/default/files/publication/161289/south-asia-wp-035.pdf

Ferdinando, S. (2023). How climate change is impacting Sri Lanka's hydropower sector, *Climate Advocates Sri Lanka*, https://www.climateadvocatessl.com/articles/article7

International Finance Corporation. (2017). *Climate Investment Opportunities in South Asia: An IFC Analysis*. Retrieved from https://www.ifc.org/content/dam/ifc/doc/mgrt/17663-ifc-srilanka-factsheet-v2.pdf

Kahandawa, S. S. K. (2017). *Sri Lanka - Climate Resilience Improvement Project (CRIP): P146314 - Implementation Status Results Report: Sequence 06* (English). Washington, D.C.: World Bank Group. Retrieved from http://documents.worldbank.org/curated/en/882801498178683167

MA (Mahaweli Authority). (n.d.). *Mahaweli Authority Act*. http://www.mahaweli.gov.lk/pdf/Mahaweli%20Act%20No%2035%20of%201976.pdf

MERE (Ministry of Environment and Renewable Energy). (n.d.). *Sri Lanka's initial national communication to the United Nations Framework Convention on Climate Change*. http://unfccc.int/resource/docs/natc/sri01.pdf

MoCPWS (Ministry of City Planning and Water Supply). (2019). *National water supply and drainage board annual report 2019*. Colombo, Sri Lanka.

NWSDB (National Water Supply and Drainage Board). (2019). *Sri Lanka national water supply and drainage board strategic business plan 2019–2023*. https://www.nwsdb.lk/images/PDFs/Strategic_Business_Plan_2019-2023.pdf

NWSDB. (2021). *Water quality monitoring system*. https://cewas.nwsdb.lk/water-quality-monitoring-system-nwsdb/

Ranathunga, U., Kumara, M. W. M. M., & Welivitiya, W. D. (2016). Assessment of rainwater harvesting systems for domestic water supply in rural Sri Lanka. *Journal of Water Supply: Research and Technology—AQUA, 65*(7), 561–573.

Ratnayake, R. M. R., & Fernando, W. J. S. (2018). Sustainable manufacturing practices in the Sri Lankan textile and apparel industry: A review. *IOP Conference Series: Materials Science and Engineering, 393*(1), 012015. https://doi.org/10.1088/1757-899X/393/1/012015

SGS. (2025). Innovation is the Future for the Textile and Apparel Industry: Sustainable dyeing and finishing practices, *SGS Consumer Compact Softlines*, https://www.sgs.com/en-lk/news/2025/07/cc-2025-q2-innovation-is-the-future-for-the-textile-and-apparel-industry

UNICEF. (2018). *Sri Lanka country programme document 2018–2022*. https://www.unicef.org/srilanka/media/1186/file/UNICEF-Sri-Lanka-Country-Programme-2018-2022.pdf

UNIDO (United Nations Industrial Development Organization). (2018). *Sri Lanka water assessment: Industry performance and technology review*. https://www.unido.org/sites/default/files/2019-4/SL%20Water%20Assessment%20Report.pdf

UNDP (United Nations Development Programme). (n.d.). *Water partnership programme*. https://www.lk.undp.org/content/srilanka/en/home/operations/projects/environmentandenergy/water-partnership-programme.html

Wipulasena, A. (2013). Rathupaswala water crisis and new moves to dam it., *Sunday Times*, https://www.sundaytimes.lk/131020/news/rathupaswala-water-crisis-and-new-moves-to-dam-it-66321.html

World Bank (2015). *Sri Lanka's Water supply and sanitation improvement project—project appraisal document*. World Bank. https://documents1.worldbank.org/curated/en/157731468188653128/pdf/PAD1223-PAD-P147827-IDA-R2015-0156-1-Box391456B-OUO-9.pdf

Prof. Chandra Embuldeniya is a multidisciplinary STEAM technocrat with a distinguished career focused on transforming organizations and fostering sustainable development. He has a proven track record in strategic management, organizational turnaround, and the re-engineering of production facilities, leveraging modern digital tools to drive mission-driven outcomes. His expertise extends to reforming education for a future-ready workforce, developing high-tech agricultural projects, and commercializing technology. As former President of the National Chamber of Commerce and founding Vice Chancellor of Uva Wellassa University, he has significantly contributed to Sri Lanka's progress. He has led innovations at the National Science Foundation, contributed to agricultural development and food security initiatives (President's Committee), and championed STEAM education transformation (Ministry of Education and NSF).

Part IV
Water Governance and Climate Change Adaptation: Policy, Institutional and Technical Perspectives

Chapter 12
Aligning Agricultural and Environmental Policies Towards Integrated Water Governance in Sri Lanka

Manoj Thibbotuwawa and Erandathie Pathiraja

Abstract The National Environmental Policy outlines the role of water and water governance across various sectors, as well as the guidelines for addressing these challenges. Agriculture is the largest water user in Sri Lanka and is highly vulnerable to climate change-induced disasters. Agriculture in Sri Lanka uses rainfall, surface water stored in reservoirs, and groundwater. Nevertheless, water availability in Sri Lanka varies spatially and temporally, and variability has been altered or induced recently due to climate change. This chapter assesses the key dimensions of water governance within Sri Lanka's environmental and agricultural policies and focuses on institutional coordination and policy integration for effective water management, water allocation mechanisms, equitable access, water pollution control measures, climate change adaptation and resilience strategies, and financial sustainability in water management.

Keywords Climate change · Agriculture and environment · Agricultural water use · Environmental sustainability · Water quality · IWRM

12.1 Introduction

Sri Lanka has implemented several policies and legal frameworks for water governance, including environmental protection and agriculture policies. Despite these efforts, the country continues to face persistent governance challenges in sustainable water management. Key issues, including pollution, climate change, over extraction, deforestation, catchment degradation, conflicts, encroachment, and illegal activities, coupled with institutional fragmentation and policy gaps, pose significant threats

M. Thibbotuwawa (✉) · E. Pathiraja
Institute of Policy Studies, Colombo, Sri Lanka
e-mail: manoj@ips.lk

E. Pathiraja
e-mail: erandathie@ips.lk

U. A. Amarasinghe et al. (eds.), *Water Governance for Climate Change Adaptation in Sri Lanka*, Global Issues in Water Policy 13,
https://doi.org/10.1007/978-3-032-04950-6_12

to agriculture, food production, economic growth, and environmental sustainability. Addressing these challenges, which have kept the country off track in achieving water security (United Nations Sri Lanka, 2023), requires a robust water governance framework that integrates policy coherence, institutional coordination, and sustainable management practices to balance agricultural demands with environmental conservation (FAO, n.d.).

Building on this context, this paper assesses the key dimensions of water governance within Sri Lanka's environmental and agricultural policies. It focuses on institutional coordination and policy integration for effective water management, water allocation mechanisms, equitable access, water pollution control measures, climate change adaptation and resilience strategies, and financial sustainability in water management. The report is structured as follows. Section 12.2 explores the role of water in Sri Lanka's agriculture and environmental systems. Section 12.3 examines the key water governance challenges arising at the intersection of these two sectors. Section 12.4 provides an overview of the policy landscape influencing water governance in agriculture and the environment. Section 12.5 examines how agricultural and environmental policies respond to these governance challenges. Section 12.6 presents future directions and offers recommendations to strengthen Sri Lanka's water governance framework, with a focus on enhancing sustainability and resilience in the face of emerging and future challenges. Section 12.7 concludes with a way forward.

12.2 The Role of Water in Sri Lanka's Agriculture and Environment

12.2.1 Agricultural Water Use

Historically, Sri Lanka has used traditional water management methods in agriculture, particularly in tanks and small-scale reservoirs. Ancient irrigation systems, such as those in the Anuradhapura and Polonnaruwa regions, are among the world's most sophisticated water management systems. These systems comprise a series of interconnected tanks, channels, and reservoirs designed to capture rainwater and store it for use during the dry season (IUCN, 2018). The system also includes catchment areas and canals that help distribute water efficiently to surrounding agricultural lands. This traditional irrigation system has supported agriculture for centuries and remains in use in many rural parts of Sri Lanka today (Satoyama Initiative, 2010).

However, modern irrigation methods have been introduced in response to increasing population pressure and the challenges posed by climate change. These methods include large-scale irrigation schemes, pumps, and drip irrigation systems that aim to use water more efficiently and effectively (Jayakody et al., 2018). Modern irrigation infrastructure, including reservoirs and dams, has been expanded to cover newly developed agricultural areas. In particular, the Mahaweli Development Project,

Sri Lanka's largest irrigation project, has transformed agriculture in the dry zone by providing reliable water sources for paddy cultivation and other crops (Mahaweli Authority, n.d.).

While irrigation has increased agricultural productivity, it also presents challenges such as high infrastructure and maintenance costs, over-extraction of groundwater, and the environmental impacts of large-scale water diversions. The tension between maintaining traditional practices and adopting modern, mechanized systems remains a significant challenge for water governance in the country. Water use in agriculture is vital not only for food production but also for environmental sustainability. Therefore, understanding the interlinkages between agricultural water use and environmental health is essential to addressing water governance challenges in Sri Lanka.

12.2.2 Agricultural Water Use, and Environmental Sustainability

Sri Lanka's rich biodiversity is directly tied to the vitality of its water systems (Mongabay, 2022). Sri Lanka's freshwater habitats support a diverse array of species, including endemic fish, amphibians, and aquatic plants, all of which are essential for maintaining ecological balance. Rivers such as the Mahaweli, Kelani, and Gin, and numerous smaller streams and tributaries provide habitats for various aquatic life forms. Wetlands, such as those in Wilpattu National Park and Bundala, further enhance biodiversity by supporting migratory birds and other wildlife (Mongabay, 2022). The nation's distinctive ecosystems—including wetlands, rivers, and reservoirs—are crucial for environmental sustainability and agricultural productivity.

In Sri Lanka, agriculture consumes the largest share of water resources, and how this water is managed significantly impacts both crop yields and the environment. As food demand rises, the expansion of irrigation systems has boosted agricultural productivity, but it often disrupts natural river flows and depletes downstream water levels (Jayawardena et al., 2021). These changes degrade aquatic habitats and diminish vital ecosystem services, such as water purification and flood regulation (Mongabay, 2022). Large-scale dams and reservoirs further fragment ecosystems, leading to biodiversity loss (ScienceDirect, 2021). Poorly managed irrigation also causes land degradation and soil erosion, reducing soil fertility and increasing sedimentation in water bodies (RWSN, 2015). Excessive water withdrawal contributes to groundwater depletion, further straining ecosystems (Chandrasekara et al., 2021).

Water quality deterioration is another major concern. Runoff from agricultural lands carries pesticides, fertilizers, and sediments into rivers and lakes, causing nutrient pollution. This can trigger eutrophication, where excessive nutrients encourage rapid algae growth, depleting oxygen levels and endangering aquatic life (ScienceDirect, 2021).

The lack of coordination between agricultural and environmental institutions in Sri Lanka has resulted in policy gaps, with agricultural policies often prioritizing crop yields without fully considering environmental impacts. This results in the degradation of water systems crucial for both agriculture and biodiversity (Chandrasekara et al., 2021; Jayawardena et al., 2021). Resolving these conflicts requires an integrated approach that engages stakeholders across sectors.

12.3 Water Governance Challenges at the Agriculture–Environment Nexus

Among the numerous governance issues in the water sector, agricultural and environmental aspects include inefficient water use, pollution, environmental degradation, climate change–induced water scarcity, deforestation, catchment degradation, over-extraction, inequitable water distribution, weak enforcement of environmental regulations, encroachment, illegal activities, lack of public participation, technological and skills gaps, institutional fragmentation, and policy gaps (Geekiyanage et al., 2015; Jayasiri et al., 2022, 2023; Shunglu et al., 2022; Sriskandarajah & Sivapalan, 2023). This section assesses how the agriculture and environment policies have addressed the key issues in the water sector.

One of the major water governance challenges in Sri Lanka's agricultural sector is inefficient water use in irrigation. The country relies heavily on traditional irrigation systems, such as tank-based irrigation and canal networks, which, despite their historical importance, are often inefficient and result in excessive water consumption. Many poorly maintained irrigation structures lead to water losses due to seepage, evaporation, and inefficient distribution. Furthermore, flood irrigation remains the dominant practice, contributing to water wastage. Due to high costs and limited awareness, farmers often lack access to modern water-efficient technologies, such as drip irrigation and sprinkler systems. As a result, a sizeable proportion of irrigation water is wasted, reducing overall water productivity in agriculture.

Water pollution in Sri Lanka arises from both natural processes and human activities, including industrial and agricultural sources (Geekiyanage et al., 2015). Water quality is considered a more serious issue than water scarcity. Pollution from agriculture, industry, and domestic waste poses a significant governance challenge. In rural areas, lakes and rivers are polluted by agricultural runoff containing pesticides and fertilizers, which poses health risks, including chronic kidney disease (CKDu) (Hettithanthri et al., 2021; Jayasumana et al., 2015).

In urban areas, untreated industrial discharges and poor waste management are the main sources of surface and groundwater pollution. Industrial effluents, including untreated chemical waste and heavy metals, pollute water bodies, particularly in regions with substantial industrial activity, such as the Western Province. Ineffective management of wastewater and solid waste in urban areas contaminates freshwater sources (Nuwanka & Gunathilaka, 2023; Shamsudduha et al., 2025). The

Mahaweli and Kelani rivers are among the most polluted water bodies. Unregulated water-intensive activities, including certain agricultural practices and domestic waste disposal, contribute to excessive pollution, making water resources unsafe for human consumption and biodiversity.

Deforestation and catchment degradation threaten water resources by disrupting natural hydrological cycles, reducing water availability, and increasing vulnerability to climate change. Deforestation leads to rapid surface runoff and reduced groundwater recharge, resulting in decreased water retention (Kumarasiri et al., 2022). Exposed soil increases erosion and reduces the capacity of rivers through siltation, resulting in poor water quality, flash floods, and landslides. Illegal logging, land encroachment, agricultural expansion in the plantation and paddy sectors, slash-and-burn agriculture, and infrastructure development (such as hydroelectric projects and roads) disturb forested watersheds (Herath & Patabandi, 2023; Hewawasam, 2010; UNR Sri Lanka, 2015).

Urbanization, illegal construction, agriculture, and tourism are all threats to wetlands and coastal areas. Wetlands support natural water filtration and absorb excess water during floods. However, encroachment, waste dumping, and illegal construction in wetlands lead to increased urban and rural flooding, as well as biodiversity loss. Unauthorized settlements and illegal sand mining near riverbanks lead to water pollution and interrupt natural water flow (Jayawardhana & Udawattha, 2022; Manage et al., 2020; Geekiyanage et al., 2015).

Changes in rainfall patterns influence water availability. Extreme weather events, leading to increased droughts and floods, are projected to be frequent with a high level of confidence, according to the recent IPCC (AR6) report (Lee et al., 2023). Rising temperatures, erratic rainfall patterns, and sea-level rise impact water resources. For instance, frequent floods damage ecosystems and pollute water, rising sea levels lead to saltwater intrusion in groundwater reserves, and prolonged droughts affect water availability for drinking, irrigation, and electricity generation (Fernando et al., 2024; Khaniya et al., 2019, 2021; Mathiventhan et al., 2022; Perera et al., 2018). Climate change has become a major threat to water availability in Sri Lanka's agriculture sector. During drought periods, water levels in reservoirs and tanks drop significantly, affecting irrigation schemes that rely on these water sources. In contrast, intense rainfall events lead to floods, which damage crops and reduce soil fertility.

Another significant challenge in Sri Lanka's water governance is the limited involvement of farmers and local communities in decision-making processes. Enhanced community engagement and information exchange can bolster water conservation initiatives and governance efficiency (Hewawasam & Matsui, 2022; Kekulandala et al., 2024). In Sri Lanka, community participation in water governance remains limited, despite the country's rich tradition of local knowledge and expertise in water management. For instance, farmers and rural communities are rarely included in water policy decision-making, except for major irrigation schemes for paddy farming, which are often managed through farmers' organizations for water allocation. Community-based solutions are more sustainable than top-down methods because they integrate traditional knowledge and promote awareness-driven responsibility. Additionally, many farmer organizations lack the necessary resources and institutional support to participate actively in water management initiatives.

Water governance in Sri Lanka is complicated by a fragmented institutional framework, where multiple agencies manage water resources without effective coordination. Various government and semi-government agencies oversee water governance, involving both direct and indirect participation from nearly 41 institutions and 52 pieces of legislation (Chandrasekara et al., 2021). The main institutions include the Ministry of Water Supply, National Water Supply and Drainage Board (NWSDB), Irrigation Department, Department of Agriculture, Mahaweli Authority, Ministry of Power, Central Environmental Authority (CEA), Water Resources Board (WRB), Ministry of Land, Ministry of Environment, and the Department of Forest Conservation. Lack of coordination among these institutions and unclear responsibilities lead to inefficiencies, delays in decision-making and implementation, and conflicts among stakeholders.

Effective water governance necessitates robust legal frameworks; however, weak policy enforcement remains a significant issue. Governance gaps emerge from inadequate monitoring, limited financial and technical resources, technological and skills deficiencies, corruption, and political interference. Despite the existence of environmental regulations in Sri Lanka, enforcement of these laws remains insufficient. Illegal waste disposal, industrial pollution, unregulated sand mining, and other mining activities continue to degrade water resources. For instance, gem mining is regarded as the primary cause of soil erosion and sedimentation in the Rathnapura district, while sand mining occurs in numerous downstream rivers (Geekiyanage et al., 2015). Penalties and monitoring for pollution are often inadequate, allowing agriculture and industries to surpass accepted pollution limits.

12.4 Mapping the Policy Landscape: Agriculture, Environment, and Water

The National Agriculture Policy (NAP) of 2022, the National Environment Policy (NEP) of 2021, and the National Environment Action Plan (Ministry of Environment, 2022) are the three primary sectoral policy documents overseeing water governance in Sri Lanka's agriculture and environment sectors. The NEP was developed to address these challenges in alignment with global policy trends, while the NEAP (2022–2030) provides a detailed set of actions for integrated water resource management, building upon the updated NEP. The Nationally Determined Contributions (NDCs) in Sri Lanka also include a specific focus on the water sector, recognizing that climate change presents significant challenges to water resources, such as more frequent and intense droughts, floods, and unpredictable rainfall patterns, which affect agriculture, drinking water supplies, and ecosystems. Additionally, the National Agriculture Policy plays a crucial role in regulating water use within the agriculture sector.

The National Agriculture Policy (NAP) of Sri Lanka

The NAP-2022 aims to transform Sri Lanka's agricultural sector into a modern, sustainable, and competitive industry that ensures food security and improves the livelihoods of farmers. The policy seeks to achieve sustainable food security, foster national prosperity, and focus on developing a socially acceptable and environmentally friendly food system. It also aims to enhance agricultural production, processing, and marketing while safeguarding the interests of producers and consumers.

The NAP-2022 outlines several strategic focus areas to achieve these goals. These include enhancing crop production and productivity, achieving food and feed self-sufficiency, and ensuring the efficient utilization of agricultural resources. The policy also emphasizes improving market access and competitiveness, both locally and internationally, as well as developing climate-resilient agricultural practices. Additionally, the policy aims to mitigate risks and uncertainties in agricultural production and marketing, promote the active participation of women and youth, and enhance collaboration between central and regional agricultural institutions. In total, the policy outlines twelve measurable goals aimed at revitalizing the agricultural sector. These goals focus on sustainability, productivity, competitiveness, and inclusivity, with the aim of ensuring food security, promoting environmental sustainability, and enhancing farmers' livelihoods.

The NAP-2022 also addresses water and irrigation issues in agriculture by prioritizing sustainable water management and the efficient use of irrigation systems. Recognizing the crucial role of water in agricultural productivity, the policy advocates for integrated water resource management (IWRM) to optimize water use while maintaining ecological balance and supporting rural livelihoods. It calls for the modernization and rehabilitation of both large-scale and small-scale irrigation infrastructure to improve water distribution and reduce waste. To combat the challenges posed by climate change, such as irregular rainfall patterns, the policy encourages the adoption of water-saving techniques like drip irrigation and rainwater harvesting systems. Additionally, the NAP stresses the development of water storage solutions, such as small-scale reservoirs and tanks, to ensure water availability during dry periods, alongside the improvement of existing storage infrastructure.

Furthermore, the policy emphasizes the need for farmer capacity building, ensuring that farmers are well-equipped with the knowledge and tools to manage water efficiently. It also highlights the importance of research and technological innovation in enhancing water use efficiency and reducing environmental impact. Through these comprehensive measures, the NAP aims to address the growing water and irrigation challenges in Sri Lanka's agriculture sector, ensuring long-term sustainability and enhanced agricultural productivity.

The National Environment Policy (NEP)—2021

The updated NEP-2021 considered global and local policy developments and environmental threats since 2003 (Ministry of Environment, 2021). The NEP is based on eleven key policy principles, namely:

- Sustainable development and green economy

- Polluter pays, beneficiary pays (such as payment for ecosystem services), and compensation
- Life cycle, circular economy, and 3R (Reduce, Reuse, Recycle)
- Safe minimum standards (SMS)
- Precautionary principle
- Rewards, stewardship, and reciprocity
- Equity, inclusiveness, and no one left behind
- Efficiency in resource use
- Planetary boundaries
- Common but differentiated responsibilities
- Public trust doctrine (right to a sound environment)

Policy statements are categorized under seven thematic areas: Land and Water Resources, Biodiversity and Ecosystems, Coastal and Marine Resources, Pollution Control and Waste Management, Built Environment and Green Development, Climate Change and Global and Regional Environmental Challenges, and Cross-Cutting Issues and Challenges.

The Land and Water Resources theme encompasses three sub-thematic areas: (a) integrated management of land and water resources, national spatial strategy and land use planning; (b) environmentally sensitive land and water resources; and (c) participatory management of land and water resources. The policy goals for land and water resources are threefold. The first goal focuses on implementing sound land use planning based on scientific criteria, including zoning and land suitability classification. The second goal promotes an integrated approach to managing land, water, and forest resources at national, provincial, and local levels. The third goal aims to increase forest cover to one-third of the country's total land area. Additionally, the water sector is addressed through other policy goals, such as pollution prevention and control in all urban and rural areas, integrated waste management at the local government level, and building climate change resilience via adaptation and disaster risk management.

The National Environment Action Plan (NEAP)—2022–2030

The NEAP will be a rolling plan that is updated periodically, with necessary arrangements in place for monitoring and evaluation. It is the primary implementation mechanism for the NEP. The NEAP for the water sector is grounded in the principles of IWRM and outlines 14 strategic actions to be implemented during the 2022–2030 period. These strategies are shaped by pressing local challenges and align with Sustainable Development Goals (SDGs) 6 and 13, as well as the Nationally Determined Contributions (NDCs) for the water sector (Ministry of Environment, 2022). The strategies aim to ensure environmental safeguards by providing safe drinking water, improving sanitation, and preventing contamination of surface and groundwater. They also seek to reduce pollution and enhance the quality of rivers, lakes, reservoirs, and groundwater, as well as promote watershed management for long-term sustainability. The strategies focus on increasing water conservation and improving water use efficiency across sectors, enhancing irrigation and groundwater management, and minimizing saltwater intrusion. Furthermore, the strategies include reducing water-induced disasters, such as floods and droughts, ensuring national water security, implementing IWRM at all levels, and introducing new policies and institutions for improved water resource planning. Additionally, the plan emphasizes public awareness, capacity building, community mobilization, and strengthening water-related research to support evidence-based policy and action.

The above strategies are used in the *Action Plan for Integrated Water Resource Management*, which describes specific activities to undertake under each strategy, including the timeline, target over baseline, budget, and implementation responsibility of each action. This water sector action plan aims to achieve Sustainable Development Goals (SDG) 6 objectives. SDG 6.1 aims to provide universal and equitable access to safe drinking water; SDG 6.3 focuses on enhancing water quality; SDG 6.4 emphasizes water-use efficiency; SDG 6.5 promotes integrated water resource management; and SDG 6.6 targets the protection and restoration of water-related ecosystems. Actions related to climate change in the water sector are explicitly discussed in the Action Plan for Climate Action *for Sustainability* of the NEAP. The NEAP has outlined nine key actions to address climate change impacts on the water sector, in line with the needs identified by the NEP, the SDGs, and the NDCs. SDG 13.1 aims to enhance resilience and adaptive capacity to climate-related hazards, while SDG 13.3 emphasizes the importance of improving education, awareness, and capacity-building regarding climate change mitigation, adaptation, impact reduction, and early warning.

12.5 Policy Approaches for Addressing Water Governance Challenges

The Role of Environmental Policies: NEP and NEAP

Water pollution and environmental degradation issues are extensively discussed under the NEP sub-theme of managing water pollution, with four policy statements. The first statement proposes enhancing the quality of surface and groundwater to a standard suitable for drinking and domestic use through consistent monitoring and real-time measurement of parameters, and effectively regulating discharges from industrial, domestic, agricultural, and waste management activities. The second statement proposes preventing sewage from entering surface and groundwater bodies through modern wastewater management techniques in all urban and suburban areas. The third statement proposes efficient river flow management, nature-based solutions, controlling sand mining at river mouths, developing necessary infrastructure, and building institutional capacities at national and subnational levels to prevent saltwater intrusion. Routine, systematic, and scientific research is also proposed to detect nonpoint source water pollution and control it in coastal ecosystems.

Related to climate change issues affecting water, the NEP has identified that adaptive governance approaches, such as IWRM, drought risk management, and mitigation plans, are essential for avoiding governance failures. Climate change affects all environmental and ecological sectors. Hence, it is considered under a common theme with strategies including adaptation, mitigation, and loss and damage. Periodic vulnerability and risk assessments, enhancing the adaptive capacity of stakeholders, climate-smart solutions, and proper institutional mechanisms are proposed. The NEAP has incorporated nine actions for water sector adaptation to climate change based on Integrated River Basin Management (IRBM) in 15 river basins. The water sector adaptation measures of the NDCs and SDGs 6 and 13 are comprehensively integrated to enhance water use efficiency and quality across sectors.

The growing demand for water resources in numerous economic sectors has made sustainable water resource management a top priority. Therefore, the NEP emphasizes managing water and land resources within an integrated framework to balance demand while optimizing economic, social, and environmental benefits in an equitable manner. Recognizing over-extraction and inequitable water distribution, the NEP introduces strategies to improve groundwater management. Such unequal distribution is primarily observed in rural areas, where industries, commercial agriculture, and urban sectors are prioritized over local communities.

Participatory management of land and water resources is considered under a subtheme, including seven policy statements to emphasize its importance in water governance. Consultation of all stakeholders, payment for ecosystem services (PES), effective tenurial reforms, promoting tank cascade systems, regulating permit systems for resource extractions, and guidelines for mining activities are included in the NEP. Regulatory measures will be implemented to ensure the sustainable management of limited groundwater resources, including both regolith and hard-rock aquifers, as specified in the NEP guidelines. This involves a scientific evaluation of aquifers, identifying different types and their potential storage capacities, monitoring extraction rates to avoid overexploitation, and optimizing the use of shallow groundwater in conjunction with surface water sources and rainfall. Moreover, natural and artificial recharge will be utilized to mitigate the impacts of floods and droughts, supply water for household and industrial use, and control salinity during dry periods. The NEP promotes the thorough restoration of historic cascade and individual tank systems to prevent floods and droughts, provide ecological benefits to humans and wildlife, and distribute resources equitably and without dispute.

Specific management plans have become mandatory for land and water resources in environmentally sensitive areas under the current NEP to prevent pollution from industry, agricultural runoff, household waste, illegal activities such as sand and gem mining, unauthorized construction, and deforestation. Similarly, regions chosen for development initiatives must be governed under specified plans. These legally binding management plans strengthen the zoning of land and water resources in ecologically vulnerable areas. As a result, legally enforceable management plans are necessary for demarcating land and water resource zones in environmentally sensitive regions and development project areas covered by the NEP. Currently, not all critical and sensitive areas have been fully identified and demarcated. Therefore, steps to demarcate, manage, protect, and sustainably use critical and sensitive areas, such as fragile areas and conservation zones, are also considered in the NEP. When development projects, such as development corridors and metro regions, are planned, water as an associated resource will be identified and included to facilitate necessary regulatory and coordination mechanisms for its sustainable use.

To address institutional weaknesses and policy gaps, the NEP recognizes the necessity of coordination across all environmental sectors. It proposes vertical and horizontal coordination of responsibilities under a central body, such as the Ministry of Environment, to monitor progress using key performance indicators (KPIs) and ensure commitments by integrating them into annual budgets, thereby minimizing coordination inefficiencies. The NEP has discussed the difficulties in implementing environmental regulations, mainly due to capacity gaps, including those related to technology and information. As a result, a real-time monitoring and reporting system has been proposed, along with a database that includes all water sources.

The NEAP is a comprehensive action plan that involves collaborative efforts to achieve specific policy goals across multiple sectors and institutions; it includes a comprehensive coordinating and monitoring mechanism. NEAP coordinating mechanisms at all levels are considered, and the actions of all lead agencies are integrated into their annual plans. The Ministry of Environment is expected to establish a dedicated unit with the necessary capacities, serving as the NEAP Secretariat within the Ministry, to facilitate and coordinate NEAP implementation and monitoring. The Ministry of Environment is designated as the lead agency for NEAP monitoring, and other key agencies have also been specified. An online platform for implementation monitoring of the NEAP is planned to ease the task and improve transparency with access for many parties. Technical support for the NEAP committees will be provided by operational NEAP Working Groups (NWGs) for each thematic and cross-cutting area. Subnational level coordination of the NEAP is arranged through Provincial Environmental Committees (PECs), District Environmental Committees (DECs), and Divisional Environmental Committees (DivEC).

Enhancing environmental outcomes and reducing environmental pressure through reciprocity, suasive tools, and community empowerment are incorporated into the NEAP. For instance, applying community-based and suasive tools for sustainable ecosystem management, identifying community-managed conservation areas, implementing projects, and developing Bio-cultural Community Protocols (BCPs) are included. This recommended action plan is based on appropriate reviews and extensive consultation with stakeholders. It identifies two sets of actions. As stated in the plan, the Ministry of Environment or institutions affiliated with the Ministry of Environment, such as the CEA and the Geological Survey and Mines Bureau (GSMB), can directly implement actions in collaboration with other cooperating agencies. The remaining activities are expected to be implemented by other leading agencies in collaboration with the assigned ministries or line agencies.

The Role of Agricultural Policies: NAP

The NAP integrates water governance into various thematic areas, focusing on improving irrigation efficiency, promoting climate resilience, and enhancing institutional coordination. One of its key achievements is the recognition of water scarcity as a significant challenge, underscoring the importance of water governance as a policy priority.

The NAP emphasizes the importance of climate adaptation strategies, including watershed conservation, drought-resistant crop varieties, and sustainable irrigation and land-use practices. Additionally, the policy promotes disaster preparedness through agro-meteorological forecasting, enabling farmers to plan their irrigation schedules more effectively. It advocates adhering to the agriculture-related actions identified in the National Adaptation Plan for Climate Change and the Nationally Determined Contributions (NDCs). Regarding sustainable groundwater and surface water management, the policy encourages the monitoring and regulation of groundwater extraction while promoting organic farming and reduced chemical fertilizer use to prevent water contamination. It also prioritizes the rehabilitation of reservoirs and irrigation canals to ensure a reliable water supply. To enhance irrigation efficiency, the policy emphasizes participatory water management, adoption of water-saving technologies such as drip irrigation and rainwater harvesting, and regulatory measures to control excessive groundwater extraction while promoting catchment area conservation.

In terms of empowering farmers in water governance, the NAP supports participatory decision-making by encouraging farmer organizations to be actively involved in water allocation processes. It also suggests training programs on efficient irrigation techniques and climate-smart water management while proposing incentives for water conservation and the adoption of efficient irrigation methods. While these efforts encourage stakeholder engagement and knowledge sharing, stronger institutional backing is needed to ensure meaningful farmer participation in decision-making processes. To strengthen institutional coordination, the NAP proposes an IWRM approach, ensuring better collaboration among key agencies, including the Department of Irrigation, the Mahaweli Authority of Sri Lanka, and the Department of Agrarian Development. It also promotes inter-agency collaboration to streamline water allocation for agricultural purposes. However, while institutional streamlining could reduce inefficiencies, the policy does not specify accountability measures to enforce coordination effectively.

12.6 Future Directions and Policy Recommendations

Despite the considerable advancements reflected in the NEP, NEAP, and NAP, several critical gaps continue to hinder effective water governance. Addressing these gaps requires a coordinated, cross-sectoral approach that strengthens institutional accountability, promotes integrated planning, and empowers stakeholders across all levels.

12.6.1 Strengthen Policy Implementation and Legal Enforcement

While the NEP and NAP articulate comprehensive objectives—ranging from pollution control and groundwater management to climate adaptation and community engagement—implementation remains weak due to limited legal enforcement, under-resourced institutions, and ambiguous monitoring and accountability mechanisms. Future policies must prioritize the operationalization of existing frameworks by:

- Enacting enforceable legal provisions for water allocation, pollution control, and groundwater regulation.
- Establishing clear accountability mechanisms among water-related agencies with designated responsibilities and performance benchmarks.
- Allocating adequate financial and technical resources to implement, monitor, and evaluate water governance targets at both national and subnational levels.

Enhance Institutional Coordination and Data Integration: Water governance in Sri Lanka suffers from institutional fragmentation, with nearly 41 institutions and over 50 pieces of legislation governing the sector. This has led to duplication, inefficiencies, and conflicts in resource allocation. Furthermore, coordination with broader national water policies is insufficient, leading to weak integration between agricultural and national water governance frameworks. Going forward, a unified institutional mechanism is essential. This includes:

- Establishing a legally mandated National Water Commission or strengthening the role of the Ministry of Environment as a central coordinating body.
- Integrating environmental, agricultural, and irrigation information systems into a centralized water data platform to ensure transparency, reduce redundancies, and support real-time decision-making.
- Creating cross-ministerial task forces for integrated land and water use planning under the umbrella of IWRM principles.

Promote Climate-Resilient Water Management: While both the NEP and NAP recognize the impact of climate change on water availability, they lack clarity in defining adaptive responsibilities and actions. Future directions should:

- Develop localized drought and flood risk adaptation plans, linked with agro-climatic zones and river basin-level governance.
- Scale up investment in climate-smart irrigation technologies such as micro-irrigation, solar-powered pumps, and water harvesting systems, supported by targeted subsidies and financial incentives.

- Link national adaptation strategies and SDG commitments with measurable climate-resilient water governance indicators.
- **Empower Local Communities and Farmer Organizations**: Although the policies acknowledge the role of participatory governance, actual community involvement in decision-making remains superficial. Stronger institutional support for local actors is required by:
- Institutionalizing community-based water governance models through legal recognition and capacity building of farmer organizations, Water User Associations, and local councils.
- Encouraging co-management arrangements for minor irrigation and tank systems with clear rights, roles, and benefit-sharing mechanisms.
- Providing incentives for water stewardship, including training on efficient irrigation, ecosystem conservation, and water-saving cultivation techniques.

Bridge the Agricultural-Environmental Policy Gap: The NAP and NEP operate in largely parallel domains, with insufficient coordination in goals, implementation timelines, and resource planning. This has led to weak integration between agricultural water use and environmental sustainability. Future reforms should:

- Establish joint planning mechanisms between the Ministries of Agriculture and Environment to align irrigation modernization with watershed protection and pollution reduction.
- Harmonize water-related targets under the National Adaptation Plan, NDCs, and SDGs 6 and 13, ensuring cohesive reporting and monitoring frameworks.
- Mainstream ecosystem-based approaches in both agricultural and environmental water governance strategies.

Mobilize Sustainable Financing for Water Governance: The effectiveness of policy implementation is undermined by chronic underfunding and heavy reliance on donor-driven pilot projects. To ensure sustainability:

- Develop a dedicated Water Governance Fund supported through domestic allocations, green bonds, climate finance, and public–private partnerships.
- Expand Payment for Ecosystem Services (PES) mechanisms in watershed areas to link upstream conservation with downstream water users.
- Explore blended finance instruments to support the adoption of community-based water infrastructure and innovation.

12.7 Conclusion

This paper examined the critical dimensions of water governance in Sri Lanka, particularly at the intersection of environmental and agricultural policy frameworks. The analysis highlights that, although substantial policy advancements have been made through instruments such as the NEP, NEAP, NAP, and the NDCs, persistent challenges continue to undermine effective water governance. Key governance deficiencies—such as institutional fragmentation, limited policy coherence, inefficient irrigation practices, pollution, and catchment degradation—pose significant threats to sustainable water management and broader socio-environmental goals.

While national policies increasingly acknowledge the importance of IWRM and climate resilience, a considerable gap remains between policy ambition and implementation. Weak enforcement mechanisms, insufficient cross-sectoral coordination, chronic underinvestment, and the lack of participatory governance structures continue to hinder the realization of sustainable water governance outcomes. Furthermore, the disconnection between agricultural productivity objectives and environmental sustainability considerations exacerbates tensions over water allocation, ecosystem degradation, and long-term water security.

Future reforms must prioritize the operationalization of integrated approaches that bridge the agricultural and environmental domains, supported by strengthened institutional mandates, robust enforcement frameworks, and inclusive governance mechanisms. The development of localized, climate-resilient water management strategies, enhanced stakeholder engagement, and the mobilization of sustainable financing are essential to overcoming current challenges. Without such coordinated and systemic interventions, Sri Lanka risks further exacerbating water scarcity, environmental degradation, and socioeconomic inequalities. Strengthening water governance is thus imperative not only for achieving Sustainable Development Goals (SDGs) 6 and 13 but also for safeguarding the resilience of agricultural systems, ecosystems, and communities in the face of emerging and future risks.

References

Chandrasekara, S. S. K., Chandrasekara, S. K., Gamini, P. H. S., Obeysekera, J., Manthrithilake, H., Kwon, H. H., & Vithanage, M. (2021). A review on water governance in Sri Lanka: The lessons learnt for future water policy formulation. *Water Policy, 23*(2), 255–273. https://doi.org/10.2166/wp.2021.152

Fernando, R., Ratnasooriya, H., Bamunawala, J., Sirisena, J., Nipuni Odara, M. G., Gunawardhana, L., & Rajapakse, L. (2024). Assessing climate-change-driven impacts on water scarcity: A case study of low-flow dynamics in the Lower Kalu River Basin, Sri Lanka. *Water, 16*(10), 1317.

Food and Agriculture Organization of the United Nations. (n.d.). Water governance. FAO. Retrieved April 2, 2025, from https://www.fao.org/land-water/water/watergovernance/en/

Geekiyanage, N., Vithanage, M., Wijesekara, H., & Pushpakumara, G. (2015). State of the environment, environmental challenges and governance in Sri Lanka. In *Environmental challenges and governance: Diverse perspectives from Asia* (pp. 106–122). Routledge. https://doi.org/10.4324/9781315717081

Herath, H., & Patabandi, K. (2023). The environmental impacts of unauthorized activities: A case of the Maragala mountain range environment protection area in Sri Lanka. *Sri Lanka Journal of Social Sciences and Humanities, 3*(2).

Hettithanthri, O., Sandanayake, S., Magana-Arachchi, D., Wanigatunge, R., Rajapaksha, A. U., Zeng, X., Shi, Q., Guo, H., & Vithanage, M. (2021). Risk factors for endemic chronic kidney

disease of unknown etiology in Sri Lanka: A retrospect of water security in the dry zone. *Science of the Total Environment, 795*, Article 148839.
Hewawasam, T. (2010). Effect of land use in the upper Mahaweli catchment area on erosion, landslides, and siltation in hydropower reservoirs of Sri Lanka. *Journal of the National Science Foundation of Sri Lanka, 38*(1), 3–8.
Hewawasam, V., & Matsui, K. (2022). Traditional water governance practices for flood mitigation in ancient Sri Lanka. *Climate, 10*(5), 69.
IUCN (International Union for Conservation of Nature). (2018). *North-South learning and sharing: Ancient water management systems in Anuradhapura.* https://www.iucn.org/news/sri-lanka/201802/north-south-learning-and-sharing-ancient-water-management-systems-anuradhapura
Jayakody, P., Amarasinghe, U. A., & Mutuwatta, L. (2018). *Water resources management in Mahaweli and adjoining river basins of Sri Lanka.* Asian Development Bank. https://events.development.asia/system/files/materials/2013/12/201312-water-resources-management-mahaweli-and-adjoining-river-basins-sri-lanka.pdf
Jayasiri, M., Dayawansa, N. D. K., & Yadav, S. (2023). Assessing the roles of farmer organizations for effective agricultural water management in Sri Lanka. *Agricultural Systems, 205*, Article 103587.
Jayasiri, M. M. J. G. C. N., Yadav, S., Dayawansa, N. D. K., & Ingold, K. (2022). Managing agricultural water resources: Addressing the complexity of innovation, social perspectives, and water governance in Sri Lanka. *Irrigation and Drainage, 71*, 71–85.
Jayasumana, C., Gunatilake, S., & Siribaddana, S. (2015). Simultaneous exposure to multiple heavy metals and glyphosate may contribute to Sri Lankan agricultural nephropathy. *BMC Nephrology, 11*(16), 103. https://doi.org/10.1186/s12882-015-0109-2
Jayawardena, M., Perera, N., & Gunasekara, A. (2021). A review on freshwater pollution in Sri Lanka: Current status and future outlook. *Environmental Challenges, 5*, Article 100278. https://doi.org/10.1016/j.envc.2021.100278
Jayawardhana, P., & Udawattha, C. D. (2022). Environmental repercussions of gem mining in Sri Lanka. *Journal of Real Estate Studies, 19*(2), 23–43.
Kekulandala, B., Jacobs, B., & Cunningham, R. (2024). The agricultural decision process in a small tank cascade system in Sri Lanka: Diagnosing options for adaptive governance. *Discover Agriculture, 2*(1), 36.
Khaniya, B., Gunathilake, M. B., & Rathnayake, U. (2021). Ecosystem-based adaptation for the impact of climate change and variation in the water management sector of Sri Lanka. *Mathematical Problems in Engineering, 2021*, 8821329.
Khaniya, B., Jayanayaka, I., Jayasanka, P., & Rathnayake, U. (2019). Rainfall trend analysis in Uma Oya basin, Sri Lanka, and future water scarcity problems in the perspective of climate variability. *Advances in Meteorology, 2019*, 3636158.
Kites Global. (n.d.). *Irrigation for coconut cultivation.* Retrieved April 2, 2025, from https://demo-smr.kites.global/irrigation-for-coconut-cultivation/
Kumarasiri, A., Udayakumara, E. P. N., & Jayawardana, J. (2022). Impacts of soil erosion and forest quality on water quality in Samanalawewa watershed, Sri Lanka. *Modeling Earth Systems and Environment, 8*(1), 529–544.
Lee, H., Calvin, K., Dasgupta, D., Krinner, G., Mukherji, A., Thorne, P.,Trisos, C., Romero, J., Aldunce, P., & Barret, K. (2023). IPCC Climate Change 2023: Synthesis report, summary for policymakers. In *Contribution of Working Groups I, II and III to the Sixth Assessment Report of the Intergovernmental Panel on Climate Change.* IPCC.
Mahaweli Authority of Sri Lanka. (n.d.). *Master Plan.* https://mahaweli.gov.lk/master%20plan.html
Manage, P., Mahagamage, Y., Ajward, R., Amaratunge, S., & Amaratunge, V. (2020). The need for proper management leading to the sustainability of the Kelani River and its lower basin. *Journal of Water and Land Development, 47*, 10–15. https://doi.org/10.24425/jwld.2020.135026
Mathiventhan, T., Gorman, D., & Jayasingam, T. (2022). Sea-level rise, coastal salinity and vegetation changes in Sri Lanka. *Philosophical Transactions of the Royal Society A, 380*(2221), 20210142.

Ministry of Environment. (2021). *National Environment Policy—2021*. Ministry of Environment. https://env.gov.lk/web/images/pdf/Policies/National_Environment_Policy_2021.pdf

Ministry of Environment. (2022). *National Environmental Action Plan 2022–2030*. Ministry of Environment.

Mongabay. (2022). *Report sums up a wealth of Sri Lanka's biodiversity and the threats it faces*. https://news.mongabay.com/2022/06/report-sums-up-wealth-of-sri-lankas-biodiversity-and-the-threats-it-faces

Nuwanka, M. W. R., & Gunathilaka, M. (2023). Complexities of water pollution: A review of surface water contamination in Sri Lanka. *International Journal of Environment, Engineering and Education, 5*(2), 72–78.

Perera, M., Ranasinghe, T., Piyadasa, R. U. K., & Jayasinghe, G. Y. (2018). Risk of seawater intrusion on coastal community of Bentota River Basin Sri Lanka. *Procedia Engineering, 212*, 699–706.

Rural Water Supply Network. (2015). *Sustainable water resources management in Sri Lanka: Present situation and way forward*. https://rwsn.blog/2015/02/20/sustainable-water-resources-management-in-sri-lanka-present-situation-and-way-forward

Satoyama Initiative. (2010). *Sri Lanka: Tank irrigation farming in dry zones*. https://satoyamainitiative.org/case_studies/sri-lanka-tank-irrigation-farming-in-dry-zones/

ScienceDirect. (2021). *Freshwater pollution trends in Sri Lanka*. https://www.sciencedirect.com/

Shamsudduha, M., Lee, J., Joseph, G., Bahuguna, A., Wijesundera, S. S., Nair, S. S., Hoo, Y. R., Wang, Q., & Ayling, S. C. E. (2025). Assessing the water quality hazard and challenges to achieving the freshwater goal in Sri Lanka. *Scientific Reports, 15*(1), 10187.

Shunglu, R., Köpke, S., Kanoi, L., Nissanka, T. S., Withanachchi, C. R., Gamage, D. U., Dissanayake, H. R., Kibaroglu, A., Ünver, O., & Withanachchi, S. S. (2022). Barriers in participative water governance: A critical analysis of community development approaches. *Water, 14*(5). https://doi.org/10.3390/w14050762

Sriskandarajah, N., & Sivapalan, M. (2023). Water security, participatory governance, and dialogue spaces for change in Northern Sri Lanka. *World Water Policy, 9*(3), 405–413.

United Nations Sri Lanka. (2023). *Integrated water resources development: A way forward for Sri Lanka to tackle the climate crisis*. Retrieved October 8, 2025 from https://srilanka.un.org/en/248878-integrated-water-resources-development-way-forward-sri-lanka-tackle-climate-crisis

UNR Sri Lanka. (2015).*Drivers of deforestation and forest degradation in Sri Lanka: Assessment of key policies and measures*. https://www.researchgate.net/profile/Nirmalie-Pallewatta/publication/314197372_Drivers_of_Deforestation_and_Forest_Degradation_in_Sri_Lanka_Assessment_of_Key_Policies_and_Measures/links/58b9790892851c471d4a8308/Drivers-of-Deforestation-and-Forest-Degradation-in-Sri-Lanka-Assessment-of-Key-Policies-and-Measures.pdf

Dr. Manoj Thibbotuwawa is a Research Fellow at the Institute of Policy Studies of Sri Lanka (IPS) with research interests in agriculture, agribusiness value chains, food security, climate change and environmental and natural resource economics. He has more than 20 years of research experience at IPS. He has collaborated and served as a consultant to several international organisations such as the World Food Programme (WFP), the Food and Agriculture Organization (FAO), the International Fund for Agricultural Development (IFAD), the International Food Policy Research Organization (IFPRI), the Australian Centre for International Agriculture Research (ACIAR), the Japan International Cooperation Agency (JICA), the United States Agency for International Development (USAID), the European Union (EU), and the World Bank. Manoj holds a BSc (Agriculture) with Honours from the University of Peradeniya, an MSc (Agricultural Economics) from the Post-Graduate Institute of Agriculture at the University of Peradeniya, and a PhD from the University of Western Australia. He has also obtained a Post-Graduate Diploma in Modelling and Accounting for Sustainable Development from the International Institute of Social Studies (ISS) in The Hague. Manoj is a recipient of the prestigious Nuffic Fellowship from the Government of the Netherlands and the Endeavour Award from the Government of Australia

Dr. Erandathie Pathiraja is a Research Fellow at the Institute of Policy Studies of Sri Lanka (IPS) with research interests in the environment and natural resource policies, climate change, sustainable energy, and analysis of industries and markets. She served as a member of the teaching panel at the Postgraduate Institute of Agriculture, University of Peradeniya, and as a visiting lecturer to the Faculty of Graduate Studies, University of Colombo. Her previous engagements are as a senior agricultural economist and an agricultural economist at the Coconut Research Institute of Sri Lanka, and research projects funded by the World Bank and the ADB. She has also served as the Secretary of the Sri Lanka Agricultural Economics Association. Erandathie holds a BSc in Agriculture from the University of Peradeniya, an MPhil in Agricultural Economics from the Postgraduate Institute of Agriculture, and a PhD in Agricultural Economics from the University of Melbourne, Australia.

Chapter 13
Water Governance Challenges in Land Use Policy

Thilani P. Munaweerage

Abstract As pressure on land and its interactions with other natural resources continue to grow, there is an urgent need for scientific land use planning to foster a sustainable balance among land, water, and the environment. This chapter examines land use changes in Sri Lanka from historical times to the present, highlighting key legislation and institutions involved in land use planning, with particular emphasis on soil conservation. To address the challenges of water governance in the context of land use, Sri Lanka must undertake a series of governance reforms.

- Prioritize the formalization of land titles and integrate traditional landholding systems into the official legal framework to reduce land conflicts and improve access to water resources.
- Establish land use regulations that are linked to clearly defined property rights, encouraging efficient resource use while minimizing conflicts and environmental degradation.
- Enhance institutional coordination among water, land, agriculture, and environmental sectors to address overlapping mandates and conflicting objectives.
- Integrate customary land rights, particularly in rural areas, into formal legal frameworks to promote inclusivity and reduce disputes.
- Develop robust climate forecasting systems and ensure access to actionable information to support informed irrigation planning, crop selection, and land-use decision-making.
- Clarify governance roles, strengthen stakeholder coordination, and establish effective accountability mechanisms for sustainable resource management.

Keywords Climate change · Governance · Land use planning · Soil conservation · Land tenure

T. P. Munaweerage (✉)
Griffith University, Gold Coast, Australia
e-mail: thilanimunaweera@gmail.com

U. A. Amarasinghe et al. (eds.), *Water Governance for Climate Change Adaptation in Sri Lanka*, Global Issues in Water Policy 13,
https://doi.org/10.1007/978-3-032-04950-6_13

13.1 Introduction

Land, encompassing all other natural resources, is fundamental to meeting the needs of food, water, and multiple ecosystem services (Bai et al., 2019; Fu et al., 2017; Runting et al., 2017). In countries like Sri Lanka, where a significant portion of the population resides in rural areas and depends on agriculture for their livelihood, scientific land use planning is essential for addressing human-environmental challenges. These challenges have become increasingly complex with demographic shifts and climate change. Poor land-use practices can have a negative impact on human health, ecosystems, and the environment (Wickramasinghe et al., 2023; Khaniya et al., 2021; Tsuchida & Takeda, 2021).

Government regulations that directly or indirectly determine land use can limit the activities undertaken on a given land or offer incentives that influence land use choices. These decisions influence industrial investment, cropping patterns, or whether to leave the land fallow (Feder, 1988). Historically, Sri Lanka's land use patterns have been shaped by various socio-economic and environmental factors, from ancient systems of communal land management to modern frameworks influenced by colonial legacies and contemporary policy initiatives (Herath, 2006). Over time, these patterns have undergone significant transformation, with notable impacts on soil quality, water resources, and ecosystem health. Approximately 40 active land laws regulate land use in Sri Lanka, and these laws are implemented by various ministries and institutions (Weerahewa et al., 2021).

Water governance is crucial in shaping sustainable agricultural practices, particularly in regions where water availability and quality significantly impact land-use planning. In the agricultural sector, land and water use decisions are crucial to achieving food security, enhancing foreign exchange earnings, improving living standards, and promoting economic development. However, the sector's vulnerability to global climate change poses significant challenges, particularly in developing countries, where agriculture is a cornerstone of food security and economic progress (Kalkuhl & Wenz, 2020; Jagustović et al., 2019; Nicholson et al., 2020). In South Asia, these challenges are magnified by the region's extensive geophysical, climatic, and ecological diversity (Pound et al., 2018), compounded by systemic issues such as poverty, malnutrition, population pressure, weak institutional frameworks, and the predominance of small-scale subsistence farming. Moreover, underdeveloped markets and limited technological adoption exacerbate the region's susceptibility to the impacts of climate change (World Bank, 2013).

In this context, the governance of water resources is integral to effective land use planning. Sustainable land and water management is essential for mitigating the degradation of natural resources, which threatens the livelihoods of millions of rural households. With finite water and land resources, future agricultural strategies must align productivity with sustainability, emphasizing integrated water governance approaches to address the complex challenges of land use planning.

This chapter examines Sri Lanka's historical land use changes, related laws and institutions, and identifies policy gaps for sustainable land use adaptation to climate change. Furthermore, the chapter examines the multifaceted challenges of water governance in the context of land use planning, emphasizing the interplay between resource management, agricultural development, and climate resilience.

The following section in this chapter reports on the changes in land use patterns in Sri Lanka over time, the key drivers of these changes, and their impact on the environment and society. The third section describes the legislation and institutions related to land use planning, with a focus on water management and soil conservation. Section four presents the challenges for water governance that contribute to sustainable land use, and the final section offers a way forward for implementation.

13.2 Land Use Changes in Sri Lanka

Land use patterns depend on socio-economic and cultural factors and can change over time and by location. In Sri Lanka, a country steeped in a rich historical legacy, land use patterns have undergone substantial transformations. Over time, distinctive land use, land tenure, and regulatory systems have aligned with changing objectives and needs, adapting to diverse circumstances and priorities. In the early stages, when plenty of land was available for a smaller population, individuals and communities determined land use based on their self-sufficiency requirements. This land use pattern had minimal environmental impact due to limited human interference. With population growth, the demand for land increased in all sectors, including housing, agriculture, industries, and other developments.

In ancient times, with a predominantly agricultural population, land use patterns were primarily agricultural, and people lived in self-sufficient communities with minimal interaction with outsiders. As noted in the *Mahawansha*, the earliest agricultural settlements emerged in the river valleys of the northern and southeastern plains, with rain-fed paddy as the main crop. Traditional farming practices and ways of life were intricately linked to water availability, with two primary farming approaches: "wet" and "dry," tailored to the available water resources. The lowland paddy was cultivated only in one season and left fallow in the dry season.

With the growing demand for food from the increasing population, Sri Lankan kings recognized the need to expand agricultural production into the dry season. This led to the construction of thousands of small irrigation tanks throughout the country's dry zone, building upon advanced irrigation technologies that showcased the rich history of Sri Lankan water management (Weerahewa et al., 2021). These cultivation practices kept the highland regions with abundant rainfall and natural montane forests as catchments for the main rivers, while rice and other field crops were cultivated on the lowland plains. Furthermore, spices and other upland crops were grown in mixed gardens in some mid-country areas, where plant canopies of varying heights absorbed solar energy more efficiently while limiting the effects of raindrops on soil erosion (Mapa et al., 2002). During this historical era, the Sri Lankan agricultural system demonstrated a remarkable level of sustainability, largely due to the practice of shifting cultivation, which played a pivotal role in maintaining soil health and land productivity (Irangani, 2020).

Subsistence agriculture remained the backbone of the country's economy until colonial rulers brought significant changes in the nineteenth century. Since 1505, the country has been ruled by three European powers: the Portuguese, Dutch, and British. These colonial rulers established a trade-based economy, with land use practices influenced by market-driven policies (Weerahewa et al., 2021). The Portuguese and Dutch controlled the coastal regions and made only a few notable alterations to the traditional land management system; however, substantial changes in land use and tenure were introduced during the British colonial period. The British neglected subsistence agriculture and focused on commercial crops for the export market, particularly plantation crops.

Unlike traditional subsistence agriculture, which prevailed in the country, plantation agriculture is characterized by monocropped large units of operation, often referred to as estates, and involves the intensive use of wage labor, focusing on export-oriented production. With the Dutch starting coffee cultivation in 1722, perennial crops such as coffee, tea, rubber, and coconut were cultivated on these plantations, which gained momentum in the nineteenth and early twentieth centuries. This shift in land use significantly altered the overall land use pattern, particularly in the upland areas, bringing substantial economic, social, and cultural changes within the country (Mapa et al., 2002).

The expansion of plantations, primarily through the clearing of forested areas, has incurred significant environmental costs, impacting the watersheds of major rivers. Soil erosion on sloping land increased siltation in primary rivers and the reservoirs they nourish (Nayakekorale, 2020). Moreover, cultivating plantation crops demanded a substantial workforce, but local Sri Lankans were hesitant to work in these areas. As a result, plantation owners brought in labor from India and established residential settlements within the plantations, often at the expense of clearing additional sections of natural reserves (Gunasekara, 2020).

British regulations altered traditional land use, restricted farmers' rights, and allocated large land areas to European investors for plantations. This shift prioritized export crops over domestic food production, which remained stagnant, meeting only household consumption needs with a small surplus for local markets (Samaratunga, 2012). Cheap food imports affected prices and local markets. Although plantation crops helped fuel GDP growth, domestic food production declined and had a diminished role by the time Sri Lanka gained independence in 1948 (Weerahewa, 2017).

After years of neglecting traditional agriculture, there was a need to enhance food production and increase land availability to meet the growing population. On a modest scale, driven by the Governor's interest, initial investments restored large, abandoned reservoirs in the Dry Zone (Shand, 2002). In the nineteenth century, broader colonial reforms were introduced to replace traditional systems viewed as obstacles to the development of a free-market economy and centralized governance. One significant reform was the abolition of *Rajakariya*. This traditional feudal labor system required individuals to provide labor or services to the king in exchange for the right to use land. *Rajakariya* was deeply ingrained in Sri Lanka's socio-economic structure, primarily linked to land ownership and agricultural practices. Its abolition disrupted traditional land management and water governance systems, particularly in farming areas that relied on communal labor for maintaining irrigation systems, such as tanks (reservoirs) and canals (Casinader et al., 2018). Consequently, the government supported the restoration of numerous village tanks. Major irrigation projects started in the 1920s and gained prominence by the 1930s. This continued until independence in 1948, with key reservoirs such as Minneriya, Parakrama Samudra, and Elahera being revitalized (Gunawardena, 1981).

The initial restoration efforts of reservoir rehabilitation focused on minimal alterations to the distribution systems, primarily benefiting major schemes and establishing new settlements. In 1970, the government initiated the Mahaweli Ganga Development Programme for irrigation and power. After a slow start, the project was accelerated in 1977 to address employment generation, food production, and power needs. With the extensive transformation of irrigation and agriculture in the dry zone, the government initiated resettlement programs to alleviate pressure in the wet zone (Abeywardana et al., 2018). This shift, however, altered the country's land use and introduced substantial environmental and social challenges that threaten the sustainable use of natural resources (Amarasekara, 1992).

Sri Lanka's total area of 6.57 million ha is primarily state-owned (83%). Agricultural land covers over three million hectares (45.5%), with 1.38 million hectares of state land farmed by private farmers under various tenure arrangements (Weerahewa et al., 2021). Key land use categories include home gardens (18%), plantation crops such as tea, rubber, and coconut (10%), and paddy fields (15%). Forest lands cover the largest share (38%); water bodies, marshlands, streams, and unutilized land make up 10%, while nonagricultural lands, consisting of built-up areas, constitute 1.3% of the total area (Table 13.1).

Table 13.2 shows the land-man ratio, indicating that the land available for a person steadily declined over time. This trend highlights the importance of effective land management in supporting a growing population, enhancing agricultural productivity, and ensuring food security. The arable land indicator, measuring arable land available per person, fluctuated between 0.075 ha/person in 1963 and 0.046 ha/person in 2004. By 2018, the indicator had slightly improved to 0.063 ha/person (World Bank, 2023). This indicator also underscores the importance of effective land management in achieving sustainable agricultural development and ensuring food security.

Legislation and Institutions Related to Land Use Planning in Sri Lanka

In Sri Lanka, oversight of agricultural lands involves about 40 operational laws, implemented by different departments and line agencies under the purview of the Ministries of Agriculture, Lands and Land Development, Mahaweli, and Irrigation (Weerahewa et al., 2021). Individuals in government-administered land settlement schemes receive rights to land use and transfer through land grants and permits. Informal arrangements of ownership and access are prevalent in rural areas, bypassing official regulations, including traditional rotational tenure, share tenancy,

Table 13.1 Land use category extent (ha)

Land Use	Extent (ha)	Percentage (%)
Nonagricultural lands- Built-up area	85,094	1.30
Agricultural lands		
Home gardens	1,192,925	18.18
Plantation crops		
Tea	228,118	3.48
Abandoned tea	5,855	0.09
Rubber	186,334	2.84
Abandoned rubber	2,359	0.04
Coconut	208,368	3.18
Paddy	983,550	14.99
Abandoned paddy	40,356	0.62
Palmyrah	15,833	0.24
Oilpalm	8,781	0.13
Cinnamon	21,015	0.32
Other Field Crops	57,041	0.87
Mixed tree and other perennials	82,228	1.25
Sparsely used croplands/chena	307,311	4.68
Forest lands		
Forest	1,848,548	28.17
Forest plantation	68,103	1.04
Scrub lands	599,787	9.14
Water bodies	374,553	5.71
Other (marsh, mangroves and streams, etc.)	85,442	1.30
Balance area (rocky and sandy area, unutilized and bare lands, etc.)	159,399	2.43
Total	6,561,000	100.0

Source Land use and policy planning department, Ministry of Lands, 2018

Table 13.2 Change in per capita land availability

Year	Population (million)	Land-man ratio (ha/per person)
1871	2.4	2.73
1901	3.5	1.88
1953	8.1	0.81
1986	16.5	0.40
2000	20.0	0.33
2022	22.2	0.30

absentee landlordism, and ownership rights through long-term use (Witharana, 2014). Encroachment onto state land, particularly in the Dry Zone, and unofficial land transactions are widespread despite government regulations on land size. Regulations exist to manage agricultural land, especially paddy, plantation crops, and forestry.

The subsequent section summarizes agricultural land regulations, focusing on ownership, land size, and land utilization. The details are in Herath (2006), Maddumabandara (1998), Marawila (2007), Gunasekara (2020), Weerahewa et al. (2021), and World Bank (2018).

Key legislation and institutions affecting land use, tenure, and size:

- The Crown Land Encroachment Ordinance (CLEO) of 1840 acquired land for plantation agriculture, particularly coffee, tea, and cash crops, by abolishing the traditional system of service tenure, resulting in the Crown owning approximately 90% of Sri Lanka's land.
- The Registration of the Temple Lands Ordinance of 1856 aimed to regulate temple lands, aligning them with British colonial economic interests.
- The Waste Lands Ordinance of 1897 facilitated the acquisition of lands deemed "unused" for plantation expansion.
- The Partition (or Estate) Ordinance of 1863 was implemented to standardize land transactions and ownership in favor of British economic activities.
- The Land Surveys Ordinance of 1863 introduced systematic surveying for land consolidation and plantation development.
- The Services Tenure Ordinance of 1870 formalized the elimination of traditional land-use rights.
- The Land Development Ordinance (LDO) of 1935 marked a shift towards government intervention in land allocation and established the Land Commissioner's Department to oversee Crown land use.
- The Irrigation Land Ordinance (1946, amended in 1951 and 1968) addressed irrigation requirements for alienated lands in the dry zone.
- The Land Acquisition Act of 1950 regulates land acquisition for public purposes.
- The Paddy Lands Act of 1958 improved tenure security for tenant farmers and regulated rental payments.
- The Agrarian Services Act of 1979 replaced the Agricultural Productivity Law and established Agrarian Services Committees to oversee the management of agricultural land.
- The National Environmental Act No. 47 of 1980 establishes the legal framework for environmental protection, including provisions for sustainable water resource management, conservation, and the prevention and control of water pollution.
- The Irrigation Ordinance No. 22 of 1900 governs the management and regulation of water resources for irrigation purposes, covering aspects like water allocation and rights, and the construction and maintenance of irrigation infrastructure.
- The Mahaweli Authority of Sri Lanka Act No. 23 of 1979 established the Mahaweli Authority responsible for planning, implementing, and managing water resources development projects, particularly in the Mahaweli River basin.

- The Coast Conservation and Coastal Resource Management Act No. 57 of 1981 focuses on coastal land use, management, and regulation, including the protection and conservation of coastal water resources.
- The Land Development Ordinance No. 19 of 1935 provides regulation and management of land development, encompassing planning, land conservation, and the protection of water resources.

Other key legislations and institutions (land use planning and soil conservation):

- The Town and Country Planning Ordinance No. 49 of 1946 established the legal framework for land use planning, empowering local planning authorities to prepare and implement development plans, control land subdivisions, and regulate land use activities.
- The National Physical Planning Policy and Implementation Framework outlined guidelines and principles for national physical planning, including land use planning and management, serving as a guiding framework for local planning authorities.
- The Soil Conservation Act No. 3 of 1951 aimed to prevent soil erosion and promote sustainable land management practices by establishing soil conservation areas, regulating land use in those areas, and implementing soil conservation measures.
- The Department of Town and Country Planning (DTCP) is responsible for implementing land use planning policies and regulations, preparing development plans, providing technical assistance to local authorities, and overseeing land use activities in designated planning areas.
- The Department of Agrarian Development (DAD) promotes sustainable land use and soil conservation in agricultural areas by implementing programs and projects related to land consolidation, soil conservation, and land use planning for agricultural purposes.
- The Forest Department plays a crucial role in the conservation and management of forest lands, protecting soil resources, as forests are essential for soil conservation and erosion prevention.
- The Central Environmental Authority (CEA) is responsible for monitoring and regulating environmental activities and overseeing the implementation of soil conservation measures, particularly in areas where land use activities can impact soil quality and erosion.

13.3 Agricultural Land Management Challenges in Sri Lanka

Agricultural land management in Sri Lanka faces several key challenges, including limited arable land, land fragmentation, land degradation, issues with land markets and ownership, competing demands from other sectors, water scarcity, climate change impacts, outdated farming practices, and the need for sustainable agricultural policies.

Limited Availability of and Competing Demand for Arable Lands

Urbanization, industrial expansion of agricultural land, and the prolonged conflicts in the North and East have been significant obstacles to food production. Despite these issues, agricultural land accounts for nearly half of the country's land area (Makumbura et al., 2022) and employs about one-third of its workforce; it contributes less than 10 percent of the country's total GDP (Central Bank of Sri Lanka, 2022). This underscores the pressure to convert agricultural land into more productive, high-income-generating sectors. Hence, effective land use management and productivity enhancement are imperative to fulfill current and future food demands and address the competing demands of other sectors.

Land Market and Ownership

Due to structural shortcomings and policy gaps, Sri Lanka faces several challenges in integrating water and land use policies. Weak land title security and the absence of an integrated land information system are correlated with increased risks in land transactions, which in turn lower the level of efficient land use, resulting in low-productivity resource management (World Bank, 2018). The large fraction of land owned by the state limits avenues for private sector participation and creates barriers to the operationalization of the land market (Bradley et al., 2016), which, in the long run, impacts the enhancement of agricultural productivity and water resource management.

The high incidence of land fragmentation, combined with existing traditional tenure systems, has an adverse impact on irrigation efficiency and exacerbates the water governance situation, ultimately facilitating land abandonment (Weerawardena and College, 1971; Moore & Wickramasinghe, 1978; Witharana, 2014; Munaweerage et al., 2021). Similarly, the delays associated with converting land permits to deeds through resettlement initiatives have led to the emergence of informal land markets, rendering the coherent planning of land and water resources more complex. Regulatory amendments aimed at controlling land use and mitigating fragmentation have, in most instances, become counterproductive, resulting in poor resource use that leads to depletion of water resources (Paranage, 2018; Bamunuarachchi, 2017). This highlights the importance of prioritizing integrated land and water governance to address these issues.

Land Fragmentation

Land fragmentation poses a challenge to the introduction of efficient farming practices and modern agricultural techniques, thereby hindering overall productivity improvements (Wickramaarachchi & Weerahewa, 2016). It hampers economies of scale, access to resources, and efficient infrastructure development. Additionally, it leads to difficulties in implementing cohesive agricultural policies and cooperative initiatives, hindering collective efforts aimed at sustainable land use and productivity enhancement (Table 13.3).

Between 1982 and 2014, Sri Lanka experienced significant growth in the number and extent of agricultural holdings, although land extent patterns varied between 2002 and 2014 (Table 13.3). Notably, there was a substantial increase in holdings of less than or equal to 1/4 of an acre, known as marginal holdings, attributed to heightened urbanization, which causes land fragmentation. Agricultural land extent increased by

Table 13.3 Agricultural holdings and extent by census of agriculture

Census year	Number of holdings (000)	Land extent (000 acres)
1982	1800	4877
2002	3253	4797
2014	4353	5643

Source Department of census and statistics

17.6 percent, primarily due to resettlement efforts following internal conflicts in the Northern and Eastern provinces. The high rate of land fragmentation has occurred across all districts in the country, but it is more pronounced in terms of absolute numbers in the districts of the Wet Zone (Department of Census & Statistics, 2018). In the wet zone districts, high population densities lead to a significant population-to-land ratio, particularly in agricultural areas. This scenario often results in fragmented land holdings, with individuals working on small plots using traditional farming methods. Unfortunately, these conditions result in limited gains and subsistence-level yields, highlighting the challenges faced by farmers in achieving sustainable livelihoods and substantial agricultural productivity in this densely populated region.

Land Degradation

One of the primary reasons for land degradation is the clearing of forest cover for alternative land uses, as well as the inappropriate land-use practices of humans. The Land Degradation Neutrality Target Setting Technical Guide, developed by the Ministry of Mahaweli Development and Environment (2017), categorizes land degradation into five major land use types: croplands, grazing lands, forests, mixed land use, and non-vegetated artificial land uses, such as mining, human settlements, and water infrastructure (Fig. 13.1). Each of these land use categories is susceptible to specific forms of land degradation, including water-induced soil erosion, wind-induced soil erosion, chemical soil deterioration, water degradation, and biological degradation.

Over time, the continuous practice of monoculture has raised significant concerns about soil quality and fertility. Modern agricultural approaches have failed to provide viable alternatives to the dominant monocropping model, which is heavily reliant on chemical inputs. More than half of Sri Lanka's agricultural lands have been degraded, as evidenced by decreased levels of organic carbon reserves, nutrient content, and increased soil acidity (Dandeniya and Dharmakeerthi 2020). Extensive portions of soil undergo degradation due to both commercial agricultural practices and natural occurrences, such as rainfall and land topography. One significant contributor to land degradation is water-induced soil erosion, exacerbated by deforestation and cultivation at higher altitudes. Over the last century, approximately 30 cm of topsoil has been lost in the mid- and upland regions (Jayakody, 2019).

Producing more food from limited land was challenging, leading to a decline in traditional cultivation methods in agricultural societies. Traditional methods of cultivation, such as shifting cultivation and long fallow periods, require a significant amount of land and time, and are not suitable for large-scale agriculture. Modern

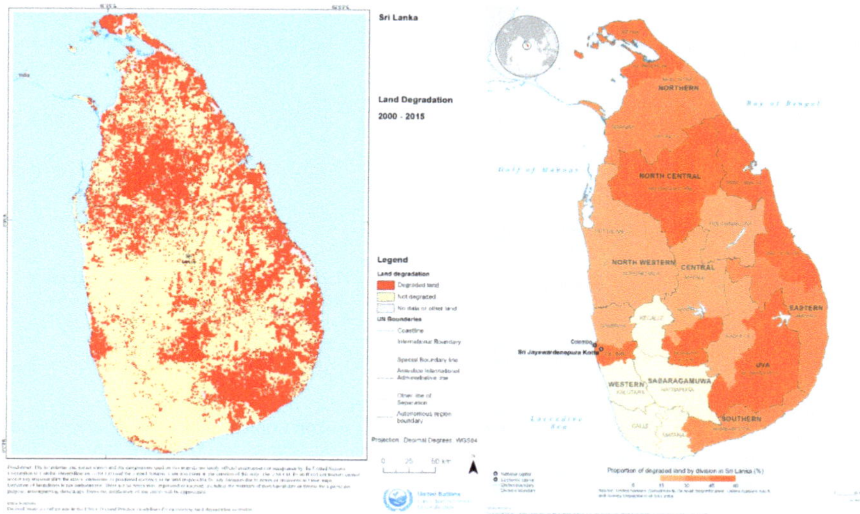

Fig. 13.1 Map from UNCCD showing the raster data on land degradation in Sri Lanka

agricultural practices, including the use of synthetic fertilizers and pesticides, have enabled farmers to produce more food from less land. However, these practices can have negative environmental impacts, such as water pollution and soil degradation. According to the Agricultural Household Survey conducted in 2016/2017 by the Department of Census and Statistics, the adoption of soil erosion control methods remains relatively low among farmers in Sri Lanka. Despite the importance of preserving soil health and productivity, many farmers have yet to fully embrace effective erosion control techniques. The choice of erosion control methods depends on factors such as the type of crop being cultivated and the land's geographical location. Two standard and effective erosion control methods are terraces and erosion control bunds (Fig. 13.2), which are widely used by agricultural households. Factors such as limited awareness, resource constraints, and traditional farming practices contribute to this low adoption rate, underscoring the need for enhanced education and support to promote sustainable land management.

The primary causes of soil degradation in the nation stem from a lack of knowledge among land users, unfavorable terms of land tenure, poverty-driven subsistence farming, land pressure, unwelcome political interventions, inadequate institutional structure, ineffective enforcement measures, conflicting economic and developmental policies, and insufficient attention to environmental issues. Consequently, farmers and other land managers adopt improper and unsustainable land-use and management approaches, leading to the degradation of land resources (Nayakekorale, 2020).

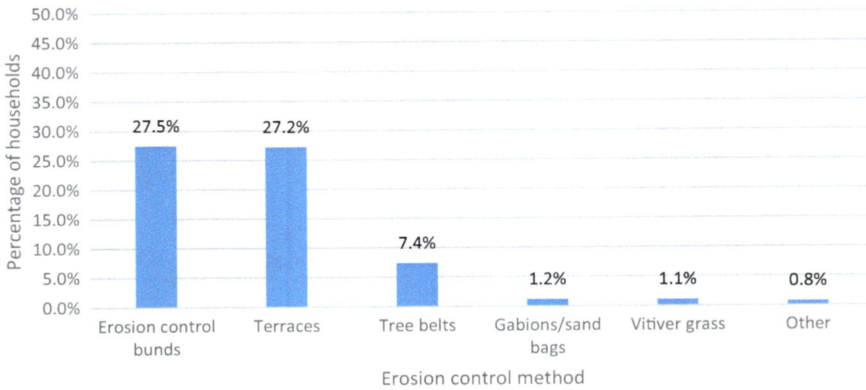

Fig. 13.2 Percentages of agricultural households under erosion control methods. *Source* Department of census and statistics

Law Enforcement and Institutional Arrangements

In the legal realm, a comprehensive set of ordinances has been established to encompass nearly all facets of land management activities, aiming to prevent land degradation effectively. However, challenges arise in the implementation of these regulations, as various acts related to land resource use, such as the Land Development Ordinance and Soil Conservation Act, face weak implementation or are not implemented. Furthermore, the long-pending enactment of the National Land Use Planning Act, drafted in 2010, and the absence of a national-level land use plan agreed upon at the national level further exacerbate the situation. Notably, land use planning is not adequately factored into decision-making processes regarding development activities. Laws designed to streamline development and allow for land use conversions may inadvertently result in adverse outcomes such as land degradation (Weerahewa et al., 2021; Nayakekorale, 2020).

On the institutional front, a solid framework exists, supported by well-qualified personnel. However, issues of overlapping responsibilities contribute to poor accountability, and a lack of coordination among different institutions responsible for land management hinders optimal functioning. Furthermore, there is a potential risk that the government may assign low priority to the activities of pertinent institutions that work towards mitigating land degradation. This is often compounded by inadequate resource allocation from the consolidated budget, specifically earmarked for comprehensive land management, further impeding effective efforts to address land degradation issues.

13.4 Water Governance Challenges in Sustainable Land Use

The strong interconnection between land and water must be a central consideration in the sustainable use of natural resources, particularly in resource allocation and governance. Growing environmental concerns and the overexploitation of resources underscore the urgent need to revise existing policies, legislation, and institutional frameworks. Although numerous policies have been introduced and implemented, few have achieved their intended outcomes, and many have fallen short (Chandrasekara et al., 2021). It is therefore crucial to examine the underlying causes of these policy failures and the gap between expectations and actual results.

Ambiguous land ownership and overlapping formal and traditional tenure systems pose significant barriers to effective water resource management and agricultural productivity. These uncertainties discourage long-term investments in irrigation and land improvements, limiting farmers' ability to optimize water use. Governance reforms should prioritize formalizing land titles and integrating traditional landholding systems into the official legal framework to reduce conflicts and improve access to water resources.

In Sri Lanka, land ownership grants individuals the freedom to use their property as they choose, often with little oversight or restriction. For example, watershed areas in the country are often held under both government and private ownership. Activities carried out on privately owned land can have significant effects on watersheds and related natural resources, including water quality, soil health, and downstream ecosystems. Therefore, implementing land use laws tied to clearly defined property rights is essential for promoting social justice, economic development, and sustainable resource management. Well-designed land use regulations help ensure that land is utilized efficiently, minimizing inefficiencies, conflicts, and environmental degradation (Rodgers, 2009).

Highly fragmented farmlands lead to inefficiencies in water distribution, making it challenging for smallholders to adopt efficient irrigation practices. This ultimately reduces both land and water productivity, thereby impacting farmers' incomes. Emphasizing the importance of proper land use in sustainable water governance, interventions should prioritize land consolidation programs. These can be implemented through various approaches, such as voluntary pooling via cooperatives, with the most appropriate method selected based on the specific context (Weerawardena and College, 1971; Munaweerage et al., 2021). Additionally, the establishment of zoning regulations and the use of GIS technology can help optimize land use and water management, aligning fragmented holdings with broader sustainability goals.

A coordinated approach among water, land, agricultural, and environmental institutions is essential to resolve overlapping mandates and conflicting objectives. Developing coherent policies that account for all relevant sectors can help align land use, water resource management, and environmental conservation efforts. For instance, revising the Soil Conservation Act to incorporate modern sustainable practices and integrating it with agricultural regulations would address soil degradation, promote efficient water use, and enhance overall productivity.

Customary norms and institutions—such as rotational land tenure and traditional water management practices—can complement formal governance systems. Since these traditional practices have evolved within localized contexts, recognizing and incorporating them into the national framework can strengthen local resource management while preserving cultural heritage. To promote inclusivity and reduce conflict, especially in rural areas where informal practices are widespread, policies must integrate customary rights with formal legal frameworks (Paranage, 2018).

Climate change exacerbates the challenges of water and land governance in Sri Lanka by increasing resource scarcity, accelerating soil degradation, and heightening the vulnerability of agricultural communities. Shifts in rainfall patterns, more frequent droughts and floods, and rising temperatures contribute to greater uncertainty in water availability and crop yields. Research highlights the importance of providing updated climate information and planning accordingly (Amaranath et al., 2023). Therefore, integrating climate data into governance frameworks is essential for building resilience. Robust climate forecasting systems and access to actionable information can support more informed irrigation planning, crop selection, and land use decisions, helping farmers adapt to changing climate conditions.

Existing land and resource management policies, though developed to address specific needs, require revision to form a comprehensive and cohesive framework. The current policy landscape is fragmented, lacking clear implementation mechanisms, and is marked by overlapping responsibilities across various levels of governance. This has led to inefficiencies, conflicting interests, and poor accountability (Chandrasekara et al., 2021). To improve the sustainable management of natural resources, it is essential to clarify governance roles, strengthen stakeholder coordination, and establish effective systems for accountability and resource use.

13.5 Conclusions

Sri Lanka's growing population has placed increasing pressure on the country's land resources, leading to a declining land-to-person ratio. This reduction in available land, combined with soil degradation and the shrinking of agricultural areas, poses a serious threat to national agricultural productivity and output. If left unaddressed, these challenges could ultimately undermine the country's food security.

Soil degradation occurs through a range of physical, chemical, and biological processes. Key contributors include soil erosion, declining fertility, acidification (dystrification), excessive nutrient buildup (eutrophication), salinization, waterlogging, and pollution. The root causes of these issues stem from various factors, such as a lack of awareness among land users, insecure land tenure, poverty-driven subsistence farming, population pressure, political interference, weak institutional structures, poor law enforcement, misaligned economic and development policies, and neglect of environmental considerations. These factors have led to the adoption of unsustainable farming methods and harmful land management practices by both farmers and land managers.

The human activities primarily responsible for land degradation include inadequate soil conservation efforts, excessive use of agrochemicals, poor soil fertility management, obstruction of drainage systems, deforestation, the release of toxic substances into the soil, improper waste disposal, and indiscriminate sand mining in rivers and streams. The consequences of these activities are severe, resulting in reduced land productivity, depletion of water resources, loss of biodiversity, downstream sedimentation, increased production costs, water pollution, diminished arable land, and heightened rural poverty.

A sustainable nexus between land and water is critical for efficient resource allocation and governance, particularly in addressing Sri Lanka's environmental challenges and resource exploitation. However, fragmented governance, overlapping jurisdictions, and inadequate accountability reduce the efficacy of policies, leading to disputes and inefficient use of resources. A cohesive governance framework, with well-defined roles and improved stakeholder coordination, is vital for achieving sustainable resource management and economic development.

References

Abeywardana, N., Bebermeier, W., & Schütt, B. (2018). Ancient water management and governance in the dry zone of Sri Lanka until abandonment, and the influence of colonial politics during reclamation. *Water, 10*(12), 1746. https://doi.org/10.3390/w10121746

Bai, Y., Ochuodho, T. O., & Yang, J. (2019). Impact of land use and climate change on water-related ecosystem services in Kentucky, USA. *Ecological Indicators, 102*, 51–64. https://doi.org/10.1016/j.ecolind.2019.01.079

Bamunuarachchi, B. (2017). Should free holding rights be granted to LDO lands? A situational analysis of dry zone in Sri Lanka. Research report no. 206. Hector Kobbekaduwa Agrarian Research and Training Institute, Colombo, Sri Lanka.

Bradley, A., Weerahewa, J., Vickers, B., Palmer, D., & Aggarawal, S. (2016). Land tenure considerations in Sri Lanka's proposed National REDD+ strategy—UN-REDD programme collaborative online workspace. Sri Lanka UN-REDD Programme, Battaramulla. https://landportal.org/library/resources/land-tenure-considerations-sri-lanka%E2%80%99s-proposed-national-redd-strategy

Casinader, N., Wijeyaratne, R. D. S., & Godden, L. (2018). From sovereignty to modernity: Revisiting the Colebrooke-Cameron Reforms–transforming the Buddhist and colonial imaginary in nineteenth-century Ceylon. *Comparative Legal History, 6*(1), 34–64.

Central Bank of Sri Lanka. (2022). *Annual report 2022.* Central Bank of Sri Lanka.

Chandrasekara, S. S. K., Chandrasekara, S. K., Gamini, P. S., Obeysekera, J., Manthrithilake, H., Kwon, H. H., & Vithanage, M. (2021). A review on water governance in Sri Lanka: The lessons learnt for future water policy formulation. *Water Policy, 23*(2), 255–273.

Dandeniya, W. S., & Dharmakeerthi, R. S. (2020). Integrated plant nutrient management in major agricultural soils of Sri Lanka: A review of the current status and the way forward. In R. P. De Silva et al. (Eds.) *Agricultural research for sustainable food systems in Sri Lanka: A pursuit for advancements*(Vol 2, pp 213–238). Springer Nature. https://doi.org/10.1007/978-981-15-3673-1_10

Department of Census and Statistics. (2019). Agricultural household survey 2016/17. Department of Census and Statistics, Sri Lanka.

Feder, G. (1988). *Land policies and farm productivity in Thailand*. Johns Hopkins University Press.

Fu, Q., Li, B., Hou, Y., Bi, X., & Zhang, X. (2017). Effects of land use and climate change on ecosystem services in Central Asia's arid regions: A case study in Altay prefecture, China. *Science of the Total Environment, 607–608*, 633–646. https://doi.org/10.1016/j.scitotenv.2017.06.241

Gunasekara, V. (2020). *Turning points in Sri Lanka's land policy: MCC and its predecessors*. Law & Society Trust and People's Alliance for Right to Land.

Gunawardena, P. J. (1981). Land policy and agrarian change in independent Sri Lanka. *Sri Lanka Journal of Agrarian Studies, 2*(2), 71–90.

Herath, J. (2006). *Distributional impacts of land policies in Sri Lanka*. Sri Lankan Agriculture Economics Association (SLEA) and Movement for National Land and Agricultural Reform (MONLAR).

Irangani, M. (2020). Sustainable principles of indigenous chena cultivation and management in Sri Lanka: lessons for contemporary agricultural problems. *International Journal of Scientific and Research Publications, 10*(11), 46–54. https://doi.org/10.29322/ijsrp.10.11.2020.p10706.

Jagustović, R., Zougmoré, R. B., Kessler, A., Ritsema, C. J., Keesstra, S., & Reynolds, M. (2019). Contribution of systems thinking and complex adaptive system attributes to sustainable food production: Example from a climate-smart village. *Agricultural Systems, 171*, 65–75.

Jogaratnam, T. (2001). Allocation of state land for peasant agriculture in Sri Lanka. In *Background Paper for Workshop: Towards an Employment Strategy Framework in Sri Lanka*.

Jayakody, A. N. (2019). Soil degradation in Sri Lanka. In E. M. Bridges (Ed.), *Response to land degradation* (pp. 130–139). CRC Press.

Kalkuhl, M., & Wenz, L. (2020). The impact of climate conditions on economic production: Evidence from a global panel of regions. *Journal of Environmental Economics and Management, 103*, Article 102360.

Khaniya, B., Gunathilake M. B., & Rathnayake, U. (2021). Ecosystem-based adaptation for the impact of climate change and variation in the water management sector of Sri Lanka. *Mathematical Problems in Engineering, 2021*(1). https://doi.org/10.1155/2021/8821329

Maddumabandara, C. M. (1998). Land use and land settlements: The last fifty years and some thought on future policy. In *Fifty Years of Sri Lanka Independence, Socioeconomic Review*. Institute of Social and Economic Studies (SLISES).

Makumbura, R. K., Samarasinghe, J., & Rathnayake, U. (2022). Multidecadal land use patterns and land surface temperature variation in Sri Lanka. *Applied and Environmental Soil Science*, Article ID 9880691.

Mapa, R. B., Kumaragamage, D., Gunarathne, W. D. L., & Dassanayake, A. R. (2002) Land use in Sri Lanka: Past, present and the future. In *Proceedings of the 17th World Congress of Social Science (WCSS)*(pp. 14–21)

Marawila, D. (2007). Revitalizing small-scale agriculture: Rental policies of alienated state lands of Sri Lanka. *Sri Lankan Journal of Agricultural Economics, 9*, 23–42. https://doi.org/10.4038/jae.v9i0.1831

Ministry of Mahaweli Development and Environment. (2017). *Final country report of the land degradation neutrality target setting programme in Sri Lanka*. Ministry of Mahaweli Development and Environment.

Moore, M. P., & Wickramasinghe, G. (1978). *Thattumaru-Kattimaru systems of land tenure*. Agrarian Research and Training Institute.

Munaweerage, T. P., Bandara, M. A. C. S., & de LW Samarasinha, G. G. (2021). *Thattumaru—Kattimaru and productivity: An assessment of traditional tenurial systems in the Wet Zone*. Research report no 2410. Colombo, Sri Lanka: Hector Kobbekaduwa Agrarian Research and Training Institute.

Nayakekorale, H. B., & Mapa, R. B. (2020). *The soils of Sri Lanka soil degradation* (pp. 103–118). Springer International Publishing.

Nicholson, C. F., Kopainsky, B., Stephens, E. C., Parsons, D., Jones, A. D., Garrett, J., & Phillips, E. L. (2020). Conceptual frameworks linking agriculture and food security. *Nature Food, 1*(9), 541–551. https://doi.org/10.1038/s43016-020-00142-3

Paranage, K. (2018). The consequences of restricting rights to land: Understanding the impact of state-land tenure policies in Sri Lanka sustainability. *Science Practice and Policy, 14*(1), 46–54. https://doi.org/10.1080/15487733.2018.1545556

Pound, B., Lamboll, R., Croxton, S., Gupta, N., & Bahadur, A. V. (2018). Climate-resilient agriculture in South Asia: An analytical framework and insights from practice. *Action on Climate Today.*

Rodgers, C. (2009). Property rights land use and the rural environment: A case for reform. *Land Use Policy, 26*, S134–S141. https://doi.org/10.1016/j.landusepol.2009.09.018

Runting, R. K., Bryan, B. A., Dee, L. E., Maseyk, F. J., Mandle, L., Hamel, P., Wilson, K. A., Yetka, K., Possingham, H. P., & Rhodes, J. R. (2017). Incorporating climate change into ecosystem service assessments and decisions: A review abstract. *Global Change Biology, 23*(1), 28–41. https://doi.org/10.1111/gcb.2017.23.issue-1. https://doi.org/10.1111/gcb.13457

Samaratunga, P. (2012). *Achieving economic efficiency and distribution of land use though reforms and/or market discipline.* VijithaYapa Publication.

Schellnhuber, H. J., Hare, B., Serdeczny, O., Schaeffer, M., Adams, S., Baarsch, F., Schwan, S., Coumou, D., Robinson, A., Vieweg, M., & Piontek, F. (2013). *Turn down the heat: Climate extremes, regional impacts, and the case for resilience.* World Bank.

Shand, R. T. (2002). *Irrigation and agriculture in Sri Lanka.* Institute of Policy Studies.

Tsuchida, R., & Takeda, S. (2021). Is resilience socially emerging or embedded? A review of "resilience" under climate change in Sri Lanka. *Journal of Safety Science and Resilience, 2*(4) 258–266 https://doi.org/10.1016/j.jnlssr.2021.11.001

Weerahewa, J. (2017). Modernizing agriculture in Sri Lanka—Status and challenges. In: P. Athukorala, E. Ginting, H. Hill, & U. Kumar (Eds.) *The Sri Lankan economy charting a new course.* Asian Development Bank.

Weerahewa, J., Hemachandra, D., & Pushpakumara, G., Tanjina, N., & Hasnat Mohammed Kamal, H. (2021). Examining international land use policies changes and conflicts land regulations tenure and land use of Sri Lanka. *A Historical Analysis on Agricultural Lands* (pp. 139–161). IGI Global.

Wickramaarachchi, N. C., & Weerahewa, J. (2016). Land fragmentation and land productivity: Empirical evidence from land distribution schemes of Sri Lanka. *International Academic Research Journal of Business and Management, 5*(1), 11–21.

Wickramasinghe, D., Prematilaka, L., Witharana, A., Halwatura, D., Perera, T., & Singh, A. (2023). *International handbook of disaster research cities and climate change: Responding to the impacts of water-related disasters in Sri Lanka* (pp. 1–23). Springer Nature Singapore.

World Bank. (2018). *Improving quality of land administration in Sri Lanka (English).* World Bank Group. http://documents.worldbank.org/curated/en/750021530107195459/Improving-quality-of-land-administration-in-Sri-Lanka

World Bank. (2023). World bank open data. [online] World Bank Open Data. Available at: https://data.worldbank.org/indicator/AG.LND.ARBL.HA.PC?locations=LK (Accessed 11 September 2023).

Ms. Thilani P. Munaweerage is currently pursuing a PhD in Economics at Griffith University Australia. She previously served for over 15 years as a Researcher at Hector Kobbekaduwa Agrarian Research and Training Institute, where she most recently held the position of Head of the Environment Division. Thilani earned her Bachelor's degree in Agriculture, specializing in Agricultural Economics, from the University of Peradeniya and obtained her Masters degree in Environmental and Resource Economics from Australian National University, Australia. Thilani's research interests focus on three broad areas: water, land use, and agricultural policy. Her specialized research areas are integrated water resource management, climate change, agricultural resource use, and environment-related issues. Her current research aims to better understand how economic incentives affect the supply of agricultural products and the interaction between agricultural production and the environment and natural resources. During her professional career, she has been engaged in more than 15+ research studies and published more than 15 research reports and four peer-reviewed journal articles.

Chapter 14
Water Governance in Traditional Tank Cascade Systems for Addressing Climate Challenges

Muditha Prassanajith Perera, S. S. K. Chandrasekara, and S. W. Goonathilake

Abstract This chapter examines the traditional land and water management practices of Sri Lanka's unique hydraulic civilization, focusing on their historical context and relevance to modern water governance in the context of climate adaptation. Key discussion areas include the construction of ancient tanks, water resource development under the monarchic system, water conservation through hydro-ecological techniques, land and water sharing practices, irrigation traditions, enforcement mechanisms for water misuse, and the sustainable use of natural resources within tank cascade systems. Particular emphasis is given to the maintenance and management of irrigation tanks as part of tank cascade systems (TCSs); to sustainable cascade management through systems such as *Kulu Wewa* and *Olagam Wewa*; and to land governance through traditional frameworks like the *Bethma* system and the *Purana Wela–Akkara Wela* system. These practices are key elements of effective water management and governance in the face of climate variability. In addition to these technical systems, the chapter also highlights hydro-ecological aspects such as cascade niches and traditional water control methods implemented by farmer organizations.

Keywords Climate change · Traditional land and water management · Tank cascade system · Kulu wewa · Olagam wewa · Bethma system · Purana wela · Akkara wela

---

M. P. Perera (✉)
Department of Geography, University of Peradeniya, Peradeniya, Sri Lanka
e-mail: mudithpras@gamil.com

S. S. K. Chandrasekara
Department of Agricultural Engineering, Faculty of Agriculture, University of Peradeniya, Peradeniya, Sri Lanka
e-mail: sewwandhich@agri.pdn.ac.lk

S. W. Goonathilake
Department of Civil Engineering and Construction, Skills College of Technology, Nugegoda, Sri Lanka

U. A. Amarasinghe et al. (eds.), *Water Governance for Climate Change Adaptation in Sri Lanka*, Global Issues in Water Policy 13,
https://doi.org/10.1007/978-3-032-04950-6_14

14.1 Introduction

Ancient hydraulic civilizations (for example, those along the Nile, Euphrates, and Tigris, Indus, Yellow, and Yangtze rivers) developed to cater to domestic and irrigation needs. For the same purpose, Sri Lanka's historical civilization emerged in the dry zone lowlands in the sixth century BC. Monarchs adopted effective surface and groundwater resource management to address variations in rainfall, due to the island's tropical location and varied geography. Ancient kings developed various-sized tanks and irrigation systems to ensure sustainable water use and protect their sovereignty.

Understanding the technical dimensions of these ancient and traditional adaptation methods is crucial for Sri Lanka to mitigate the impacts of climate change, especially in the dry zone lowlands. This chapter discusses traditional water-conserving methods based on hydro-ecological techniques and water-sharing technologies used during water-deficient seasons.

14.2 Historical Context of the Ancient Irrigation System in Sri Lanka

Ownership of ancient irrigation systems, including the Tank Cascade System (TCS) and its associated technical structures, was predominantly held by individuals. Other irrigation schemes were allocated to royals, elite groups, prominent local families such as the *Paramuka Vapihamika*, and wealthy households (Abeywardana et al., 2018). Kings enacted laws rooted in customs, norms, and traditions—after consulting with ministers and monastic authorities—and inscribed these laws in stone.

Ancient inscriptions in Sri Lanka reveal that water governance included grants for irrigation, declarations of ownership and profession, official announcements, and provisions related to income generation that supported irrigation. The compulsory labor system known as *Rajakariya* (along with hired labor) was employed for the operation and maintenance of the TCS. The state also allocated grants for the maintenance of irrigation infrastructure, including tanks, canals, and fields (Abeywardana et al., 2018). The labor force, known as Vari Sal, was supported by the state treasury.

Numerous irrigation professionals were responsible for operating and maintaining these schemes.

If individuals or communities violated these maintenance or damaged irrigation structures, their actions were considered theft, and they were required to pay compensation to the king.

The communal operation and maintenance of the TCS was known as the Greater Tradition, or *Mahasammatha*, and was carried out by villagers (Kenyon et al., 2004). This tradition encompassed the enforcement of state laws and the administration of the reservoirs. Participatory decision-making took place in village assemblies under the leadership of a local administrative officer (the *Gamika* or *Gamladda*), while the *Wel Vidane* managed water distribution in command areas.

The tanks from Sri Lanka's ancient irrigation infrastructure, initiated by monarchs in the 5th to sixth centuries BC (Hewawasam & Matsui, 2022), still exist and contribute actively to the effectiveness of the water sector in Sri Lanka. Furthermore, the dry zone's water-related environment sustainably met the basic needs of human, plant, and animal communities. In 2017, the minor/small TCS was identified by the United Nations Food and Agriculture Organization (FAO) as a Globally Important Agricultural Heritage System (The World Bank, 2023). These communal and ecosystem services helped rural communities in Sri Lanka's dry zone survive through frequent political transitions in the past.

Archaeological evidence of major irrigation works dates back to the third century BC. However, evidence for minor tanks from that period is limited. During the Anuradhapura period (377 BC–1017 CE) (Zubair, 2005), construction and renovation of minor tanks peaked from 200 to 950 CE (Gilliland et al., 2013). The establishment of TCSs was recognized as the most important development in historic urban societies in Sri Lanka's north-central dry zone because (a) rainfed agriculture was insufficient to meet food requirements at the time (Amarasinghe & Nguyen, 2010), and (b) storing water for dry seasons was necessary.

However, frequent transitions of kingdoms and epidemics in the fourteenth century led to the abandonment of many tank systems (Dharmasena, 2019; Zubair, 2005). Although tank cascades suffered significant damage during the 1818 freedom struggle in Uva Wellassa, British colonial governance later recognized the importance of Sri Lanka's irrigation works. It focused on rehabilitating irrigation structures and reforming governance by: (a) introducing the first water-resources legislation—the Irrigation Ordinance No. 32 of 1946—and (b) abolishing the compulsory labor system (*Rajakariya*). Later, the Agrarian Services Act No. 4 of 1991 was enacted to legalize Farmer Organizations (FOs) and mandate participatory maintenance of medium and minor irrigation schemes, coordinated by the Divisional Officer of the Department of Agrarian Development. Sri Lanka undertook numerous irrigation projects to construct large-scale multipurpose reservoirs, but it neglected the development of minor tanks. Furthermore, low farmer incomes, youth migration to urban areas, deforestation for population growth, and ecosystem degradation have contributed to the abandonment of tank systems in Sri Lanka in recent decades (FAO, 2023).

14.3 Technical Dimensions of Tank Cascade Management

Climate variability is not a phenomenon that emerged abruptly in the twenty-first century, although its increased severity in recent times warrants due recognition. Hence, assessing water conservation through tanks—an ancient strategy for mitigating climate variability—would be a timely measure for climate change adaptation in Sri Lanka. This approach can integrate traditional principles with modern technologies into existing water resource management strategies. Consequently, the restoration of tanks has become an important adaptation strategy for the Sri Lankan government to mitigate the impacts of climate change on the country's water resources. The government is focusing on investigating and adapting traditional approaches to sustain existing tanks under a changing climate, restoring system structures, building community capacity to conserve water and optimize its use during the dry season, and incorporating novel methodologies such as climate-smart agriculture.

14.3.1 Managing Water Through Hydrological Techniques

Efficient utilization of water resources was essential for long-term agricultural sustainability in ancient Sri Lanka. According to Brohier (1935), man-made tanks were constructed to store and discharge water efficiently to irrigable areas in Sri Lanka's dry zone. As Geekiyanage and Pushpakumara (2013) note, "These ancient irrigation structures evolved with the advancement of traditional knowledge over two millennia and still act as essential elements of water management for agriculture in the dry zone of Sri Lanka." Thus, cultivation faced no hindrance even during periods of minimal rainfall in the dry zone, because adequate water was available.

These tanks were interconnected. In fact, Madduma Bandara (1985) reported that more than 85% of Sri Lanka's small tanks in the dry zone are organized into cascade systems. A Tank Cascade System (TCS) is a connected series of tanks organized within a meso-catchment of the dry zone landscape, storing, conveying, and utilizing water from an ephemeral rivulet (Madduma Bandara, 1985). Tennakoon (1995) noted that a local term for these cascades is "*Ellangawa.*" Panabokke et al. (2002) verified this and found that the North Central dry zone (Rajarata) contains the core of these cascades, identifying 457 cascade systems nationwide. Perera (2010) further described the complexity of these cascades by introducing a local term, ***Wew Mandulla***, and categorizing tank interlinkages as follows:

- **Linear system of tanks**
- **Dendritic localization of tanks**
- **Parallel-network behavior of tanks.**

Perera (2010) explained that these tanks are interlinked within mini-basins and clarified that cascades are not confined to tanks along a single stream. They encompass entire catchments, command areas, and their biophysical and human environments, which are inherently part of the cascade system. Accordingly, Perera (2019) defined a tank cascade as a "connected series of tanks with its micro-catchments and command areas, organized within a small valley/basin" (Fig. 14.1).

The importance of TCSs lies in collecting rainfall during the main rainy season and utilizing that water multiple times along a valley in the dry season. Once collected, water in the uppermost tank is distributed to its command area. The return flow,

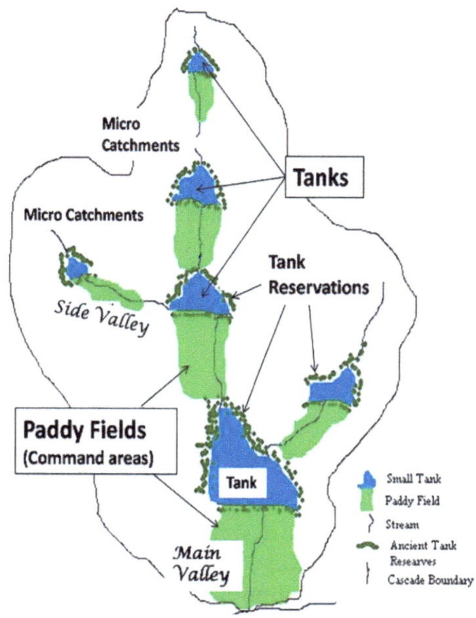

Fig. 14.1 A schematic diagram of a tank cascade. *Source* Perera (2019)

along with seepage water, is then collected by the lower tank, allowing for reuse downstream.

The International Union for Conservation of Nature (IUCN) in Sri Lanka identified the main functions of TCSs as: (a) storing rainfall during the main rainy season (the Northeast monsoon from November to February, which provides about 80% of the annual rainfall for Sri Lanka's dry zone); (b) recycling and reusing water by incorporating natural ecosystems and drainage patterns into the TCS; and (c) minimizing and mitigating flood and drought impacts by transporting and distributing the stored water. Small and medium-scale reservoirs were built upstream of the cascades, and larger reservoirs were built downstream to store additional water.

TCSs were integrated with social and environmental aspects for sustainability, such as: (a) community participation in upper catchment management through forest cover maintenance, water sharing, and conservation; (b) aligning administrative boundaries with cascade boundaries to minimize conflicts; (c) maintaining buffer zones for future settlements and natural disaster mitigation; (d) developing food and medicinal resources from the system; and (e) incorporating spiritual reverence toward the TCS. These tanks are typically located in villages and are referred to as *village tanks*. However, some tanks are not near villages; these are called Kulu Wewa, which are small tanks or ponds located in the immediate catchment of medium or large tanks to trap sediment carried by ephemeral streams. They do not include irrigation infrastructure such as sluices or spillways. Rainwater is stored in Kulu Wewa for a short period and then gradually released downstream. Under heavy rainfall conditions, *Kulu Wewa* acts as a control structure: when rainfall intensity is high, sudden inflows are first captured by the *Kulu Wewa* before reaching the main tank, preventing abrupt surges.

In every medium- to large-sized tank, an additional spillway, known as *Pita wana*, was provided outside the main spillway to release excess water during heavy rainfall. This reduces hydraulic pressure on the main spillway, preventing water from overtopping the tank bund and thereby protecting the bund from damage.

For water control, tanks had two types of sluice gates: an upper tower sluice gate (*Goda Sorowwa*) and a lower sluice gate (*Mada Sorowwa*). When the tank is full, the *Goda Sorowwa* is opened to release water. In the dry season, water from the bottom of the tank is released through the *Mada Sorowwa* to sustain irrigation in the command area.

14.3.2 Managing Water Through Ecological Techniques

Ancient reservoir design included ecological features that play a vital role in tank hydro-ecology. *Gasgommana*, *Perahana*, and *Kattakaduwa* were key components of a tank's conservation buffer zone (Fig. 14.2). The *Gasgommana* is a strategically placed belt of trees within the tank bed near the water's edge, functioning as a windbreak to reduce evaporation. *Gasgommana* covers between 10 and 20% of the tank's surface area, significantly conserving water during the dry season.

Perahana refers to the peripheral strip of grasslands and scattered shrubs encircling the tank. This ecological buffer zone acts as a natural filter, preventing topsoil

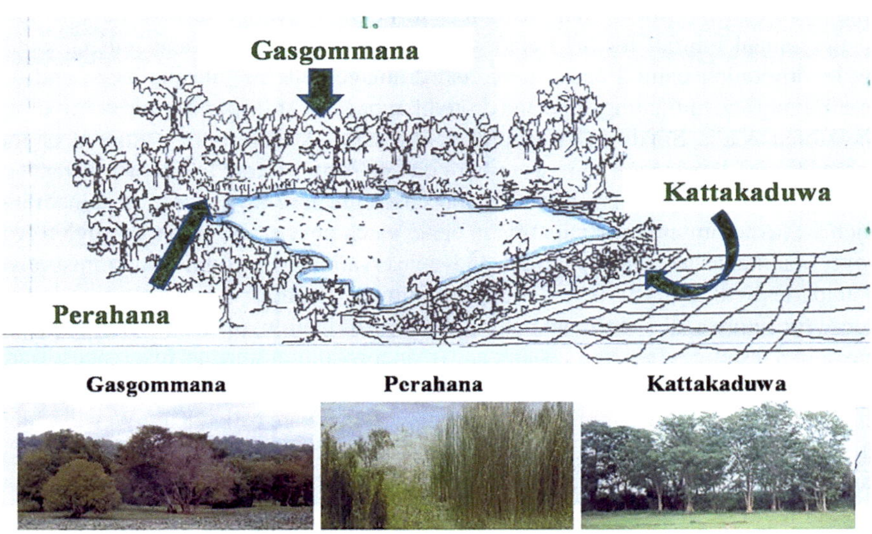

Fig. 14.2 Main tank reservations. *Source* Perera (2010)

from adjacent agricultural lands from eroding into the tank. This is crucial for mitigating sedimentation and preserving the tank's water storage capacity.

Kattakaduwa is a designated strip of land between the tank bund and the paddy fields, planted with well-established trees. It serves multiple purposes: it minimizes water seepage through the bund and acts as a natural barrier protecting the command area from salinity intrusion (Fig. 14.2).

In analyzing ecological and human interactions, *cascade niches* can be identified not only in downstream relationships along cascades but also in horizontal relationships within a TCS (Fig. 14.3). This concept emphasizes the interconnection between ecological elements and human activities across various levels of the cascade, particularly the movement of fauna among components under varying climatic conditions. For example, birds and reptiles may migrate among parallel tanks depending on water availability. These niches offer insights into the complex dynamics that shape both the ecological landscape and the socio-economic fabric of surrounding communities. This interdisciplinary perspective highlights the intricate web of relationships and dependencies within cascade systems under intra-annual climate fluctuations.

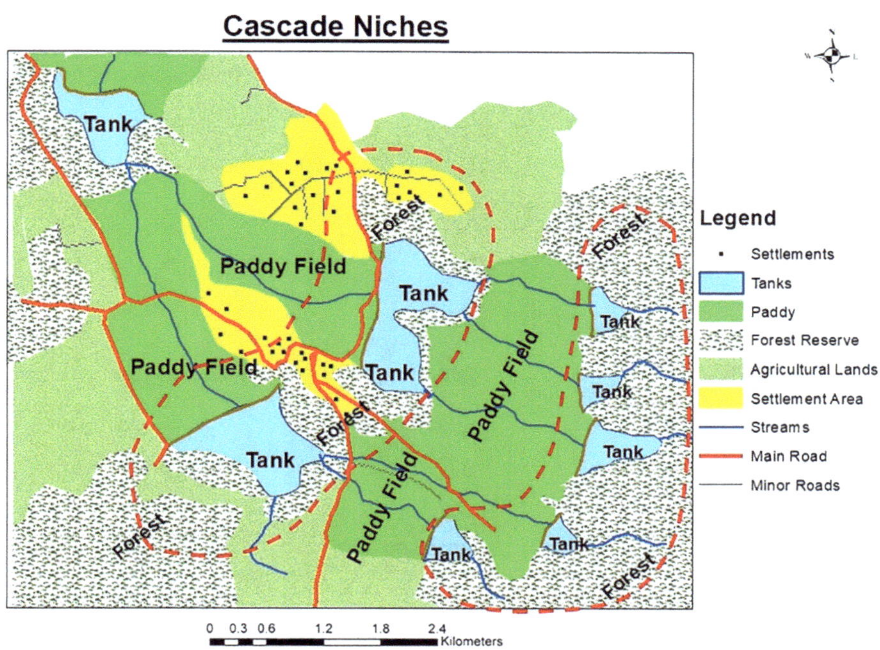

Fig. 14.3 Cascade niches in Anuradhapura district, Sri Lanka's dry zone. *Source* Perera and Madduma Bandara (2025)

14.3.3 Land and Water Sharing Techniques: Bethma *System*

The dry season in Sri Lanka's dry zone poses significant challenges for tank irrigation systems. With limited water availability, ancient farmers devised a unique solution for sharing scarce water resources equitably among all farmers within the same command area: the *Bethma system.*

The *Bethma* system operated on the principle of proportional allocation. Roughly half of the available tank water was designated to irrigate 50% of the total command area. This ensured that at least some cultivation was possible for all farmers while fostering cooperation in sharing the limited water supply. Land allocation under *Bethma* followed three main approaches, allowing farmers to choose the method best suited to local conditions (Fig. 14.4):

- **Selected Canal Approach**: Farmers choose a single major irrigation canal on one side of the command area. The lands served by this canal are cultivated in rotation, ensuring that every farmer connected to the canal receives water for their crops.
- **Proximity-Based Allocation**: The portion of the command area closest to the tank is prioritized for irrigation. All farmers share this nearest portion, which maximizes water use efficiency by focusing on lands with the least water loss due to distance.

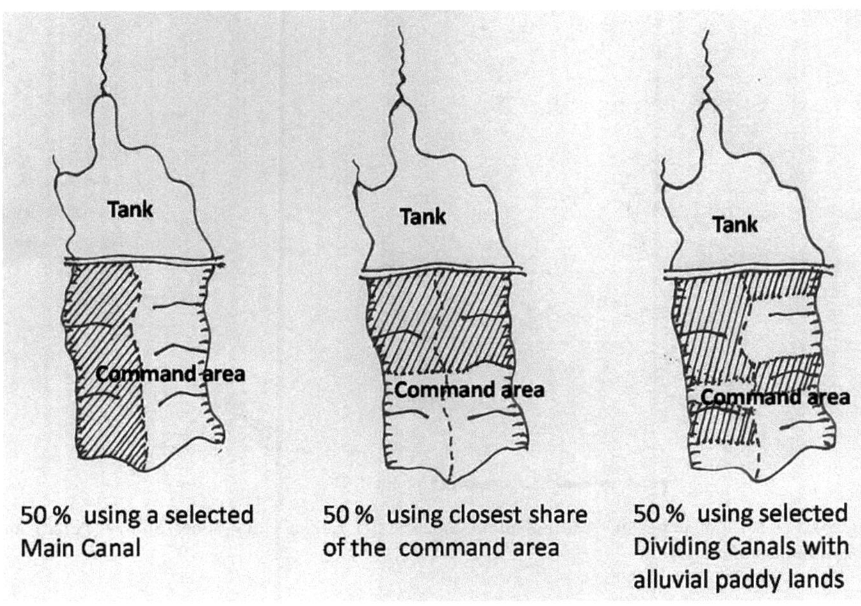

Fig. 14.4 *Bethma* cultivation methods—**a** Selected Canal approach; **b** Proximity-based allocation; **c** Alluvial land selection. *Source* Perera (2010)

- **Alluvial Land Selection**: This method focuses on the most fertile, alluvial soils in the command area. Canals are used to divide this fertile land into sections, allowing each farmer to cultivate a portion. This approach leverages both water availability and high soil quality, making it particularly effective for larger tanks, as fertile land generally requires less water to achieve good yields.

In addition, subdivisions of the command area may be important for water allocation under different seasonal conditions. The command area of a tank irrigation system is the total land designated for cultivation using tank water. Within this area, variations in cultivation practices can occur depending on seasonal water availability. Two such sub-areas are: (1) the **main (old) paddy field—*Purana Wela***, which refers to the upper area of the command (nearest the tank) with a long history of cultivation; and (2) the **partial (uncultivated) paddy field—*Akkara Wela***, which includes lands where paddy cultivation is not consistent. Purana Wela fields are typically closest to the tank, receive the most reliable water supply, and are usually the most fertile. In contrast, *Akkara Wela* fields may experience inconsistent cultivation due to limited water resources, poorer soil quality, or changing land use practices. When water is abundant, farmers may cultivate both types of fields. Understanding these subdivisions within the command area is crucial for effective water management, as it enables targeted water allocation strategies and informed land-use decisions to optimize agricultural productivity under varying water conditions.

14.3.4 Water Sharing Technique with the Nearby Tank (Budding Tank)

Budding tanks are small tanks situated closely together (Fig. 14.5). Two main theories explain how budding tanks contribute to water sharing within a cascade during climatic extremes. The first theory is that during dry periods, when one tank's water is insufficient, a neighboring budding tank can release additional water, sustaining irrigation in the command area. The second theory is that budding tanks can act as storage for overflow water from the main tank. This helps regulate water levels in the main tank and prevents overflow and potential damage to the tank bund. By capturing excess water, budding tanks may also help reduce downstream flooding.

14.3.5 Irrigation Traditions to Control the Misuse of Water

In Sri Lanka's dry zone, several traditional ceremonies are associated with tank water, though they are now rarely practiced. One ritual occurs after the tanks are filled by rain. Villagers carry pots of water to the Bo tree at the local temple—the most sacred tree in Buddhism—and pour it at its base to invoke blessings for a successful irrigation season. Historically, this offering was made with milk, but today water is

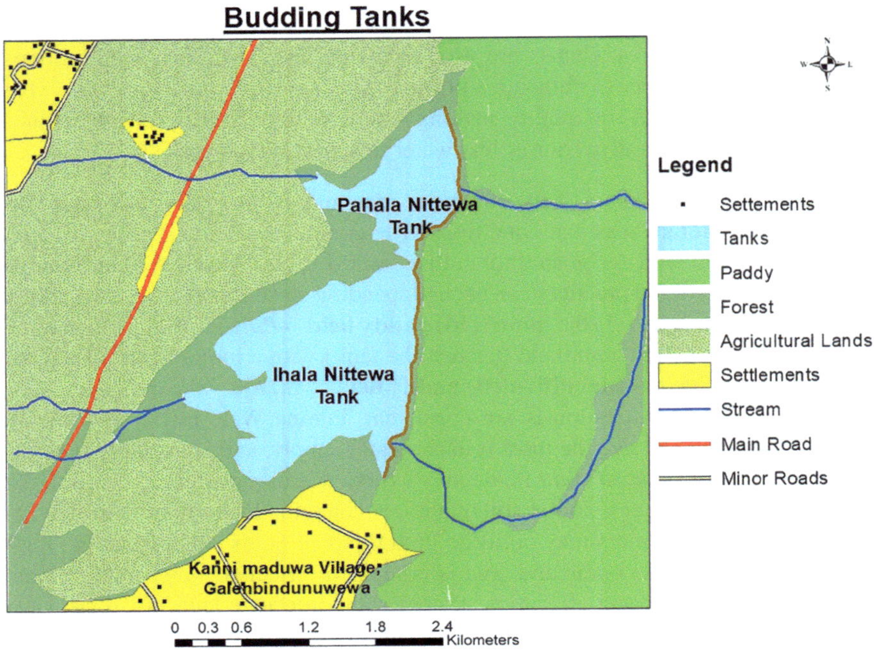

Fig. 14.5 Budding tanks in Galenbindunuwewa, Anuradhapura district, Sri Lanka's dry zone. *Source* Authors's (Modified from Metrics Maps)

used symbolically. This act is a tribute to the sacred Bo tree, which once sheltered the Buddha.

When a tank reaches full capacity and begins to spill, and if the overflow is excessive, village elders gather at the tank. They offer a ceremonial salute to the gods by firing two gunshots, followed by a *Yathika* (ritual chant or declaration) and a vow by the village chief. This tradition, known as *Pūjā Wedilla*, is associated with *Panduru Bendeema*, a prayer seeking divine protection for the tank. After the harvest, another ceremony called *Mutti Mangalle* is held, in which villagers offer milk rice in remembrance of the *Panduru Bendeema*. A final ceremony is often held near the tank to honor a local deity, further emphasizing the spiritual connection between water, agriculture, and community.

Another traditional practice is *Ana Bola Bendeema*, which prohibits fishing during severe droughts when aquatic life is threatened. The head of the farmer organization enforces this ban by placing a tree branch in the tank as a marker. Once the tank is refilled with sufficient rain, the branch is removed and fishing is permitted again.

Additionally, accountability systems were in place to prevent water misuse. The *Wel Vidane*, the official responsible for village water management, had the authority to punish those who wasted water, ensuring the responsible and equitable use of water among farmers.

14.3.6 Environmental Equilibrium Within the Tank Cascade for Addressing Climate Challenges

In addition to village tanks, some tanks were built without nearby villages; these are known as ***Olagam Wewa*** (Olagam tanks) (Fig. 14.6). As populations expanded and village tank harvests became insufficient, farmers would enter the forest and build a tank without disturbing the environment or wildlife. These *Olagam Wewas* were cultivated only during the Maha season under the condition that wildlife be protected. After the Maha harvest, the tank was left full of water for wildlife consumption during the rest of the year.

Olagam Wewas also help sustain downstream flows through seepage and return flows. These flows maintain baseflow in downstream areas, sustain riverine forest strips, and preserve biodiversity of both flora and fauna, thus contributing to environmental equilibrium in the system.

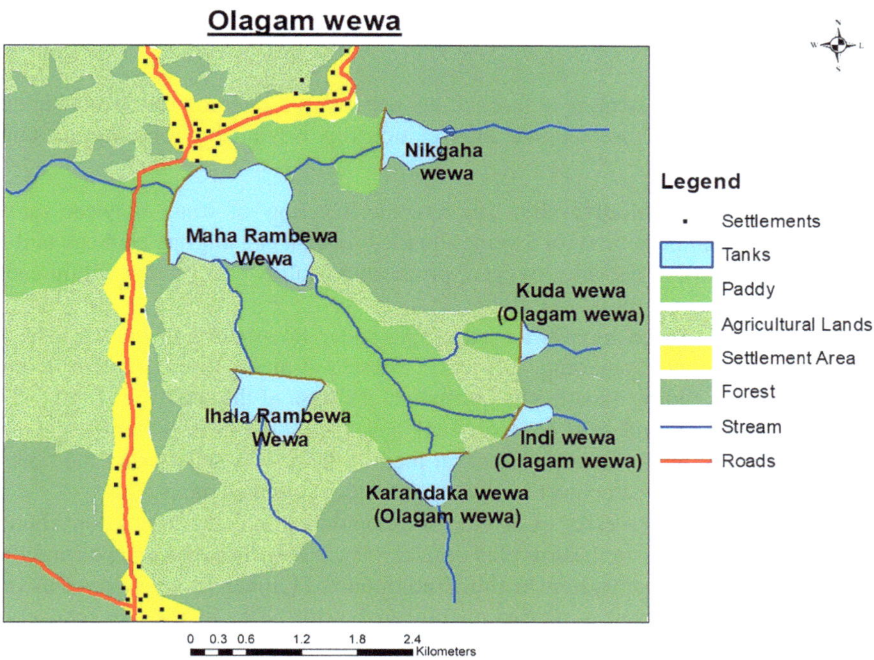

Fig. 14.6 An *Olagam Wewa* in a tank cascade at Palugaswewa, Anuradhapura district, Sri Lanka's dry zone. *Source* Perera and Madduma Bandara (2025)

14.3.7 Groundwater Availability in the Vicinity of the Tanks to Address Drought

Minor TCSs in the dry zone play a crucial role in groundwater management. During the rainy season, water stored in tanks contributes to groundwater recharge. As water percolates through the soil beneath the tank bed and surrounding areas, it replenishes the underlying aquifers. This groundwater storage becomes a vital source of water during dry periods.

Understanding groundwater dynamics around tanks is essential for sustainable water resource management, especially in regions reliant on tank irrigation systems. Factors such as geology, topography, land use, and tank management practices play pivotal roles in shaping groundwater availability.

14.3.8 Hydro-Ecological Interrelationship Within a Tank Cascade Toward a Sustainable Environment

The tank cascade system is a fascinating network of interconnected ecological processes—a true hydro-ecological interrelationship. Key interactions within a tank cascade include:

- **Water flow and biodiversity**: The cascade-like flow of water between tanks creates a dynamic habitat for diverse aquatic life. Fish can migrate between tanks, and flowing water provides oxygen, promoting healthy ecosystems within each tank.
- **Sedimentation and nutrient cycling**: As water flows through the cascade, it carries sediments containing nutrients. These sediments often fertilize downstream tanks, aiding aquatic plant growth at the base of the food web.
- **Bird migration and pest control**: Tank cascades serve as important stopovers for migratory birds. These birds often feed on insects in and around the tanks, acting as natural pest control for the crops in the command areas.
- **Maintaining biodiversity**: A diverse bird population in paddy fields contributes to a balanced ecosystem. Birds play crucial roles in seed dispersal and pollination, promoting overall ecosystem health. Traditional Sri Lankan paddy systems incorporated *Kurulu Paluwa*—narrow uncultivated strips along field edges that serve as sanctuaries for birds. This simple practice demonstrates a harmonious relationship between agriculture and nature, supporting sustainable food production while benefiting bird biodiversity.

These ancient strategies help maintain hydro-ecological relationships even in a dry zone. Keeping a healthy ecological balance within the system ensures the long-term viability of these vital water resources for both humans and other life forms.

14.4 Conclusion

Many traditional technical elements are embedded in Sri Lanka's unique hydraulic civilization, contributing to effective water management and governance in the face of climate change. Notable examples include hydro-ecological techniques such as the maintenance and management of irrigation tanks arranged in cascades; the sustainable use of auxiliary tanks like *Kulu Wewa* and *Olagam Wewa*; and the implementation of traditional tank reservation systems.

In addition, equitable land and water distribution systems—such as the *Bethma* system and the *Purana Wela/Akkara Wela* irrigation traditions—played crucial roles. These systems included mechanisms to penalize water misuse and promote the sustainable use of natural resources within the cascade system.

Together, these practices reflect the distinctiveness of Sri Lanka's traditional water management and governance approaches. They offer valuable insights and models that can be adapted for use in other tropical countries with similar environmental and social conditions to address the challenges posed by climate change.

References

Abeywardana, N., Bebermeier, W., & Schütt, B. (2018). Ancient water management and governance in the dry zone of Sri Lanka until abandonment, and the influence of colonial politics during reclamation. *Water, 10*, 1746. https://doi.org/10.3390/w10121746

Amarasinghe, U. S., & Nguyen, T. T. T. (2010). Enhancing rural farmer income through fish production: Secondary use of water resources in Sri Lanka and elsewhere. In S. S. De Silva, & F. B. Davy (Eds.), *Success stories in Asian aquaculture*. Springer. https://doi.org/10.1007/978-90-481-3087-0_6

Brohier, R. L. (1935). *Ancient irrigation works in Ceylon*. Ministry of Mahaweli Development.

Dharmasena, P. B. (2019). Chapter 3—cascaded tank-village system: Present status and prospects. In *Agricultural research for sustainable food systems in Sri Lanka, vol. 1: A historical perspective*. Springer Nature.

FAO (Food and Agricultural Organization). (2023). *Sri Lanka among globally important agricultural heritage systems*. Retrieved October 18, 2023, from https://www.fao.org/srilanka/news/detail-events/en/c/1118377/

Geekiyanage, N., & Pushpakumara, D. K. N. G. (2013). Ecology of ancient tank cascade systems in island Sri Lanka. *Journal of Marine and Island Cultures, 2*, 93–101. https://doi.org/10.1016/j.imic.2013.11.001

Gilliland, K., et al. (2013). The dry tank: Development and disuse of water management infrastructure in the Anuradhapura hinterland, Sri Lanka. *Journal of Archaeological Science, 40*(2), 1012–1028. https://doi.org/10.1016/j.jas.2012.09.034

Hewawasam, V., & Matsui, K. (2022). Traditional water governance practices for flood mitigation in ancient Sri Lanka. *Climate, 10*(5), 69. https://doi.org/10.3390/cli10050069

Kenyon, P., Pollett, C., & Wills-Johnson, N. (2004). Sustainable water management practices: Lessons from ancient Sri Lanka, No. 14(5). http://ssrn.com/abstract=1002254

Madduma Bandara, C. M. (1985). Catchment ecosystems and village tank cascades in the dry zone of Sri Lanka. In J. Lundqvist, U. Lohm, & L. Falkenmark (Eds.), *Strategies for river basin development* (pp. 99–113). Reidel Publishing. https://doi.org/10.1007/978-94-009-5458-8_11

Panabokke, C. R., Sakthivadivel, R., & Weerasinghe, A. D. (2002). *Evolution, present status, and issues concerning small tank systems in Sri Lanka*. International Water Management Institute (IWMI).

Perera, M. P. (2010). *Irrigation heritage of Sri Lanka* (in Sinhala). Center for Environmental Justice.

Perera, M. P. (2019). *Watershed management: Theories and practices*. Godage International Publishers (ISBN: 978-624-00-0392-6)

Perera, M. P., & Madduma Bandara, C. M. (2025). Tanks and tank cascade systems of Sri Lanka: Concepts, components, and complexity. In Geekiyanage et al. (Eds.), *Tank cascade systems of Sri Lanka: A globally important agricultural system* (pp. 26–38). Taylor and Francis Group. https://doi.org/10.1201/9781003516019

The World Bank. (2023). *Sri Lanka's ancient treasure trove: Rehabilitating tank cascades to champion climate resilience*. Retrieved October 18, 2023, from https://www.worldbank.org/en/news/feature/2023/03/22/sri-lanka-s-ancient-treasure-trove-rehabilitating-tank-cascades-to-champion-climate-resilience

Zubair, L. (2005). Modernization of Sri Lanka's traditional irrigation systems and sustainability. *Science, Technology and Society, 10*(2), 161–195. https://doi.org/10.1177/097172180501000201

Prof. Muditha Prasannajith Perera graduated from the University of Peradeniya and continued his postgraduate studies and research on the hydro-ecology of Tank cascades, Hydraulic Civilization, and Watershed Management. He has working experience at the International Irrigation Management Institute, Forest Department, University of Ruhuna, and is currently a Professor in Geography at the University of Peradeniya. He has published more than 60 publications in International refereed journals, Books, and International Book Chapters. He is an Editorial Board Member and Reviewer for several national and international journals. Occasionally, he contributes to the Forest Department, Irrigation Department, Education Department, and the Ministry of Mahaweli Development and Environment. Further, he has earned experience in Watershed Management Projects in Sri Lanka and studied the Kalipur Water Management project in Bangladesh. He was the national coordinator (Training) of the Sri Lanka Environmental Management Programme (SLIMP) and currently coordinating the short course on "Tank Rehabilitation and Cascade Development" in the University of Peradeniya. Further, he is serving as the Coordinator, of the "Water Resources Management Postgraduate Programme", Post Graduate Institute of Science, University of Peradeniya.

Dr. Sewwandhi Chandrasekara is a lecturer at her alma mater—the Department of Agricultural Engineering, University of Peradeniya. Recently, she received her Ph.D. in Hydraulic Engineering at Sejong University, South Korea. She has experience as a research scientist in local and international institutions, including APCC Climate Center, South Korea. Her expertise focuses on water resources management, including hydro-climatological assessments, agricultural water management, and analysis of irrigation and water policies. She has many publications on her expertise. Currently, she serves as the country coordinator for Cap-Net Lanka, a network dedicated to capacity development in sustainable water management in Sri Lanka and worldwide.

Ms. Sharmila Wanamali Goonathilake received her first degree in Civil Engineering from the Institution of Engineers Sri Lanka (IESL). After completing her degree, she has worked as a demonstrator in Civil Technology at University College, Kuliyapitiya. Currently, she is a lecturer (probationary) at the Department of Civil Engineering and Construction, Skills College of Technology, Nugegoda. Her research interests include sustainable water management practices, hydro-ecology, and Irrigation technology. She is also pursuing a master's degree in water resources management at the Postgraduate Institute of Science (PGIS), University of Peradeniya.

Chapter 15
Role and Need for Climate Information Services in Climate Change Adaptation

Shiromani Jayawardane, Niranga Alahacoon, and Surajit Ghosh

Abstract Climate Information Services, a key platform for reducing the impacts of climate vulnerability, is an important source of information access and knowledge sharing. The current climate information services in Sri Lanka provide forecasted climatic variables, primarily rainfall, wind, and river water levels, to various end-users at daily to weekly temporal scales. However, the dissemination of timely and reliable information needs significant improvements for climate risk reduction and adaptation activities. Moreover, data-driven water governance can facilitate more informed and efficient decisions on water allocation by leveraging data and analytical tools that balance the needs of different stakeholders and promote sustainable water management. This chapter describes the current status and gaps in climate information services and data-driven water governance, and assesses the challenges and prospects for future adaptation activities.

Keywords Climate change · Climate adaptation · Climate information services · Meteorological · Climate resilience

15.1 Introduction

Over the past few decades, climate change has emerged as one of the most critical global challenges of the twenty-first century, profoundly affecting both natural and human systems. Extreme weather events—such as heatwaves, droughts, floods,

---

S. Jayawardane (✉)
Climate Change Secretariat, Ministry of Environment, Battaramulla, Sri Lanka
e-mail: shirojaya2000@yahoo.com

N. Alahacoon · S. Ghosh
International Water Management Institute (IWMI), Colombo, Sri Lanka
e-mail: n.alahacoon@cgiar.org

S. Ghosh
e-mail: s.ghosh@cgiar.org

U. A. Amarasinghe et al. (eds.), *Water Governance for Climate Change Adaptation in Sri Lanka*, Global Issues in Water Policy 13,
https://doi.org/10.1007/978-3-032-04950-6_15

cyclones, and wildfires—underscore the vulnerability of ecosystems and communities to climate variability (IPCC, 2021). These events have disrupted food and water supplies, damaged infrastructure, increased health risks, and affected mental well-being. The impacts are especially severe in regions with high climate variability, where livelihoods and development are closely tied to climate-sensitive sectors like agriculture, water, and energy. Adaptive capacity varies significantly across regions, with developing countries often facing greater limitations (Smit & Wandel, 2006).

Population growth has intensified water demand, increasing pressure on already stressed resources (Cosgrove & Loucks, 2015). Urbanization reduces the area of natural water recharge, while excessive water extraction for agriculture, industry, and domestic use leads to the depletion of aquifers and surface water sources (Bilgiç & Baba, 2023). Climate change exacerbates these challenges by altering rainfall patterns, increasing the frequency of extreme events, and shifting the distribution of water resources (Konapala et al., 2020). These trends require a resilient and adaptive water governance framework to ensure sustainable management.

Climate Information Services (CIS) are essential tools for addressing these challenges. They convert climate data into user-focused information and services, such as forecasts, trends, and risk assessments tailored to specific sectors (Sivakumar & Lucio, 2018). CIS enhances decision-making in agriculture by informing planting schedules, irrigation planning, and crop management. They support efficient water use, flood control, and drought preparedness in water resource management. For disaster risk reduction, CIS improves early warning systems and response planning. In the health sector, they help mitigate climate-related health risks by guiding interventions against diseases and extreme heat events.

By delivering timely, accurate, and actionable information, CIS strengthens adaptive capacities and supports evidence-based planning across sectors, contributing to more resilient societies in a changing climate.

CIS have gained global importance due to technological advances and rising climate risks; however, developing countries still face challenges in accessing and implementing them. In South Asia, where climate-induced disasters are common, CIS play a critical role in supporting adaptation. Sri Lanka, for example, has developed a CIS guide to address floods, droughts, and monsoon variability, aiming to enhance water security and resilience (Amaranath et al., 2019). By translating scientific data into user-friendly information, CIS bridges the gap between knowledge and action, supporting informed decision-making across climate-sensitive sectors such as agriculture, water management, disaster risk reduction, and energy (Graham et al., 2015).

The Global Framework for Climate Services (GFCS), the South Asian Seasonal Climate Outlook Forum (SASCOF), and the Monsoon Forum (MF) play key roles in strengthening Climate Information Services (CIS) at global, regional, and national levels in South Asia.

The GFCS aims to enhance climate resilience by improving the availability, accessibility, and practical use of climate services (Daron et al., 2022). It offers a structured approach to delivering climate information, supports capacity building, and promotes the integration of climate services into policies and planning (Hewitt et al., 2020).

SASCOF facilitates regional collaboration in producing and disseminating seasonal climate outlooks for South Asia. Building consensus around seasonal forecasts, particularly monsoon predictions, enhances their accuracy and relevance, facilitating proactive planning and preparedness for climate-related events (Daron et al., 2022; Jayawardena et al., 2022).

The Monsoon Forum focuses on the national application of climate services. It serves as a collaborative platform where stakeholders evaluate forecasts, propose improvements, and maintain ongoing dialogue between forecasters and end-users. This feedback loop strengthens the relevance and usability of climate information, supporting effective planning and risk management in climate-sensitive sectors.

Together, these initiatives significantly enhance the quality, accessibility, and application of climate services, contributing to more effective climate adaptation and resilience strategies.

This chapter examines the crucial role and need for Climate Information Services (CIS) in adapting to climate change. Section 2 highlights the role of CIS in managing risks and building resilience. Section 3 outlines the essential components of effective CIS frameworks, types, products, and services for climate adaptation. Furthermore, the chapter discusses the key components necessary for creating and delivering climate services. Section 4 discusses the importance of enhancing meteorological service capacity, the role of national hydro-meteorological institutions, and the future of CIS, considering emerging technologies and the necessity for better integration with the Sustainable Development Goals. By providing a comprehensive understanding of CIS, this chapter aims to equip policymakers, practitioners, and researchers with the knowledge to leverage these services for effective climate adaptation.

15.2 CIS Frameworks for Enhancing Climate Resilience

Effective climate services play a critical role in various sectors by providing vital information for strategic planning and risk management. In agriculture, agrometeorological advisories based on seasonal weather outlooks are essential for optimizing crop production and mitigating adverse weather impacts (Hansen et al., 2019). Coordination among agricultural research and development institutes further enhances the application of these advisories. In the health sector, climate services help reduce health impacts from extreme weather conditions, and collaboration with the Disaster Management Center (DMC) ensures preparedness and response.

Disaster Risk Reduction (DRR) benefits from climate services through preparedness measures at both national and local levels, as well as through emergency planning and awareness programs. In water management, rainfall forecasts are crucial for mitigating floods and droughts, maximizing hydropower generation, and optimizing irrigation services. The energy sector relies on accurate climate forecasts to minimize impacts on the power supply, enabling precise forecasting of system demand and renewable energy generation. These integrated climate services enhance resilience and efficiency across multiple sectors, supporting sustainable development and climate adaptation efforts.

Effective climate services for water security encompass a range of products designed to support water managers, policymakers, and communities in making informed decisions. By leveraging meteorological data, these services provide insights and forecasts that help mitigate the impacts of adverse weather conditions and optimize water resource management. The following sections outline the types of climate service products available, each serving a specific purpose and time frame.

15.3 Products and Services in CIS for Adaptation

CIS involves transforming climate-related data and other information into customized products, including projections, trends, economic analyses, best practices advice, and the development and evaluation of solutions. These and any other climate-related services benefit society. These services include data, information, and knowledge that support adaptation, mitigation, and disaster risk management.

Effective climate services for water management encompass a diverse range of products designed to empower stakeholders with the information they need to make informed decisions. These products utilize meteorological data to provide insights and forecasts, helping to mitigate the adverse effects of weather conditions and optimize water resource management. Additionally, data-driven water governance plays a crucial role in decision-making, ensuring efficient water allocation and sustainable management practices. This section details the various climate service products available to water managers, each tailored to address specific needs and time frames (Table 15.1). The subsequent sections elaborate on the types of climate service products accessible, highlighting their unique purposes and applicable periods while emphasizing the importance of integrating climate information services with water governance strategies for comprehensive climate change adaptation.

Figure 15.1 presents a comprehensive overview of the seamless forecasting services offered by the Department of Meteorology, Sri Lanka. These services encompass a range of timescales for climate and weather predictions, tailored to the country's specific needs. Effective climate services for water management involve a range of products designed to equip stakeholders with the information necessary for informed decision-making. By leveraging meteorological data, these products deliver valuable insights and forecasts that help mitigate the negative impacts of weather events and enhance water resource management. Furthermore, data-driven water governance plays a key role in supporting informed decisions, ensuring efficient water allocation, and promoting sustainable management practices.

Future Climate Change Projections offer long-term insights by forecasting the impacts of climate change over decades to centuries. These projections include visual representations of expected changes in rainfall and temperature patterns in Sri Lanka.

Table 15.1 The role of weather forecasts (short, medium, and long), climate outlooks, and climatological projections for Climate Information Services

Products	Description	Applications
Early warning and alerts	Early warning systems are crucial for providing timely alerts about imminent weather events that could potentially harm water resources and infrastructure. These systems use real-time data to predict extreme weather conditions such as storms, floods, droughts, and heat waves. By receiving early warnings, water managers and communities can proactively protect water supplies and infrastructure	• Flood warnings: enable the relocation of assets and implementation of flood defenses • Drought alerts: prompt water-saving techniques and allocation of water resources • Heatwave alerts: help manage water demand and protect vulnerable populations
Short-range forecasts (up to 3 days)	Short-range forecasts provide detailed weather predictions for the immediate future, typically covering up to three days. These forecasts are crucial for daily water management operations, aiding in decisions about water allocation, reservoir management, and emergency preparedness	• Water allocation: optimize water use based on predicted rainfall • Reservoir management: manage reservoir releases to prevent flooding • Emergency preparedness: ensure adequate water supply during dry spells
Medium-range forecasts (up to 10 days	Medium-range forecasts extend the prediction period to ten days, offering a broader outlook for planning water resource activities. These forecasts are valuable for mid-term decision-making, allowing water managers to prepare for changes in weather patterns and adjust their operations accordingly	• Water distribution scheduling: align distribution with expected weather conditions • Maintenance planning: schedule maintenance activities based on forecasts • Demand anticipation: plan for changes in water demand

(continued)

Table 15.1 (continued)

Products	Description	Applications
Weekly forecasts	Weekly forecasts offer an extended outlook covering four weeks into the upcoming month. They serve as a bridge between medium-range predictions and monthly/ seasonal outlooks	• Water storage management: plan for changes in storage levels • Demand anticipation: plan for changes in water demand, ensuring adequate supply during expected high usage or conserving water during anticipated low demand periods • Hydropower generation planning: optimize hydropower production schedules by aligning with predicted inflow variations, ensuring efficient energy generation
Monthly climate outlooks	Monthly climate outlooks offer predictions for the upcoming month, providing a longer-term perspective on weather trends. These outlooks help water managers anticipate seasonal changes and plan their activities over a more extended period	• Water storage planning: anticipate and plan for storage needs • Irrigation management: adjust irrigation schedules to optimize water use • Drought/flood preparedness: prepare for potential drought or flood conditions

(continued)

Table 15.1 (continued)

Products	Description	Applications
Seasonal climate outlooks	Seasonal climate outlooks predict the upcoming season, typically covering three to six months. These outlooks are crucial for strategic planning, helping water managers make decisions that affect their operations over an entire season	• Water allocation planning: developing strategies to allocate water resources efficiently based on seasonal weather predictions • Risk management: creating strategies to mitigate the impact of unfavorable seasonal weather on water availability and quality • Long-term resource allocation: planning the allocation of resources such as labour and budget for the season, ensuring preparedness for anticipated conditions

(continued)

Table 15.1 (continued)

Products	Description	Applications
Tailored information	Tailored climate services provide customized information specific to a region's water resources and management practices. These services combine local knowledge with meteorological data to offer actionable advice	• Location-specific forecasts: providing weather predictions that consider local hydrological conditions • Water management advice: offering guidance on best practices for water conservation and usage based on current and predicted weather conditions • Personalized risk assessments: identify potential risks specific to the water management system and suggest mitigation strategies

(continued)

Table 15.1 (continued)

Products	Description	Applications
Observed climate trends and impacts	Understanding observed climate trends and their impacts on water resources is crucial for effective climate adaptation planning. Water governance and security depend on informed decisions that consider current and future risks	• Analyzing historical climate data to identify trends and assess impacts on water resources • Supporting water resource management, risk assessment, and adaptation strategy development
Future climate projections based on different emission scenarios	Future climate projections provide insights into potential changes in climate variables under various emission scenarios. These projections are essential for planning and implementing long-term water management strategies	• Scenario planning and risk management for future climate conditions • Helping policymakers and water managers develop adaptation strategies for sustainable water governance and resource allocation

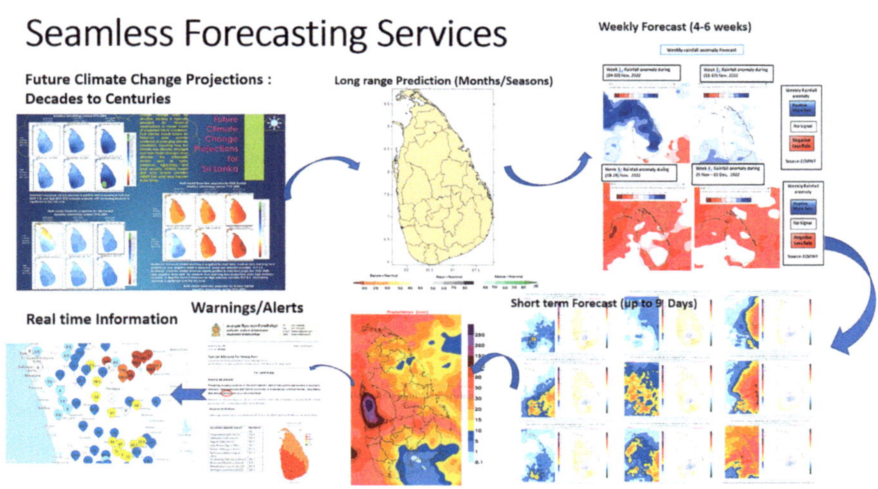

Fig. 15.1 Seamless forecasting services provided by the Department of Meteorology

- **Long-Range Forecasts**: Provide monthly and seasonal timescales, offering a deeper understanding of climate variability through detailed maps.
- **Weekly Forecasts**: Provide rainfall anomaly predictions for 4 to 6 weeks ahead, indicating whether rainfall is expected to be above normal, normal, or below normal.
- **Short-Term Forecasts**: Cover periods of up to 9 days, delivering detailed daily maps of predicted precipitation and temperature.
- **Real-Time Information**: Features current weather data, including maps that display ongoing rainfall and other meteorological parameters.
- **Warnings and Alerts**: Deliver critical information on severe weather events, offering advisory notices and maps that highlight affected areas.

The following sections provide further details on each of these climate service products, outlining their specific purposes and timescales. They also emphasize the importance of integrating climate information services into water governance strategies to support effective adaptation to climate change.

15.4 Key Components Necessary to Create and Deliver CIS

Developing and delivering climate services that are both relevant and practical for water management requires a comprehensive, multi-dimensional approach. Key elements include strengthening the capacity of meteorological services to provide accurate and timely information; fostering partnerships and co-production processes to integrate diverse stakeholder perspectives; and leveraging scalable communication channels to effectively reach a wide audience (WMO, 2018). Moreover, actively engaging vulnerable populations ensures that those most at risk from climate impacts receive the necessary support. Striking a balance between scalable solutions and context-specific interventions further enhances the relevance and effectiveness of these services. Collectively, these components help make climate services more accessible, actionable, and impactful for sustainable water resource management.

15.4.1 Enhanced Meteorological Service Capacity

To deliver effective climate services, meteorological agencies must strengthen their capacity through infrastructure upgrades, improved data collection and analysis, and provision of timely, accurate, and actionable information. This requires investment in modern weather monitoring technologies, such as automated weather stations and satellite systems, to obtain precise, real-time data. Equally important is the development of advanced data processing tools and techniques to analyze climate and weather data efficiently. Building the skills of meteorologists and support staff to interpret data and produce targeted advisories for water management is also essential.

Furthermore, the creation of digital platforms and mobile applications that present weather information in a clear and user-friendly format will significantly enhance the accessibility and impact of these services (WMO, 2018).

15.4.2 Partnerships and Processes for Co-production of Services

Collaborating with various stakeholders, including water management authorities, research institutions, and government agencies, is essential for developing relevant and effective climate services. Co-production ensures that services meet specific needs and are grounded in local realities. This involves engaging stakeholders in designing and developing climate services to ensure their needs and preferences are taken into account. Establishing partnerships between meteorological services, water management experts, and local organizations is crucial to co-creating solutions (Vincent et al., 2018). Implementing mechanisms for regular feedback from water managers helps continuously improve and adapt services based on their experiences and suggestions. Testing new services in pilot regions allows for gathering and refining insights before broader implementation.

15.4.3 Scalable Communication Channels

To effectively reach a broad audience, climate services must be disseminated through multiple communication channels that cater to varying literacy levels, technological access, and preferences. This involves utilizing media such as radio, television, SMS, mobile apps, and social media to distribute weather information and advisories. Ensuring information is available in local languages enhances comprehension and relevance. Leveraging existing community structures and networks facilitates more effective information dissemination. Developing interactive platforms where users can ask questions, provide feedback, and receive personalized advice further enhances the accessibility and usefulness of climate services (WMO, 2018).

15.4.4 Engaging Vulnerable Populations

Climate services must prioritize the needs of vulnerable populations most affected by climate risks, as these groups often lack access to resources and information. This makes targeted engagement crucial (Photiadou et al., 2021). This involves mapping and identifying communities and individuals most susceptible to climate impacts.

Designing specific outreach programs ensures these groups receive timely and relevant information. Providing training and resources empowers vulnerable populations to understand and use climate information effectively. Additionally, ensuring that marginalized groups are actively involved in developing and disseminating climate services fosters inclusive participation and addresses their unique needs.

15.4.5 Balancing Scalable Services with Context-Specific Needs

Effective climate services must strike a balance between scalability and customization. While scalable solutions are necessary to reach a broad audience, context-specific needs must also be addressed to ensure relevance and effectiveness (Hansen et al., 2019). This involves blending traditional practices and local knowledge with scientific data to create services that resonate with users. Developing modular services tailored to specific regions and practices allows for greater customization. Deploying extension agents to provide personalized support and advice based on local conditions enhances the practical application of these services. Additionally, regularly updating and adapting services to reflect changing climate patterns and emerging challenges ensures they remain relevant and effective.

15.5 Challenges in Climate Information Services

Despite the benefits, climate information services encounter several challenges that must be addressed to enhance their effectiveness and reliability. These challenges include uncertainty in climate information, limited data quality, imperfect models, inherent complexity, and gaps in understanding and communication.

15.5.1 Uncertainty in Climate Information

One of the most significant challenges inherent to climate services is uncertainty in climate information. This uncertainty stems from several factors:

- **Limited and Poor-Quality Data**: In many regions, especially developing countries, there are insufficient weather stations and data collection points. This leads to gaps in data coverage and affects the accuracy of weather and climate models (Daron et al., 2022).
- **Inconsistent Data Quality**: Variations in data collection methods and technologies can result in inconsistencies, reducing the reliability of climate predictions (Daron et al., 2022).

- **Historical Data Gaps**: A lack of comprehensive historical climate data makes it difficult to establish accurate baseline conditions and identify long-term trends (Daron et al., 2022).

15.5.2 Imperfect Models

Current climate models have limitations in accurately simulating complex climate systems. These limitations can lead to discrepancies between predicted and actual weather patterns. Many models operate at coarse resolutions, which may fail to accurately capture local climate and weather variations. Higher-resolution models are necessary for more precise predictions. Uncertainty in the values of model parameters can propagate through the modeling process, causing uncertainties in the final predictions (Kniveton et al., 2015).

15.5.3 Inherent Complexity

The climate system is inherently complex, involving interactions among the atmosphere, oceans, land surface, and biosphere. This complexity makes it challenging to predict climate behavior accurately. Many climate processes are non-linear, meaning that small changes in one component can lead to significant and sometimes unpredictable changes in the overall system. Feedback mechanisms, such as the melting of polar ice or changes in vegetation cover, can amplify or dampen climate responses, adding another layer of complexity to predictions (Kniveton et al., 2015).

15.5.4 Lack of Understanding of Climate Services

There is often a lack of awareness among users regarding the potential and limitations of climate services. This can result in unrealistic expectations or missed opportunities for utilizing available services. Effective communication of climate information is essential. However, technical jargon and complex data presentations can impede understanding and usage by non-experts.

15.5.5 Lack of Proactive Dialogues

Proactive dialogues between climate service providers and users are often lacking. Engaging stakeholders in developing and refining climate services is essential for ensuring their relevance and usability. Establishing robust feedback mechanisms that allow users to provide input on the usefulness and accuracy of the services they receive is critical for continuous improvement.

15.6 Role of National Meteorological and Hydrological Services (NMHS)

National Meteorological and Hydrological Services (NMHS) play an essential role in providing climate services at the national level, supporting the National Adaptation Plan (NAP) process. NMHS enhance societal safety and well-being, reduce poverty, sustain development and economic growth, and improve access to clean drinking water. They also enhance food production, improve health outcomes, mitigate and help societies adapt to climate change, exploit renewable energy sources, and increase the prosperity of populations (WMO, 2018, 2019).

NMHS play a crucial role in providing effective and timely climate services, including the analysis of current climate conditions and future climate change scenarios. They assess climate vulnerabilities and identify adaptation options, enhancing the capacity for planning and implementing adaptation measures, which are essential for informed decision-making at various governmental and societal levels. NMHS actively collaborate with academia, government departments, international organizations, non-governmental organizations (NGOs), and the private sector to improve data coverage, enhance model resolution, and create specialized products that benefit society. These partnerships ensure that climate services are accurate, relevant, and valuable for a broad range of stakeholders.

By providing weather and climate information, NMHS assist water managers in anticipating, preparing for, and responding to water-related risks, managing both short-term climate extremes like droughts and heatwaves and long-term trends such as the increased frequency of cyclones and desertification. They play a significant role in disaster risk reduction by providing historical and real-time data on loss and damage, contributing to multisectoral plans, offering climate information for informed investment decisions, and supporting risk financing and transfer strategies. NMHS work with the public and private sectors to implement multi-hazard early warning systems to reduce fatalities caused by weather-, water-, and climate-related disasters, enhance societal resilience, sustain productivity, and minimize property damage. The health sector benefits significantly from NMHS weather and climate information, helping health decision-makers understand how climate impacts disease transmission and occurrence, identify at-risk populations, and use decadal climate projection maps for vulnerability assessments. NMHS also provide long-term climate data for hydrological modeling, aiding in better water resource management and prioritized allocation. Their input is crucial in developing robust water and flood management policies and strategies, identifying extreme weather and climate hazards, and implementing effective risk management and emergency preparedness practices. Additionally, the energy sector relies on detailed weather and climate information from NMHS for resource assessment, infrastructure development, site selection, and maintenance, thereby helping to balance energy supply and meet demand, ensuring the sustainable development of energy systems, which is vital for both economic growth and environmental protection (WMO, 2018).

15.7 Conclusion

Climate change poses significant challenges to natural and human systems, increasing vulnerability through extreme weather events such as heat waves, droughts, floods, cyclones, and wildfires. These events disrupt ecosystems, food and water supplies, infrastructure, health, and mental well-being, with developing countries often having lower adaptive capacity. Climate Information Services (CIS) are crucial in addressing these challenges by transforming climate data into tailored products and empowering stakeholders to make informed decisions.

Effective CIS requires enhanced meteorological service capacity, including modern monitoring equipment and skilled meteorologists, partnerships for co-production with stakeholders, scalable communication channels, and targeted engagement with vulnerable populations. However, CIS faces data limitations, model imperfections, inherent complexity, and communication gaps. Addressing these challenges involves increasing data coverage and quality, developing higher-resolution models, improving understanding of climate processes, and simplifying the communication of technical information.

NMHS plays a vital role in providing accurate and timely climate information, supporting adaptation plans, societal safety, and sustainable development. By tackling these challenges and enhancing CIS capacity, stakeholders can improve resilience and adaptive capacity, mitigate risks, optimize resource management, and support sustainable development in the context of climate change. Climate services can significantly contribute to global climate adaptation and resilience efforts through continued collaboration, innovation, and engagement.

References

Amarnath, G., Ghosh, S., Alahacoon, N., Ravan, S.K., Taneja, P.K., Dave, N. & Srivastava, S.K. (2019). Insurance as an agricultural disaster risk management tool: Evidence and lessons learned from South Asia. In *CGIAR research program on water, land and ecosystems*, WLE Briefing Series No. 27. Colombo, Sri Lanka: International Water Management Institute (IWMI).

Bilgiç, E., & Baba, A. (2023). Effect of urbanization on water resources: Challenges and prospects. In *Groundwater in arid and semi-arid areas: Monitoring, assessment, modelling, and management* (pp. 81–108).

Brief, U. W. P. (2019). *Climate change and water*. UN Water.

Cosgrove, W. J., & Loucks, D. P. (2015). Water management: Current and future challenges and research directions. *Water Resources Research, 51*(6), 4823–4839.

Daron, J., Soares, M. B., Janes, T., Colledge, F., Srinivasan, G., Agarwal, A., Hewitt, C., Richardson, K., Nepal, S., Shrestha, M. S., & Rasul, G. (2022). Advancing climate services in South Asia. *Climate Services, 26*, Article 100295.
Graham, R., Ticehurst, H., Leathes, B., Wade, S., Visman, E., Bayley, S., Kane, C., Shongwe, M., Ferreira, T., Amato, R., & Bain, C. (2015). Scoping, options analysis and design of a 'Climate Information and Services Programme' for Africa (CIASA).
Hansen, J., Furlow, J., Goddard, L., Nissan, H., Vaughan, C., Rose, A., Fiondella, F., Braun, M., Steynor, A., Jack, C., & Chinowsky, P. (2019). *Scaling climate services to enable effective adaptation action.* www.gca.org
Hewitt, C. D., Allis, E., Mason, S. J., Muth, M., Pulwarty, R., Shumake-Guillemot, J., Bucher, A., Brunet, M., Fischer, A. M., Hama, A. M., & Kolli, R. K. (2020). Making society climate resilient: International progress under the global framework for climate services. *Bulletin of the American Meteorological Society, 101*(2), E237–E252.
IPCC. (2021). *Climate change 2021: The physical science basis.* In V. Masson-Delmotte et al. (Eds.), *Contribution of working group I to the sixth assessment report of the IPCC.* Cambridge University Press. https://doi.org/10.1017/9781009157896
Jayawardena, I. M. S. P., Punyawardena, B. V. R., & Karunarathne, M. D. R. K. (2022). Importance of integration of subseasonal predictions to improve climate services in Sri Lanka—case study: Southwest monsoon 2019. *Climate Services, 26*, Article 100296.
Kniveton, D., Visman, E., Tall, A., Diop, M., Ewbank, R., Njoroge, E., & Pearson, L. (2015). Dealing with uncertainty: Integrating local and scientific knowledge of the climate and weather. *Disasters, 39*(s1), S35–S53.
Konapala, G., Mishra, A. K., Wada, Y., & Mann, M. E. (2020). Climate change will affect global water availability through compounding changes in seasonal precipitation and evaporation. *Nature Communications, 11*(1), 3044.
Photiadou, C., Arheimer, B., Bosshard, T., Capell, R., Elenius, M., Gallo, I., Gyllensvärd, F., Klehmet, K., Little, L., Ribeiro, I., et al. (2021). Designing a climate service for planning climate actions in vulnerable countries. *Atmosphere, 12*, 121. https://doi.org/10.3390/atmos12010121
Sivakumar, M. V., & Lucio, F. (2018). Climate services for sustainable development. In *Bridging science and policy implication for managing climate extremes* (pp. 81–100).
Smit, B., & Wandel, J. (2006). Adaptation, adaptive capacity and vulnerability. *Global Environmental Change, 16*(3), 282–292.
Vincent, K., Daly, M., Scannell, C., & Leathes, B. (2018). What can climate services learn from theory and practice of co-production? *Climate Services, 12*, 48–58.
WMO. (2018). *Step-by-step guidelines for establishing a national framework for climate services.* WMO-No. 1206.
WMO. (2019). *The future of climate services* (Vol. 68). World Meteorological Organization.

Dr. Shiromani Jayawardena Regional Integrated Multi Hazard Early Warning System (RIMES), Thailand - Dr. Shiromani Jayawardena is a highly experienced meteorologist with over 25 years in weather and climate research. She currently serves as Senior Adviser for Weather and Climate Services at RIMES (Regional Integrated Multi-Hazard Early Warning System), where she focuses on climate resilience, disaster preparedness, and early warning systems across the region. Previously, Dr. Jayawardena held senior leadership roles as Director of Meteorology (Forecasting and Decision Support) and Deputy Director (Climate Change and Research) at the Department of Meteorology, Sri Lanka, where she led initiatives to strengthen national early warning systems and climate services. Her research focuses on climate variability, intraseasonal oscillations, monsoon dynamics, and the impacts of extreme weather events on Sri Lanka's climate resilience and early warning systems.

Dr. Niranga Alahacoon has a strong background in remote sensing and disaster risk management. He holds a PhD in Remote Sensing, awarded by the University of Colombo in 2021. He received the prestigious "Astronomer Royal Sir Arnold Wolfendale Memorial Gold Medal" for the Best Postgraduate Student in Physics at the University of Colombo. He completed an MSc in Geoinformation and Earth Observation (Geo-Hazard) in 2010 at the ITC in the Netherlands and obtained his BSc (Hons) in Geology in 2007 at the University of Peradeniya. Currently, he works at the International Water Management Institute Headquarters as a Regional Researcher, contributes significantly with application of remote sensing data and image processing, geographic information systems (GIS), hydrological and hydrodynamic modelling, geology, and rapid emergency response mapping, and has a publication record of more than 40 scientific publications and 22 training manuals. Prior to joining IWMI, Niranga served as a lecturer at the Faculty of Fisheries and Marine Sciences and Technology (FMSC), University of Ruhuna, Sri Lanka.

Dr. Surajit Ghosh is an experienced researcher in the environmental services industry. His skill set ranges from Machine learning and Geospatial Big Data analytics to cloud computing, and he has experience in Hydrological and Ecological modelling. Currently, he is a researcher at the International Water Management Institute (IWMI). Previously, Surajit has worked for the International Centre for Agricultural Research in the Dry Areas (ICARDA) and IORA Ecological Solutions Pvt. Ltd, and the Indian Institute of Remote Sensing (IIRS). His current research focuses on integrating multi-sensor data to develop spatial decision support systems that mitigate Water and Climate Risks.

Chapter 16
Water Governance for Climate Change Adaptation in Sri Lanka: Policy Options

Upali A. Amarasinghe, Amarnath Giriraj, and Hasula Wickremasinghe

Abstract The water governance gaps and challenges for climate adaptation are only emerging now, with the increased number of foreign and locally funded adaptation projects. The governance research is imperative in meeting the sustainable development goals of Integrated water resources management. This chapter summarizes the major findings and policy recommendations regarding the water governance challenges identified in the previous chapters of the book and discusses the policy options and future research needs for tackling these challenges in the context of climate change.

Keywords Climate change · Water governance · Storage · Major irrigation · Minor irrigation · Land · Pollution · Water quality

16.1 Introduction

Sri Lanka will face unprecedented challenges due to climate change. Climate change-related issues, such as increasing rainfall variability, heavy rainfall events, extreme temperatures, changing wind patterns, and associated floods and droughts, are already causing significant damage. Being a small nation, Sri Lanka can only make a small contribution to mitigating climate change. However, with accelerated climate adaptation programs, the country can make significant progress in mitigating climate change risks.

U. A. Amarasinghe (✉) · A. Giriraj
International Water Management Institute, Colombo, Sri Lanka
e-mail: U.amarasinghe@cgiar.org

A. Giriraj
e-mail: a.giriraj@cgiar.org

H. Wickremasinghe
Climate Change Secretariat, Ministry of Environment, Battaramulla, Sri Lanka

U. A. Amarasinghe et al. (eds.), *Water Governance for Climate Change Adaptation in Sri Lanka*, Global Issues in Water Policy 13,
https://doi.org/10.1007/978-3-032-04950-6_16

Climate change has a direct impact on water, land, and the environment. Climate-induced disasters—such as floods, droughts, landslides, and sea-level rise—likewise impact these resources. The 16 chapters of this book highlight various water governance challenges across different sectors and scales. Although the list of challenges identified is not exhaustive, it serves as a set of issues requiring urgent attention to enhance adaptation outcomes and build climate resilience in the water sector. This chapter identifies the way forward for addressing the water governance challenges outlined in the book. It summarizes policy actions on 11 governance aspects, including agricultural digital transformation, managing various storage, managing the major and medium irrigation systems, minor irrigation systems, including cascade systems, and land, empowering the private sector, water quality and pollution management, institutional coordination and policy framework, financing for operations and management, and demographic changes.

16.2 Policy Options

1. **Agricultural Digital Transformation**

There is a clear need to refine climate services, primarily by increasing stakeholders' access to existing digital tools.

- **Develop reliable weather forecasts at finer spatial scales (at least to the Grama Niladhari (GN) division level)**. These services are crucial for enhancing the efficiency of water management, agricultural input planning, risk mitigation through insurance, and related activities. The forecasts should be communicated effectively through easily accessible channels such as smartphones, enabling direct access by local communities and farmers. Short messages in regional languages, sent to the phones of farmers, women, and other community leaders, have proven highly beneficial in helping farmers optimize the timing of irrigation and the application of agricultural inputs.
- **Promote private sector participation** through policies and incentives to expand the hydro-meteorological data collection network and provide climate advisory services. These advisories would go a long way toward planning and implementing local adaptation measures required to address climate variability-related issues.
- **Adapt and deploy existing digital tools**, such as AWARE and the Climate Smart Governance (CSG) Dashboard, to support anticipatory disaster planning. Tools like AWARE help national and regional planners and communities proactively prepare for disasters, minimizing damage to life and property. The CSG Dashboard reduces duplication of climate investments across multiple locations and prioritizes investments for climate-vulnerable communities with targeted adaptation interventions.

2. **Managing the Storage**

Given the importance of water storage for irrigation, fisheries, livestock, domestic and industrial water supplies, hydropower generation, and ecosystem services, give high priority to storage management. Storage management systems should adopt an integrated approach to understand interlinkages within a river basin or between different types of storage, including ponds, large and small tanks, lakes, wetlands, groundwater, and soil moisture. A proper understanding of these interlinkages can help different storage types complement each other's services, contributing to a reduction in seasonal and annual water stress in river basins. In particular:

- **Promote efficient use of rainfall** through accurate rainfall forecasts. This will increase rainwater use efficiency, thereby reducing irrigation requirements.
- **Enhance soil moisture storage** by mulching or no-till farming to reduce irrigation requirements.
- **Promote groundwater use** with appropriate policies to ensure that crop water requirements at critical growth periods are met through supplemental groundwater irrigation, while safeguarding against overexploitation.
- **Value the services provided by maintaining minimum river flows and by storage in lakes and wetlands, as well as the value of water for domestic and industrial uses.** Such assessments highlight the trade-offs between different water uses.

3. **Managing Major Irrigation Systems**

Major irrigation systems play a significant role in ensuring national food security. Key actions for these systems include:

- **Introduce flow measurement systems** in canal networks at least down to the distributary level. This enables canal automation and bulk water allocation, and increases farmers' confidence in the equitable distribution of water (water duty), thereby improving on-farm water management. It also encourages private investment in high-tech irrigation and diversification into high-value crops, along with other measures to enhance water-use efficiency and both physical and economic water productivity.
- **Enhance the technical skills of irrigation personnel and establish minimum capacity requirements** to effectively manage high-tech canal irrigation systems.
- **Promote conjunctive management of surface water and groundwater** in major irrigation schemes *by design* rather than by default. Introduce the necessary policies and regulations to facilitate planned groundwater use and crop diversification in canal command areas.

4. **Managing Minor Irrigation Systems**

Many minor irrigation systems are scattered throughout the dry-zone landscape. They should receive special attention due to their widespread contribution to mitigating floods and droughts in many river basins, as well as their provision of multiple services to rural communities. Managing minor irrigation systems effectively requires a mix of traditional knowledge and modern technology, a community-centered institutional mechanism, strong policy and legal support, and the linking of tank cascade systems with river basin management plans. In particular, in tank cascade systems:

- **Provide scenarios of surface and groundwater availability and demand**, and illustrate cascade interconnectivity through return flows, to increase communities' awareness of the value of tank storage management.
- **Establish cascade management committees (CMCs)** at various levels to increase the participation of women, youth, and other water user groups in managing different tanks. The CMCs can include:
- a *tank cascade-level management committee* with all stakeholders (farmers, fishers, women's organization leaders, and other community leaders), featuring a business plan and cooperative models within the cascade for the economic empowerment of the community;
- an *administrative divisional-level (Divisional Secretariat) cascade management committee*, comprising the Divisional Secretary, representatives from various departments (irrigation, agriculture, forestry, wildlife, marketing, etc.), private sector representatives, and a representative from the tank cascade-level committee, to oversee operations at the cascade level.

5. **Managing Rainfed Agriculture Systems**

Rainfed seasonal and annual cropping areas often feature high-value crops, and these areas (excluding tea, rubber, and coconut) generate more value per cubic meter of water use than paddy cultivation. In rainfed areas, promote the following measures:

- **Better management of soil moisture** to increase the efficiency of rainwater use.
- **Adoption of high-yielding, drought-tolerant crop varieties** suited for rainfed conditions.
- **Supplemental groundwater irrigation** to meet water needs during critical periods of crop growth.
- **Use of solar-powered pumps** for groundwater extraction in remote or off-grid locations.
- **High-efficiency irrigation techniques**, such as drip irrigation, sprinkler systems, and underground piping, especially in tea and coconut plantations.
- **Intercropping** in tea and coconut plantations to maximize land and water productivity

6. **Managing Land**

  - Introduce land reforms that prioritize formalizing land titles and integrating traditional landholding systems into the official legal framework, along with enforcing land-use regulations, to prevent unsustainable practices that threaten land, water, and ecosystem services.
  - Consolidate small land parcels through voluntary pooling (e.g., via cooperatives or similar mechanisms) by:
  - involving farmers in decision-making,

- using GIS and remote sensing technologies and land records to map and document land holdings,
- establishing a transparent mechanism to value land parcels based on size, location, soil, and farming practices,
- establishing a clear legal procedure for land ownership and registration, and
- ensuring strong institutional coordination and appropriate legal mechanisms to address accountability issues and farmers' grievances.

7. **Empowering the Private Sector**

- Empower the corporate sector, including rural service providers, by establishing clear policy guidelines that enable them to provide services, thereby reducing the financial burden on government departments. It is imperative to remove constraints that hinder the private sector's ability to participate in planning and implementing adaptation initiatives.

8. **Water Quality and Pollution Management**

Water quality deterioration is caused by the increasing use of chemical fertilizers, herbicides, and pesticides in agriculture, as well as by waste discharges from the service and industrial sectors. This issue requires thorough attention.

Agriculture sector:

- **Conduct a proper assessment of chemical use** in agriculture and its contribution to water and land quality deterioration.
- **Promote traditional pest and weed control practices** (e.g., biological and integrated pest management) that can reduce reliance on agrochemicals.
- **Promote the use of bio-fertilizers in farming**, providing incentives if necessary (such as support for livestock rearing) to encourage adoption.

Industrial and service sectors:

- **Enhance water quality monitoring** in major industrial zones.
- **Strengthen pollution regulations** and enforcement for industries.
- **Promote wastewater treatment and its reuse** in agriculture, especially in peri-urban areas.

Urban Water Supply and Sanitation

Inadequate and poor-quality drinking water supplies are major contributors to multidimensional poverty. Because of rapid urbanization, it is essential to:

- **Manage water service delivery in coordination with other sectors**, ensuring that urban water supply is integrated with overall water resource management.
- **Promote demand management and reduce water waste** through measures such as public awareness, metering, and leak reduction.
- **Monitor the quality of water supplied** by drinking-water facilities to safeguard public health.

9. **Institutional Coordination and Policy Frameworks**

There are numerous institutions involved in the water sector with overlapping responsibilities. To improve institutional coordination:

- **Develop a coherent national water policy** to minimize overlapping responsibilities among various institutions.
- **Set up an inter-agency body** to coordinate and monitor the activities of different water-related institutions.
- **Develop strong legal mechanisms** to hold institutions accountable for their actions.

10. **Financing for Operations and Management**

Financing for the operation and management of the water sector, especially in canal irrigation, remains unclear except at the level of farmer organizations.

- Develop **financing instruments** that take into account all stakeholders in water use, including farmers, fishers, women, the domestic sector, and industries.

11. **Demographic Changes**

- Demographic trends are major factors influencing long-term investments in the water sector. It is important to focus on recent demographic trends. The most recent population census shows a decreasing population in many districts. However, future planning is based on projections that are significantly higher than the current population The changing rural–urban population distribution, the aging farming demographic, the evolving economic structure (with significant contributions from the service and industrial sectors), and changing lifestyles should be key considerations in determining adaptation investments in the water sector.

Dr. Upali A. Amarasinghe Emeritus Scientist at the International Water Management Institute, has over 33 years of experience as a researcher and project leader in water sources management, as well as ten years of teaching experience as a university lecturer. He holds a Ph.D. degree in Statistics from the University of Illinois, Urbana-Champaign, USA. His research focused on water management issues at different spatial scales; which included estimating water productivity and footprints of crops to river basins; food and water supply and demand and projections of river basins to countries; benchmarking of performance of canal irrigation systems; poverty and vulnerability assessment; poverty mapping using small area estimation; water security assessment of industries; impact assessments of interventions in water resources management. He has extensive experience managing research projects in India and Sri Lanka, as well as short-term work experiences in other countries, including Pakistan, China, Malaysia, Thailand, and Vietnam. Dr. Amarasinghe has published over 130 research articles and edited four books.

Dr. Amarnath Giriraj has a Ph.D. in Remote Sensing. He is the Research Group Leader for Water Risks and Disaster at IWMI, Sri Lanka, a Steering Committee member of Sentinel Asia by JAXA, and Chair of the Working Group on Water-related Disasters at the Asia Pacific Space Agency Forum. Dr. Giriraj's primary expertise includes disaster risk assessment, hydrological process research, agricultural disaster risk management, climate services, and food security, with a focus on developing solutions for pressing climate risk challenges. Many of his research and development projects have demonstrated the impact of climate risks, utilizing advanced science and technology in various hydro-climatic regions worldwide. He is keen on translating science-based knowledge into effective policies and strategies contributing to environmental, economic, and societal sustainability. He established the Water Risks and Disasters mission at IWMI and developed regional and national water-risk data systems and reports to inform policy and decision-makers on key disaster, agricultural, and water challenges and solutions.

Ms. Hasula Wickremasinghe works at the Ministry of Environment. She has worked at the Biodiversity Secretariat for ten years and coordinated the preparation of the National Red List (fauna groups) in 2012. She has worked on the Mitigation and Adaptation sectors at the Climate Change Secretariat for the past seven years. She assisted in submitting the Updated Nationally Determined Contributions (NDCs) in 2021 and is currently involved in developing the NDC Implementation Plan. She was also instrumental in designing a Climate Smart City for Kurunegala and assisted in implementing the National Adaptation Plan Readiness Support Project. Hasula has a degree in Biosciences from the Open University of Sri Lanka, an MSc in Environment Sciences from the University of Colombo, and a Postgraduate Diploma in Archaeology from the University of Kelaniya

MIX
Papier aus verantwortungsvollen Quellen
Paper from responsible sources
FSC® C105338

If you have any concerns about our products,
you can contact us on
ProductSafety@springernature.com

In case Publisher is established outside the EU,
the EU authorized representative is:
Springer Nature Customer Service Center GmbH
Europaplatz 3, 69115 Heidelberg, Germany

Printed by Libri Plureos GmbH
in Hamburg, Germany